Evolution of Sleep

Research during the past two decades has produced major advances in understanding sleep within particular species. Simultaneously, new analytical methods provide the tools to investigate questions concerning the evolution of distinctive sleep state characteristics and functions. This book synthesizes recent advances in our understanding of the evolutionary origins of sleep and its adaptive function, and it lays the groundwork for future evolutionary research by assessing sleep patterns in the major animal lineages.

DR. PATRICK MCNAMARA is an Associate Professor of Neurology at Boston University School of Medicine and Veterans Administration (VA) Boston Healthcare System. He is based in the Department of Neurology at Boston University School of Medicine. He is the director of the Evolutionary Neurobehavior Laboratory and was awarded a National Institutes of Health (NIH) grant to study the phylogeny of sleep. Dr. McNamara is the recipient of a Veterans Affairs Merit Review Award for the study of Parkinson's disease and several NIH awards for the study of sleep mechanisms. He is also the author of *Mind and Variability: Mental Darwinism, Memory and Self*; *An Evolutionary Psychology of Sleep and Dreams*; and *Nightmares: The Science and Solution of Those Frightening Visions During Sleep*.

DR. ROBERT A. BARTON is a Professor at Durham University and Director of the Evolutionary Anthropology Research Group. He has published numerous papers on the topic of brain evolution, and, in addition to an NIH-funded project on the phylogeny of sleep, he has collaborated with Dr. Charles L. Nunn on the application of comparative methods to questions in mammalian biology and physiology.

DR. CHARLES L. NUNN is an Associate Professor in the Department of Anthropology at Harvard University. Dr. Nunn completed his Ph.D. at Duke University in biological anthropology and anatomy, and he conducted postdoctoral research on primate disease ecology at the University of Virginia and University of California Davis. He has had academic appointments in the United States (University of California Berkeley) and Germany (The Max Planck Institute for Evolutionary Anthropology). He is an author of *Infectious Diseases in Primates: Behavior, Ecology, and Evolution*, and his current research focuses on phylogenetic methods, disease ecology, and the evolution of primate behavior.

Evolution of Sleep

Edited by
PATRICK MCNAMARA
Boston University

ROBERT A. BARTON
Durham University

CHARLES L. NUNN
Harvard University

*Phylogenetic and Functional
Perspectives*

CAMBRIDGE
UNIVERSITY PRESS

CAMBRIDGE UNIVERSITY PRESS
Cambridge, New York, Melbourne, Madrid, Cape Town, Singapore,
São Paulo, Delhi, Dubai, Tokyo

Cambridge University Press
32 Avenue of the Americas, New York, NY 10013-2473, USA

www.cambridge.org
Information on this title: www.cambridge.org/9780521894975

First published 2010

Printed in the United States of America

A catalog record for this publication is available from the British Library.

Library of Congress Cataloging in Publication data

Evolution of sleep: phylogenetic and functional perspectives / edited by Patrick
McNamara, Robert A. Barton, Charles L. Nunn.
  p.   cm.
Includes bibliographical references and index.
ISBN 978-0-521-89497-5 (hardback)
1. Sleep.   2. Evolution.   I. McNamara, Patrick, 1956–   II. Barton, Robert
A. (Robert Alexander), 1959–   III. Nunn, Charles L.   IV. Title.
QP425.E96   2009
612.8′21–dc22   2009011589

ISBN 978-0-521-89497-5 Hardback

Contents

v

Contributors

Mourad Akaârir
Institut Universitari de Ciències de la Salut
Universitat de les Illes Balears

Charles J. Amlaner
Department of Biology
Indiana State University

Sanford Auerbach
Sleep Disorders Center
 Boston University School of Medicine

Robert A. Barton
Evolutionary Anthropology Research Group
Durham University

Isabella Capellini
Evolutionary Anthropology Research Group, Department of Anthropology
Durham University

Subimal Datta
Sleep and Cognitive Neuroscience Research Laboratory, Department of Psychiatry
Boston University School of Medicine

Susana Esteban
Institut Universitari de Ciències de la Salut
Universitat de les Illes Balears

Antoni Gamundí
Institut Universitari de Ciències de la Salut
Universitat de les Illes Balears

Kristyna M. Hartse
Sonno Sleep Centers
El Paso, Texas

J. Lee Kavanau
Department of Ecology and Evolutionary Biology
University of California

James M. Krueger
Programs in Neuroscience
Washington State University

Patrick McNamara
Department of Neurology
Boston University School of Medicine

M. Cristina Nicolau
Institut Universitari de Ciències de la Salut
Universitat de les Illes Balears

Patrick M. Nolan
Mammalian Genetics Unit
Medical Research Council, Harwell

Charles L. Nunn
Department of Anthropology
Harvard University

Brian T. Preston
Department of Primatology
Max Planck Institute for Evolutionary Anthropology, Leipzig

Niels C. Rattenborg
Sleep and Flight Group
Max Planck Institute for Ornithology

Ruben V. Rial
Institut Universitari de Ciències de la Salut
Universitat de les Illes Balears

Mahesh M. Thakkar
Department of Neurology University of Missouri
Harry Truman Memorial VA Hospital

Valter Tucci
Department of Neuroscience and Brain Technology
Italian Institute of Technology

I. V. Zhdanova
Laboratory of Sleep and Circadian Physiology
Department of Anatomy and Neurobiology
Boston University School of Medicine

Acknowledgments

This book is a consequence of our recent phylogenetic comparative studies of mammalian sleep. As we learned more about variation in mammalian sleep, we were naturally drawn toward broader patterns of sleep across different organisms. Several questions formed in our minds, such as: Would patterns that we documented in mammals hold in other groups of organisms, and which other organisms should be studied? How would we be able to identify sleep, and thus test hypotheses comparatively, in fish, reptiles, and insects? And are the hypotheses that we focused on in mammals even relevant to nonmammals?

Mammalian sleep itself is remarkably variable, with aquatic mammals exhibiting specializations for sleep that are not found in terrestrial mammals, and marked variation in the expression of rapid-eye-movement (REM) and non–rapid-eye-movement (NREM) sleep, sleep cycles, and the organization of sleep into one or multiple bouts per 24-hour period. As we stepped outside the world of mammals, we found that sleep is pervasive phylogenetically, and we discovered that it is even more varied than we expected. This book summarizes what is currently known about variation in sleep patterns and presents some new data and analyses. We hope that the chapters herein will inspire others to collect datasets similar to those now available for birds and mammals. Further research along the lines described by the chapters in this volume will only deepen our understanding of this fundamental behavior, and is sure to lead to deeper understanding of the function—or functions—of sleep.

We have many people to thank for their time, encouragement, and inspiration. First, we would like to thank Chris Curcio from Cambridge University Press for his advocacy of this project. He played a key role in seeing this project through to the end, and we appreciate his guidance as we navigated the many hurdles of a book project. We would also like to thank our many collaborators who have played a role in our comparative research on mammals, especially Isabella Capellini, Brian Preston, Alberto Acerbi, and Patrik Lindenfors.

Erica Harris helped out on all aspects of this project, from communication with the authors to overseeing the final formatting of the book manuscript. Her organizational help has meant all the difference throughout and we are grateful for her unflagging assistance. We would also like to thank Emily Abrams, Donna Alvino, Andrea Avalos, Catherine Beauharnais, Emily Duggan, Patricia Johnson, Deirdre McLaren, and Alexandra Zaitsev for their help with editing and formatting the references for all of the chapters in the book. These assistants worked both conscientiously and carefully.

We would also like to thank Aleksandra Vicentic, the National Institutes of Health (NIH) Program Officer on our grant "Phylogeny of Sleep (5R01MH070415–01)," and NIH itself for supporting our work.

Lastly, we would also like to thank all of the authors who contributed chapters to this volume. This book would have been impossible without their combined knowledge, and they all went the extra mile to provide up-to-date reviews of sleep expression in their target taxa and an evolutionarily informed evaluation of sleep characteristics in those species.

Introduction

PATRICK MCNAMARA, CHARLES L. NUNN, AND ROBERT A. BARTON

Why do we and other animals sleep? When we are asleep, we are not performing activities that are important for reproductive success, such as locating food, caring for offspring, or finding mates. In the wild, sleep might make an animal more vulnerable to predation, and it certainly interferes with vigilance for predators. Sleep is found across the animal kingdom, yet it varies remarkably in its most fundamental characteristics across species. And for almost every pattern associated with sleep, exceptions can be found. For all of these reasons, sleep continues to be an evolutionary puzzle. Fortunately, sleep also has attracted much scientific interest, with many significant findings in the past 10 years.

The aim of this volume is to summarize recent advances in our understanding of the diversity of sleep patterns found in animals. Many of the chapters that follow examine sleep in different taxonomic groups, including insects, fish, reptiles, birds, and mammals. We take this "comparative approach" because it is one of the key ways in which biologists investigate the evolution of a trait (Harvey & Pagel, 1991). Indeed, the comparative method has long been used to investigate the evolution of sleep, particularly in mammals (e.g., Meddis, 1983; Zepelin, 1989). More recent comparative studies have capitalized on advances in the study of phylogenetic relationships to test hypotheses on the evolution of sleep (Capellini, Barton, Preston, et al., 2008a; Lesku, Roth, Amlaner, et al., 2006; Preston, Capellini, McNamara, et al., 2009; Roth, Lesku, Amlaner, et al., 2006). In mammals, these studies have revealed that species experiencing greater risk of predation at their sleep sites sleep less, that sleep duration correlates with immunocompetence across species, and that evolutionary increases in metabolic rate relative to body mass are associated with reductions in sleep. By incorporating phylogeny, a recent study also demonstrated that an apparent association between body mass and sleep is in fact a phylogenetic artifact (Capellini et al., 2008a; see also Lesku et al., 2006).

1

Other chapters provide syntheses of new advances in our understanding of the physiology and genetics of sleep as well as advances in phylogenetic analysis and informatics. These chapters are essential for uncovering sleep functions because evolution works on the genome, and many aspects of animal biology constrain the types of physiological patterns of sleep that are found across species. For example, marine mammals must continuously come to the water's surface to breath air, and this limits the kind of sleep in which they can engage. Similarly, animals that lack highly developed forebrains will be unable to exhibit classically defined sleep, which includes both behavioral and electrophysiological criteria for mammals and birds. Importantly, the study of interspecies variation requires careful compilation of data collected under diverse conditions as well as the application of comparative methods that use phylogeny to study evolutionary patterns. All of these components are essential for making sense of the variation in sleep patterns across species, and thus also for uncovering the function – or functions – of sleep.

In most cases, chapters in this volume have integrated taxonomic perspectives and details on sleep physiology, natural history, and genetics. Such integration is essential to understand sleep and to stimulate future comparative and evolutionary studies of sleep. We see the need for new comparative studies in a broader phylogenetic perspective – as well as experimental research – as a way to assess the generality of sleep patterns and the factors that influence sleep. Much of this effort will require laboratory and fieldwork to obtain new quantitative data on sleep in relatively unstudied animals, such as fish, insects, and reptiles. Even in the case of mammals and birds, sleep has been quantified in remarkably few species and often on the basis of the availability of particular species rather than in relation to specific questions concerning sleep and its evolution. We hope that this volume will spur more research along these lines.

To help set the stage for what follows, it is helpful to briefly review basic characteristics of sleep that are essential for studying sleep in comparative perspective. An important starting point involves the definition of sleep. As summarized in Table I.1, sleep is composed of behavioral, physiological, and electrophysiological characteristics as well as evidence for homeostatic regulation (i.e., sleep rebound). Behavioral measures of sleep vary according to the biology of the species involved. These measures can include a species-specific body posture and sleeping site, reduced physical activity (quiescence), reduced muscle tone (especially neck/nuchal muscle tone in rapid-eye-movement [REM] sleep), and increased arousal threshold. To distinguish the quiescent state from other states, such as coma or hibernation, it is usually required that the animal shows rapid reversibility to wakefulness upon arousal. Electrophysiological measures of REM include low-voltage fast waves, rapid eye movements, theta rhythms in the hippocampus, and pontine-geniculo-occipital (PGO) waves. Electrophysiological measures of

Table I.1. *Criteria for the definition of sleep[a]*

1. Behavioral
   - Typical body posture
   - Specific sleeping site
   - Behavioral rituals before sleep (e.g., circling, yawning)
   - Physical quiescence
   - Elevated threshold for arousal and reactivity
   - Rapid state reversibility
   - Circadian organization of rest–activity cycles
   - Hibernation/torpor
2. Electrophysiological
   EEG
       NREM: high-voltage slow waves (quiet sleep)
         - spindles in some animals
         - K-complexes in some primates
       REM: low-voltage fast waves (REM, Paradoxical sleep or AS [active sleep])
         - hippocampal theta; PGO waves
   Electro-oculogram (EOG)
       NREM: absence of eye movements or slow, rolling eye movements
       REM: rapid eye movements
   EMG
         - Progressive loss of muscle tone from Wake→NREM→REM
3. Physiological
   - REM: instabilities in heart rate, breathing, body temperature, etc.; penile tumescence
   - NREM: reduction in physiologic/metabolic processes; reduction of about 2°C in body temp
4. Homeostatic regulation
   - enhancement of sleep time
   - intensification of the sleep process (e.g., enhanced EEG power in the Delta range)

[a] Adapted from Moorcroft, 2003; Campbell & Tobler, 1984.

non-rapid eye movement (NREM) include high-voltage slow waves (HVSW), spindles, and K-complexes. Functional indices of sleep include increased amounts of sleep after sleep deprivation, and increased sleep intensity after sleep deprivation. Physiologic indices of sleep include significant reductions in temperature and metabolism during NREM and significant lability in autonomic nervous system (ANS), cardiovascular, and respiratory measures during REM, along with increases in metabolism. Lastly, as noted earlier, sleep typically involves a rebound effect, in which a sleep-deprived animal must make up for lost sleep by sleeping longer or more deeply.

For most animals, sleep can be identified only via measurement of its behavioral and functional sleep traits, as their nervous systems do not support what has become known as full polygraphic sleep – that is, electrophysiological measures of both REM and NREM sleep identified via the electroencephalogram or EEG. It has become common, however, to use the term "full polygraphic sleep" to refer to an animal that exhibits most or all of the other three major components of sleep in addition to the electrophysiologic measures. When an animal exhibits all four major components of sleep – including the behavioral, electrophysiological, physiological, and functional components – then it is said to have full polygraphic sleep. Full polygraphic sleep in this sense has so far been documented only in mammals and in birds. Although REM and NREM have been identified in 127 mammalian species representing 46 families across 17 orders (McNamara, Capellini, Harris, et al., 2008), NREM in most of these species cannot be differentiated into distinct "light" and "deep" stages as it is in several primate species. We estimate that REM and NREM sleep states have also been documented in about 36 avian species.

Overview of the volume

Krueger's chapter focuses on the neural basis of sleep. He suggests that core sleep characteristics are a property of small groups of neurons, and he summarizes the accumulating evidence that sleep is a network-emergent property of any viable group of interconnected neurons. Many biochemical sleep-regulatory events are shared by insects and mammals, suggesting that they evolved from metabolic regulatory events and that sleep is a local use-dependent process. Relationships between sleep and tumor necrosis factor (TNF) are used to examine the local use-dependent sleep hypothesis. Krueger argues that the need for sleep is derived from the experience-driven changes in neuronal microcircuitry that necessitate the stabilization of synaptic networks to maintain physiological regulatory networks and instinctual and acquired memories.

Hartse provides an overview of sleep in insects. Her work necessarily probes the definition of sleep while also giving some context to natural sleeping patterns in insects. An important discovery in the past two decades is that insects can serve as a model organism for studying sleep. She reviews the literature on sleep in *Drosophila* and the role of such studies in understanding sleep as a general phenomenon. Many insects, in fact, display all of the standard behavioral phenomena of sleep, such as periodic reduction in activity, increase in arousal threshold when quiescent, and rebound or increased rest–sleep durations after sleep deprivation.

Tucci and Nolan review the genetics of sleep in mice. They highlight the importance of understanding the genetic mechanisms of sleep – for example, by identifying functional genes. Mouse models of sleep disorders are also extremely useful for probing potential functional effects of sleep-related genes. Current progress in mouse functional genetics promises to increase the rate of discovery of sleep-related genes. There can be little doubt that basic sleep processes are influenced by genes, and it may be that separate sets of genes regulate expression of REM and NREM in mammals.

Chapters by Zhdanova and Kavanau review the literature on sleep in fishes. Fish are an ancient lineage and exhibit extensive variation in behavior and ecology. Resting behavior in fish shares several similarities with mammalian sleep. The behavioral criteria for sleep, such as periodic reduction in activity, increase in arousal threshold, and rebound after sleep deprivation are common in fish. Similarly, the principal neuronal structures involved in mammalian sleep, with the notable exception of the cerebral cortex, are conserved in fish and have neurochemical composition similar to that of higher vertebrates. In her studies of zebra fish, Zhdanova demonstrated both increased duration of sleep and changes in plasticity and behavioral performance following sleep deprivation.

Kavanau focuses on the phenomenon of schooling in fishes and the effects of schooling on sleep. Kavanau points out that by virtue of the rich variety and great permissiveness of aquatic habitats, some fish appear never to have encountered selective pressures for sleep. It is remarkable that three continuously active states of perpetual vigilance exist in these fishes, in which they achieve comparable, and even greater, benefits than accrue to animals that sleep. Even some continuously active but nonschooling fishes (some "pelagic cruisers") probably achieve highly efficient brain operation at all times, illustrating the exceptional demands of pelagic environments (open oceans).

Rial et al. review sleep processes in reptiles. While behavioral signs of sleep are clearly observable in reptiles, correlations between these behavioral signs of sleep and selected EEG indices are difficult to evaluate, given the complexities of recording sleep EEGs from the reptilian scalp and brain. Early studies of reptilian sleep reported an association between behavioral sleep and intermittent high-voltage spikes and sharp waves recorded from various brain structures in crocodilians, lizards, and turtles. Other investigators found no such association between behavioral sleep and high-amplitude spikes and sharp waves in the same animals. Rial et al. propose that mammalian sleep is a residual of reptilian waking states that were shunted aside when new cortical-based waking states became possible in early mammals.

Because birds and mammals exhibit electrophysiological signs of both REM and NREM while reptiles do not, sleep processes in birds and mammals may reflect common descent from a reptilian ancestor with similar sleep patterns. Alternatively, similar sleep processes of birds and mammals may be due to convergent evolution. Convergent evolution would suggest that similar sleep patterns of birds and mammals occur because these animals developed a similar solution to a common problem. Both birds and mammals are endothermic species. Sleep processes are implicated in temperature regulation, at least in mammals, and therefore the evolution of similar REM and NREM sleep processes in birds and mammals may be due to the emergence of the need for complex thermoregulatory processes to support endothermy in these animals.

Rattenborg and Amlaner review the literature on sleep in birds. As in mammals, birds can either sleep with a monophasic pattern (one consolidated period of sleep per day) or a polyphasic pattern (several short episodes of sleep per day). Birds also appear to exhibit a special form of slow-wave activity (SWA) and very little REM-like sleep. As in aquatic mammals, unilateral eye closure and unihemispheric slow-wave sleep (USWS) also occur in birds. Rattenborg and Amlaner first describe the basic changes in brain activity and physiology that accompany avian SWS and REM sleep. The unihemispheric nature of avian sleep is emphasized and reduction in sleep expression in migratory birds is considered. Rattenborg and Amlaner note that SWS-related spindles and hippocampal spikes, and the hippocampal theta rhythm that occurs during mammalian REM sleep, have not been observed in birds, even though they are readily detectable in epidural EEG recordings from the mammalian neocortex. They propose that the evolution of similar sleep states in mammals and birds is linked to the convergent evolution of relatively large and highly interconnected brains capable of complex cognition in each group.

Thakkar and Datta review the evolution of REM sleep. There is no evidence to suggest that REM sleep is present in invertebrates. Within the vertebrates, there is no evidence that supports the presence of REM sleep in fishes or amphibians. Some weak evidence exists to indicate the presence of REM sleep in reptiles, but further detailed studies are necessary before it can be concluded with any certainty that REM sleep is present in reptiles. REM sleep is definitely found in birds, marsupials, and mammals. However, major differences exist between avian and mammalian REM sleep. As compared to mammals, for example, REM bouts are shorter and the total amount of time spent in REM sleep is much smaller in birds than in mammals. These differences between birds and mammals may provide clues about the function of REM sleep.

The chapters by Capellini et al. and Nunn et al. utilize recent advances in phylogenetic methods in their analyses of the adaptive function of sleep in mammals and primates, respectively. Phylogenetic comparative analyses provide a means to reconstruct ancestral states, examine correlated evolution, and identify variation in how traits change over time. Capellini et al. review their work on the links between ecology and sleep in mammals. They show that predation pressure, trophic niche, and energy demands can, in part, explain patterns of interspecific variation in mammalian sleep architecture. Thus, the ecological niche that animals inhabit can exert significant evolutionary pressure on sleep durations as well as on how sleep is organized across the daily cycle. Nunn et al. focus on primate sleep, using a taxonomic subset of data that was analyzed by Capellini et al. They reconstruct the evolutionary history of primate sleep, use the data to investigate the function of sleep in primates, and pinpoint species in need of further research. In one new finding, Nunn et al. show that nocturnal species have longer sleep durations than do diurnal species.

McNamara and Auerbach discuss evolutionary medicine as a relatively new field of inquiry that attempts to apply findings and principles of evolutionary anthropology and biology to medical disorders. Although several medical disorders have been explored from the perspective of evolutionary medicine (see the collection of papers in Trevathan, Smith, & McKenna, 1999, 2007), sleep disorders have not so far been among them. This gap should be seen as an opportunity, as application of evolutionary theory to problems of sleep disorders may yield significant new insights into both causes and solutions of major sleep disorders. McNamara and Auerbach note that natural selection operates on the intensity dimension of sleep and thus that insomnia can be construed as resistance to homeostatic drive. Disorders involving excessive amounts of sleep, on the other hand, appear to be the result of chronic immune system activation.

Lacunae

A single volume cannot possibly cover all the dimensions of sleep across the tree of life or in the context of new advances in understanding sleep genetics and physiology. It is worth mentioning two areas that are not covered in this book: sleep in aquatic mammals and the phenomena of hibernation and torpor.

Sleep in aquatic mammals was recently the focus of a comprehensive review (Lyamin, Manger, Ridgeway, et al., 2008) and so is not covered here. Aquatic mammals include cetaceans (dolphins, porpoises, and whales), carnivores (seals, sea lions, and otters), and sirenians (manatees). These species are important because they depart from the typical patterns of mammalian sleep, for the obvious reason that they must come to the surface to breathe. Cetaceans exhibit a clear form of unihemispheric SWS (USWS). EEG signs of REM are absent, but cetaceans show other behavioral signs of REM, including rapid eye movements, penile erections, and muscle twitching. The two main families of pinnipeds, Otariidae (sea lions and fur seals) and Phocidae (true seals), show both unihemispheric and bihemispheric forms of sleep. Phocids sleep underwater (obviously holding their breaths) while both hemispheres exhibit either REM or SWS. Amazonian manatees (*Trichechus inunguis*) also sleep underwater, exhibiting three sleep states: bihemispheric REM, bihemispheric SWS, and USWS. Both hemispheres awaken when the animal surfaces to breathe.

Departures from the typical mammalian pattern provide an opportunity to test specific functions of sleep. For example, sleep deprivation in an animal exhibiting unihemispheric sleep has been shown to result in unihemispheric sleep rebound, prompting some authorities to claim that sleep serves a primary function for the brain rather than the body. The data on unihemispheric sleep in marine mammals also suggest that REM and NREM serve distinct functions, as animals without full polygraphic REM can survive. In addition, when REM occurs in marine mammals, it is always bihemispheric. The bilateral nature of REM may be considered one of its costs, and the brain structure of certain marine mammals, apparently, cannot bear these costs.

Hibernation and torpor are not typically considered part of the definition of behavioral sleep – yet intuitively most investigators feel that hibernation and torpor are states closely related to sleep. Several orders of mammals contain hibernating species or species that enter torpor, including the monotremes (echidna), the marsupials (mouse opposum), insectivores (hedgehog), bats (brown bat), primates (dwarf lemur), and some rodents (Kilduff, Krilowicz, Milsom, et al., 1993). Contrary to popular belief, bears are not true hibernators. During winter their body temperature does not decrease beyond the level of normal sleep, and the bear remains alert and active in its den. Typically it is the pregnant female who retires to the den for the entire winter. She gives birth to her cubs and nourishes them, often while in a state of sleep. To accomplish this feat, she bulks up during the feeding season and lives off fat reserves during the winter.

Interestingly, a hibernation bout is entered through slow-wave sleep (SWS), which thus suggests that some links exist to physiological processes involved in sleep. Body temperature drifts to ambient temperature until it is below 10°C. Metabolism shifts to lipid catabolism in a kind of slow starvation. Both REM sleep and wakefulness are suppressed. Interestingly, animals arouse from hibernation and promptly go into SWS, suggesting to some investigators that the hibernating animal is in fact sleep-deprived! Whatever the function of hibernation, the fact that the hibernator regularly arouses to go into SWS suggests that the function of SWS may not simply be to conserve energy, as hibernation would be a more efficient way to conserve energy.

Future directions

Further comparative and field research are needed to improve our understanding of sleep. In particular, it remains unclear whether ecological correlates of sleep durations found in well-studied groups, such as mammals, also account for patterns of sleep in other groups, such as birds, insects, reptiles, and fish. Similarly, more studies are needed on the links between sleep cycles, number of sleep bouts per day, and ecology as well as whether consolidating sleep into a single uninterrupted time period provides more efficient acquisitions of the benefits of sleep (Capellini, Nunn, McNamara, et al., 2008b). Other gaps in our knowledge include the effects of environmental seasonality on circadian rhythms and sleep, the links between sleep and infection in wild animals, quantification of the "opportunity costs" of sleep, and better understanding of how ecological factors constrain sleep. In the latter case, for example, could it be that the great energy requirements of some of the largest dinosaurs would have eliminated their opportunity for sleep? Models of sleep ecology coupled with digestive physiology could help to shed light on this question.

Another critical area for future research involves measures of sleep intensity. This could be achieved by tabulating those studies that provide quantitative data on SWA. Intensity indexes physiological need and is thus a target of natural selection. Avian sleep is similar to mammalian sleep in many ways except that SWA alone may not index sleep intensity in avian species as accurately as it does in mammalian species. Thus, a comparison of intensity expression in mammals versus birds may reveal potential additional sleep factors (e.g., depth or length of the sleep cycle) that are required for restorative effects of sleep in birds. Similarly, there is currently little understanding of what can be termed the evolutionary architecture of sleep: how variations in the physiological intensity of sleep, the length of sleep cycles, the length of sleep bouts and daily sleep durations, all interrelate. The determination of this architecture should lead to greater understanding of how constraints on overall sleep durations are accommodated at a physiological level.

Sleep function remains an enigma of modern biology. This is especially surprising in view of the substantial time animals and humans spend in this distinct physiological state, major similarities in its behavioral manifestations observed in different species, and typically deleterious effects of sleep deprivation on behavioral, autonomic, and cognitive functions. Although all this attests to sleep being a basic necessity, the question of whether sleep function is single and universal among diverse taxa remains to be determined. To reveal such common function requires in-depth investigation of the sleep processes in phylogenetically distant organisms that are adapted to different environments.

The study of variation in sleep expression among human populations also needs attention. It is likely that sleep duration, sleep phasing, and sleep expression varies dramatically across cultures, yet very few reliable data exist on this matter. Sleep of hunter-gatherers likely differs substantially from sleep of city dwellers in industrialized nations, for example. Surely ecologic conditions of a culture impacts sleep expression in that culture.

One last critical area for future research involves the collection of new data on sleep from wild mammals and birds. Most of the data in existing comparative databases comes from laboratory animals subjected to conditions different from those in the wild. Just as we might imagine that our own sleep would vary considerably if we were forced to sleep in the wild without shelter, easy access to food, or clothing, so can we imagine that animals will sleep differently when brought into conditions that are both more stressful (e.g., in terms of restraints or constant lighting) and less stressful (e.g., with constant access to food). Recent advances in EEG data loggers are providing new opportunities to collect data from wild animals that are ranging freely in their natural habitats (Rattenborg, Martinez-Gonzalez, & Lesku, 2009; Rattenborg, Martinez-Gonzalez, Lesku, et al., 2008; Rattenborg, Voirin, Vyssotski, et al., 2008). As these breakthrough methods are applied to more species of animals, we are likely to code at least some species as having different sleep durations. It will be interesting to see if new estimates of sleep from wild animals lead to different conclusions in comparative tests.

In summary, the study of sleep is at an exciting stage. Together with advances in the genetics and physiology of sleep, our understanding of sleep in different taxonomic groups is finally providing some answers to the question: Why do we sleep? Future research will undoubtedly build on the research synthesized here and elsewhere, and perspectives on functional aspects of sleep expression will change as this field of research develops.

References

Campbell, S. S., & Tobler, I. (1984). Animal sleep: A review of sleep duration across phylogeny. *Neuroscience and Biobehavioral Reviews, 8*, 269–300.

Capellini, I., Barton, R. A., Preston, B., McNamara, P., & Nunn, C. L. (2008a). Phylogenetic analysis of the ecology and evolution of mammalian sleep. *Evolution, 62*(7), 1764–1776.

Capellini, I., Nunn, C. L., McNamara, P., Preston, B. T., & Barton, R. A. (2008b). Energetic constraints, not predation, influence the evolution of sleep patterning in mammals. *Functional Ecology, 22*(5), 847–853.

Harvey, P. A., & Pagel, M. (1991). *The comparative method in evolutionary biology.* Oxford: Oxford University Press.

Kilduff, T. S., Krilowicz, B., Milsom, W. K., Trachsel, L., & Wang, L. C. (1993). Sleep and mammalian hibernation: Homologous adaptations and homologous processes? *Sleep, 16*(4), 372–386.

Lesku, J. A., Roth, T. C., II, Amlaner, C. J., & Lima, S. L. (2006). A phylogenetic analysis of sleep architecture in mammals: The integration of anatomy, physiology, and ecology. *American Naturalist, 168*(4), 441–453.

Lyamin, O. I., Manger, P. R., Ridgway, S. H., Mukhametov, L. M., & Siegel, J. (2008). Cetacean sleep: An unusual form of mammalian sleep. *Neuroscience and Biobehavioral Reviews, 32*(8), 1451–1484.

McNamara, P., Capellini, I., Harris, E., Nunn, C. L., Barton, R. A., & Preston, B. (2008). The phylogeny of sleep database: A new resource for sleep scientists. *The Open Sleep Journal, 1*, 11–14.

Meddis, R. (1983). The evolution of sleep. In A. Mayes (Ed.), *Sleep mechanisms and functions in humans and animals: An evolutionary perspective* (pp. 57–106). Berkshire, England: Van Nostrand Reindhold.

Moorcroft, W. H. (2003). *Understanding sleep and dreaming.* New York: Springer.

Preston, B. T., Capellini, I., McNamara, P., Barton, R. A., & Nunn, C. L. (2009). Parasite resistance and the adaptive significance of sleep. *BMC Evolutionary Biology*, *9*, 7.

Rattenborg, N. C., Martinez-Gonzalez, D., & Lesku, J. A. (2009). Avian sleep homeostasis: Convergent evolution of complex brains, cognition, and sleep functions in mammals and birds. *Neuroscience and Biobehavioral Reviews*, *33*(3), 253–270.

Rattenborg, N. C., Martinez-Gonzalez, D., Lesku, J. A., & Scriba, M. (2008). A bird's-eye view of sleep. *Science*, *322*(5901), 527.

Rattenborg, N. C., Voirin, B., Vyssotski, A. L., Kays, R. W., Spoelstra, K., Kuemmeth, F., et al. (2008). Sleeping outside the box: Electroencephalographic measures of sleep in sloths inhabiting a rainforest. *Biology Letters*, *4*(4), 402–405.

Roth, T. C., Lesku, J. A., Amlaner, C. J., & Lima, S. L. (2006). A phylogenetic analysis of the correlates of sleep in birds. *Journal of Sleep Research*, *15*, 395–402.

Trevathan, W. R., Smith, E. O., & McKenna, J. (Eds.). (1999). *Evolutionary medicine and health: New perspectives.* New York: Oxford University Press.

Trevathan, W. R., Smith, E. O., & McKenna, J. (Eds.). (2007). *Evolutionary medicine and health: New perspectives.* New York: Oxford University Press.

Zepelin, H. (1989). Mammalian sleep. In M. H. Kryger, T. Roth, & W. C. Dement (Eds.), *Principles and practices of sleep medicine* (pp. 30–49). Philadelphia: W. B. Saunders.

1

Ecological constraints on mammalian sleep architecture

ISABELLA CAPELLINI, BRIAN T. PRESTON, PATRICK MCNAMARA,
ROBERT A. BARTON, AND CHARLES L. NUNN

Introduction: sleep and ecology

All mammals so far studied experience some form of sleep. When mammals are sleep-deprived, they generally attempt to regain the lost sleep by exhibiting a "sleep rebound," suggesting that sleep serves important functions that cannot be neglected (Siegel, 2008; Zepelin, 1989; Zepelin, Siegel, & Tobler, 2005). When sleep deprivation is enforced on individuals, it is accompanied by impaired physiological functions and a deterioration of cognitive performance (Kushida, 2004; Rechtschaffen, 1998; Rechtschaffen & Bergmann, 2002). In the rat, prolonged sleep deprivation ultimately results in death (Kushida, 2004; Rechtschaffen & Bergmann, 2002). Together, these observations suggest that sleep is a fundamental requirement for mammalian life, and much research has focused on identifying the physiological benefits that sleep provides (Horne, 1988; Kushida, 2004).

Are there also costs associated with sleep? If so, what are the selective pressures that constrain the amount of time that individuals can devote to sleep? Sleep is probably associated with "opportunity costs" because sleeping animals cannot pursue other fitness-enhancing activities, such as locating food, maintaining social bonds, or finding mates. Sleeping animals may also pay direct costs. For example, sleep is a state of reduced consciousness, and thus sleeping individuals are less able to detect and escape from approaching predators (Allison & Cicchetti, 1976; Lima, Rattenborg, Lesku, et al., 2005). These ecological factors are likely to be important constraints on sleep durations and may also affect how sleep is organized over the daily cycle.

In this chapter, we review the evidence for how ecological factors, including predation risk and foraging requirements, might shape patterns of sleep among mammals. We also highlight the need for more research on the degree to which animals can exhibit flexibility in their sleep requirements, as such plasticity could provide a means to overcome constraints, particularly when the costs associated with sleep vary on daily or seasonal time scales. We begin by discussing if the available data are informative and appropriate for studying the role of ecology in the evolution of sleep architecture. We then move on to review how different characteristics of sleep have evolved alongside one another, as these traits form the foundation for our discussion of ecological constraints that follows.

We restrict our discussion to terrestrial mammals and exclude monotremes, such as the platypus (*Ornithorhynchus anatinus*) and echidna (*Tachyglossus aculeatus* and *Zaglossus* sp.). Aquatic mammals (Cetacea, Pinnipedia, and Sirenia), in fact, exhibit a different sleep architecture (with facultative or obligatory unihemispheric sleep; Rattenborg & Amlaner, 2002; Siegel, 2004), and it is still uncertain whether monotremes possess two distinct sleep states – rapid-eye-movement (REM) and non–REM (NREM) sleep – as is observed in most other mammals (Zepelin et al., 2005). We note, however, that the dramatic differences in sleep characteristics of terrestrial and aquatic mammals provide evidence for the claim that ecology influences sleep architecture. In aquatic environments, mammals appear to forego REM sleep – or at least REM indices are truncated in aquatic species relative to the range of values seen in terrestrial species – and unihemispheric NREM sleep is found (Zepelin et al., 2005). Some authors argue that the evolution of unihemispheric sleep and suppression of REM sleep, with its associated paralysis, allows cetaceans and eared seals to maintain the motor activity necessary to surface and breathe (Mukhametov, 1984, 1995), while others suggest unihemispheric sleep might facilitate predator detection (reviewed in Rattenborg, Amlaner, & Lima, 2000) or help balance heat loss to the water by constantly swimming (Pillay & Manger, 2004).

Sleep and laboratory conditions

The large majority of sleep estimates have been obtained from laboratory animals, mostly because of the difficulties associated with recording sleep times using electroencephalographic (EEG) equipment in the wild. This raises two potential challenges for comparative studies that aim to understand the evolution of sleep architecture. First, different laboratory conditions and procedures may impact sleep times, creating error in comparative datasets composed of data from different research groups. Second, it is possible that sleep times in a laboratory setting do not reflect sleep times in the wild (Bert, Balzamo, Chase, et al., 1975; Campbell & Tobler, 1984; Rattenborg, Voirin, Vissotski, et al., 2008). In addition to these concerns about data quality, comparative studies must consider the possibility that more closely related species exhibit more similar trait values, which can inflate rates of type I errors (Felsenstein, 1985; Garland, Bennett, & Rezende, 2005; Harvey & Pagel, 1991; Martins & Garland, 1991; Nunn & Barton, 2001). Thus, comparative studies on any biological trait need to assess whether there is a "phylogenetic signal" in the data (Blomberg & Garland, 2002; Blomberg, Garland, & Ives, 2003; Freckleton, Harvey, & Pagel, 2002), and if so, to control for the resulting nonindependence statistically. In this section, we address the first two issues, while the importance of accounting for species' shared evolutionary history is discussed by Nunn et al. in Chapter 6 of this volume.

First, concerning data quality, the data collected on different species must be comparable for cross-species evolutionary studies to be informative. This is particularly important in the case of sleep studies, given that different housing conditions and measurement procedures have the potential to influence sleep duration estimates (Berger, 1990; Bert et al., 1975; Campbell & Tobler, 1984; Siegel, 2005). For example, total daily sleep was twice as high in guinea pigs (*Cavia porcellus*) that were habituated to laboratory conditions as compared to nonhabituated animals (Jouvet-Monier & Astic, 1966). Other factors that might influence the comparability of data in different studies include the amount of time over which sleep is examined, ad libitum feeding conditions, photoperiod, ambient temperature, whether experimental animals were restrained during the recording session, and finally whether EEG methods were used (Campbell & Tobler, 1984).

Using an updated comparative dataset on adult sleep quotas for 127 mammals (McNamara, Capellini, Harris, et al., 2008), we assessed how laboratory procedures influence estimates of sleep quotas and total sleep time in terrestrial mammals by comparing sleep durations from the same species that were obtained under different conditions (Capellini, Barton, McNamara, et al., 2008a). We found that studies that recorded sleep for less than 12 hours significantly underestimated sleep times; similarly, EEG estimates of sleep duration were higher than behavioral estimates (Capellini et al., 2008a). Surprisingly, there was no statistically significant difference between studies in relation to habituation and restraint. However, a small sample size (n = 5) might explain the lack of significance; further tests should be carried out when more data for these and other variables become available.

Importantly, when we investigated the evolution of mammalian sleep architecture with a "restricted" dataset of sleep estimates collected under consistent laboratory conditions (EEG estimates with at least 12 hours recording time), the pattern of association between sleep and several variables of interest changed greatly (Capellini et al., 2008a), thus casting doubt on a number of inferences that had been drawn from previous analyses. For example, a previously reported positive relationship between mammalian brain sizes and REM sleep durations (Lesku, Roth, Amlaner, et al., 2006) became nonsignificant when data collected under consistent procedures were used.

Second, concerning the ecological validity of sleep estimates collected in the laboratory, comparative studies assume that sleep durations recorded in the laboratory reflect sleep times in the wild. This assumption is justified by claims that either sleep is a functional requirement, and therefore has little variability in its expression, or that sleep in the laboratory is an estimate of the optimal sleep need for a species (Campbell & Tobler, 1984). Under the latter scenario, wild animals may sleep less than is recorded in the laboratory, because ecological and social factors might disrupt and reduce the time available for sleep. In the wild, sleep times may more closely represent the minimal sleep requirement of an individual (Rattenborg et al., 2008).

Until more data on sleep durations of different species in the wild have been recorded, this issue cannot be resolved, as the evidence that is currently available is conflicting. Saarikko and Hanski (1990), for example, have shown that total sleep time did not vary between laboratory and wild conditions in three species of shrews (*Sorex araneus, S. isodon, S. caecutiens*). They found differences in the overall activity level, however, with wild shrews spending more time traveling to and from foraging sites at the expense of time spent resting quietly. In contrast, a recent landmark study on sloths (*Bradypus variegatus*) (Rattenborg et al., 2008) found that wild sloths appeared to sleep less (9.63 h/day) than sloths in a laboratory setting (15.85 h/day) (Galvão de Moura Filho, Huggins, & Lines, 1983). This finding was obtained by fitting minimally invasive EEG recorders on wild animals, and the authors concluded that this disparity was caused by differing conditions in the laboratory and in natural settings. The same study found that the EEG structure of REM and NREM sleep did not vary between laboratory and wild animals.

While the ability to record sleep in the wild is a major advance, the interpretation of the findings of Rattenborg et al. (2008) must be treated with some caution. The total sleep time estimated in the laboratory was based on the average sleep durations of adults and an unspecified number of juveniles (Galvão de Moura Filho et al., 1983) and, because sleep times in mammals can be much higher in juveniles than in adults (Zepelin et al., 2005), it is unclear to what extent the greater sleep durations recorded in the laboratory study were due to the inclusion of these younger animals. Thus, it remains an open question whether laboratory procedures provide appropriate estimates of sleep times in the wild.

Sleep architecture: Correlated evolution of sleep durations, sleep cycle length, and phasing of sleep

Mammalian sleep is composed of two distinct states – REM sleep and NREM sleep – and these states alternate in cycles over a sleep bout (Zepelin, 1989; Zepelin et al., 2005). REM and NREM sleep exhibit contrasting physiological

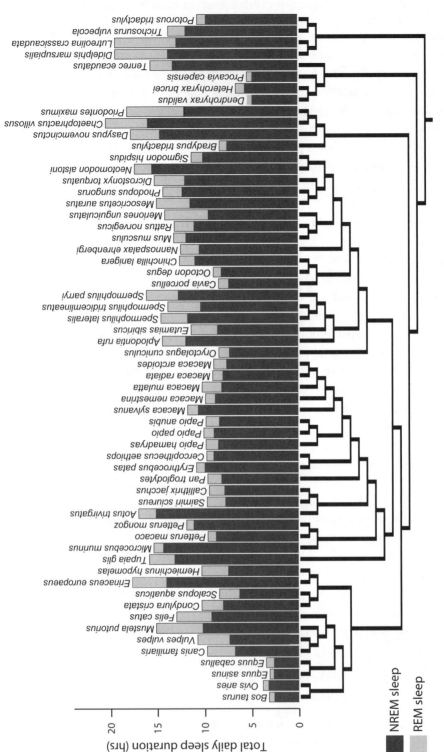

Figure 1.1. Interspecific variation in total sleep time in terrestrial mammals. Although more distantly related species might still have similar sleep durations by convergent evolution, closely related species are more similar to one another in sleep durations than expected by chance (REM sleep shown in light grey and NREM sleep in dark grey). Statistical analyses have proven that sleep times thus exhibit a strong "phylogenetic signal" (see Capellini et al., 2008a), i.e., phylogeny correctly predicts the pattern of variation in sleep architecture among the species. This suggests that some of the variance in sleep architecture is due to common ancestry. (Phylogenetic tree and sleep quotas from Capellini et al., 2008a.)

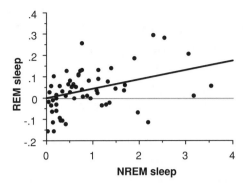

Figure 1.2. Correlated evolution of REM and NREM sleep durations. Phylogenetically independent contrasts analysis showed that REM and NREM sleep times are positively associated in terrestrial mammals ($t_{58} = 4.47$, $R^2 = 0.26$, $P < 0.0001$). Only species with EEG estimates and a recording time of at least 12 hours were included in the analysis. (From Capellini et al., 2008a.)

characteristics, which have led scientists to suggest that the two sleep states have distinct functions (Rechtschaffen, 1998; Siegel, 2005; Zepelin, 1989; Zepelin et al., 2005). The term "sleep architecture" encompasses how much time is spent in REM and NREM sleep (sleep quotas), the duration of the REM–NREM sleep cycles, and how sleep is organized and distributed across the daily cycle (phasing of sleep).

Mammals vary extensively in all these sleep traits (see Figure 1.1). For example, average total daily sleep duration ranges from 3 hours in the donkey (*Equus asinus*) (Ruckebush, 1963) to 20 hours in armadillos (*Chaetophractus villosus*) (Affani, Cervino, & Marcos, 2001), while average sleep cycles vary from 6 minutes in the chinchilla (*Chinchilla lanigera*) (Van Twyver, 1969) to 90 minutes in humans and chimpanzees (*Pan troglodytes*) (Tobler, 1995). Finally, there is great interspecific variation in how sleep time is organized within the activity budget. Sleep can be concentrated mostly in one bout per 24 hours (monophasic sleep) or divided into multiple bouts interrupted by waking phases (polyphasic sleep) (Ball, 1992; Stampi, 1992). This remarkable diversity in sleep architecture is probably due to interspecific differences in both the benefits and the costs of sleep.

How do these different characteristics of sleep architecture evolve with one another and what can we infer from these patterns? Across terrestrial mammals, we found that NREM and REM sleep quotas increase with one another (see Figure 1.2), and most of total sleep time was composed of NREM sleep (Capellini et al., 2008a). This pattern of correlated evolution between the two sleep states is in agreement with the results of physiological studies showing that REM and NREM sleep are physiologically integrated (Ambrosini & Giuditta, 2001; Benington & Heller, 1994, 1995; Steiger, 2003; Van Cauter, Plat, & Copinschi, 1998). For example, some authors suggest that REM partially reverses some of the processes occurred during NREM sleep (such as neural activation/deactivation of different brain regions or regulation of hormone release) (Benington & Heller, 1994, 1995), while others focus on the integration of NREM and REM sleep in memory processing (Ambrosini & Giuditta, 2001). Our results reveal that this integration remains even when examining patterns at the cross-species level, at least in terms of correlations among sleep state durations. Consistent with the "constraints" framework presented here, these results also suggest that when animals have more time available for sleep, they increase both sleep states.

Our comparative tests revealed that mammals that sleep polyphasically and in short REM–NREM sleep cycles have longer NREM (but not REM) sleep quotas than those that sleep monophasically or with longer sleep cycles (see Figure 1.3a to d) (Capellini, Nunn, McNamara, et al., 2008b). In addition, polyphasic sleep and short sleep cycles are associated with each other (see Figure 1.3e) and with smaller body size, and polyphasic sleep is the ancestral state in mammals (Capellini et al., 2008b).

Laboratory studies have shown that both monophasic sleepers and polyphasic sleepers exhibit "light" and "deep" NREM stages (with some groups having up to four different NREM stages – such as primates; e.g., Berger & Walker, 1972; Bert, Pegram, Rhodes, et al., 1970; Lesku, Bark, Martinez-Gonzalez, et al., 2008; Ursin, 1968; Wauquier, Verheyen, Van Den Broeck, et al., 1979). Therefore we proposed that monophasic sleep and sleeping with longer REM–NREM sleep cycles may be favored evolutionarily because they represent a more efficient way to gain the benefits of sleep. Organizing sleep into longer cycles across one daily bout would reduce the amount of time that animals spend in the lighter stages of sleep, which appears to be necessary to achieve the deeper and probably more beneficial sleep phase (e.g., slow-wave sleep, or SWS, during NREM sleep). This effect arises because partitioning one SWS phase into more bouts or cycles would require a phase of light sleep for each additional deep sleep bout or cycle (Figure 1.4).

Therefore monophasic sleepers and species with long sleep cycles may be able to gain more benefits from the same overall time asleep, as compared to polyphasic or short-cycle sleepers (Ball, 1992; Capellini et al., 2008b). This hypothesis could be investigated by examining the efficiency of monophasic and polyphasic sleep in the laboratory and by testing the degree to which light sleep stages can be skipped or compressed in time. A recent study on the plastic response of sleep architecture to predation in rats, however, showed that both REM and NREM sleep times were reduced after encounters with predators, while the time in light sleep stages was unaffected (Lesku et al., 2008, and see below).

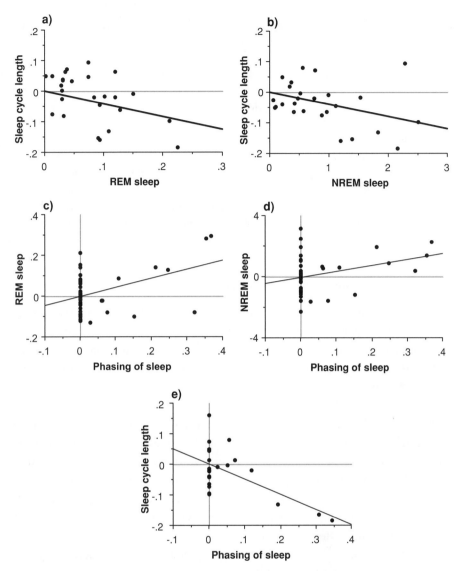

Figure 1.3. Correlated evolution of sleep durations with phasing of sleep and sleep-cycle length. Phylogenetically independent contrasts of sleep-cycle length with (a) REM sleep ($t_{25} = -2.93$, $R^2 = 0.26$, $P = 0.007$) and (b) NREM sleep ($t_{25} = -3.33$, $R^2 = 0.31$, $P = 0.003$). Phasing of sleep with (c) REM sleep ($t_{43} = 3.56$, $R^2 = 0.23$, $P = 0.001$; after bootstrapping: p = 0.132), (d) NREM sleep ($t_{43} = 2.35$, $R^2 = 0.11$, $P = 0.024$), and (e) sleep-cycle length ($t_{22} = -4.07$, $R^2 = 0.43$, $P = 0.001$; after bootstrapping: $P = 0.054$). Only species with EEG estimates and a recording time of at least 12 hours were included in the analysis. Phasing of sleep was coded and treated as a dummy variable in the comparative tests (0 = monophasic sleep; 1 = polyphasic sleep). (From Capellini et al., 2008b.)

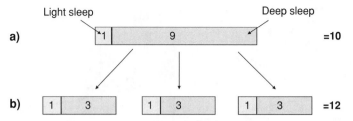

Figure 1.4. Sleep durations and efficiency. Total sleep time increases when deep sleep is fragmented into more bouts or shorter sleep cycles because for each new bout or episode of "deep sleep," a new "episode" in transitional light sleep is required. Thus, assuming that sleep intensity is held constant, even though the overall time in deep sleep is equivalent in (a) and (b), total sleep time is greater when sleep is fragmented (b). This might explain why monophasic sleep and sleeping in longer cycles are associated with shorter NREM sleep durations.

Sleep architecture and predation

Predation is believed to be among the most influential factors shaping mammalian sleep, but the nature of its influence is still debated. Some authors have argued that sleep may have evolved to protect animals from predators by making them less conspicuous when other activities are dangerous or unprofitable (the "immobilization hypothesis"; Meddis, 1975; Zepelin et al., 2005). However, we agree with the alternative view that predation – in combination with the safety level of the sleep site – is likely to represent a constraint on how much time individuals can spend asleep. Responsiveness to external stimuli is reduced during sleep (Zepelin et al., 2005); thus a sleeping animal is less aware of potential threats than an animal that is quietly resting (Tobler, 2005). Sleep should therefore be associated with a *greater* risk of predation relative to quiet resting (Allison & Cicchetti, 1976; Lima et al., 2005), particularly when an animal is sleeping in an open area with no shelter (see below).

If sleep is a dangerous state, sleep time is predicted to be reduced in species that face higher predation risk; for example, (1) in species that sleep in more exposed sleeping sites (e.g., on the ground in open grassland) as compared to species that sleep in fully enclosed sleeping sites (e.g., tree holes or dens), and (2) in "prey" relative to "predators." These predictions have been supported by various studies that developed indices of animals' vulnerability while sleeping and diet-based indices as surrogates of trophic level (Allison & Cicchetti, 1976; Capellini et al., 2008a). Both REM and NREM sleep durations are lower when animals sleep in more exposed and vulnerable sites and have a more herbivorous diet (see Figure 1.5; but see next section for the relationship between diet and sleep time). These findings indicate that total sleep time is constrained in species

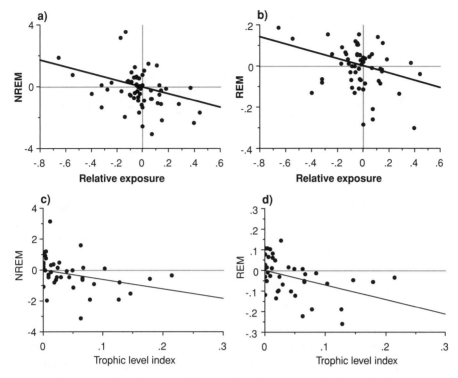

Figure 1.5. Sleep durations, diet, and sleep site exposure. Phylogenetically independent contrasts of NREM (a and c) and REM sleep (b and d) with contrasts of sleep site exposure index after controlling for body mass (NREM: $t_{57} = -2.76$, $R^2 = 0.12$, $P = 0.008$; REM: $t_{57} = -2.57$, $R^2 = 0.10$, $P = 0.013$) and a diet-based trophic level index (NREM: $t_{39} = -2.61$, $R^2 = 0.15$, $P = 0.013$; REM: $t_{39} = -3.71$, $R^2 = 0.26$, $P < 0.0001$). Sleep site exposure was a three-states variable coded as 1 = fully enclosed sleeping sites (e.g., burrows and tree holes); 2 = partially exposed sites (e.g., vegetation on the ground or in trees); 3 = fully exposed sites (e.g., in open habitats with no protection). Trophic level was a diet-based index (data from Lesku et al., 2006) coded as 1 = diet based exclusively on vertebrates; 2 = small insects; 3 = large insects; 4 = entirely herbivorous diet. (From Capellini et al., 2008a.)

that experience higher predation risk (Allison & Cicchetti, 1976; Capellini et al., 2008a).

The impact of ecological factors on the evolution of sleep architecture in mammals appears to be more complex than has so far been appreciated. Consider, for example, the expectation that an animal's predation risk while sleeping should decrease as a function of group size, owing to detection and dilution effects (reviewed in Caro, 2005). One might therefore predict that individuals that commonly sleep in groups should suffer lower predation risk than those sleeping alone and should thus be less constrained in their opportunity to sleep. Contrary to this

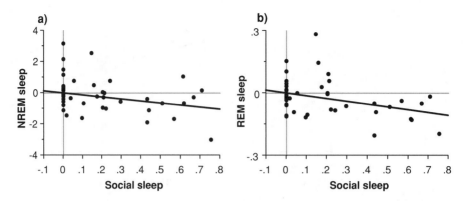

Figure 1.6. Sleep durations and social sleep behavior. Phylogenetically independent contrasts of NREM and REM sleep times are shorter when the degree of social sleep behavior is greater (NREM: $t_{42} = -2.39$, $R^2 = 0.12$, $P = 0.021$; REM: $t_{42} = -3.09$, $R^2 = 0.19$, $P = 0.004$). Social sleep behavior was coded as: 1 = both sexes sleep alone; 2 = females but not males with socially (sleeping with the offspring was not considered social sleep unless it was prolonged into adulthood); 2 = both males and females sleep socially. (From Capellini et al., 2008a.)

prediction, however, both REM and NREM sleep quotas are significantly lower in species that sleep socially as compared to those in which individuals sleep alone (Capellini et al., 2008a) (see Figure 1.6).

This result suggests that social species face a trade-off between socializing and sleeping, raising the intriguing possibility that sociality might have influenced the evolution of sleep architecture. Alternatively, individuals that sleep socially may perceive their immediate surrounding as safer and could therefore increase the intensity of sleep, thus gaining the benefits of sleep more rapidly (Capellini et al., 2008a). Further studies are needed to test the idea that social species sleep less but more efficiently and to evaluate if and to what extent sociality constrains the time available for sleep.

In addition to constraining sleep durations, predation may influence how the benefits of sleep are obtained and how sleep is organized; specifically, predation may influence the length of the REM–NREM sleep cycle and the number and duration of sleep bouts per day (Lima et al., 2005; Van Twyver & Garrett, 1972; Voss, 2004). Based on the observation that episodes of REM sleep at the end of a cycle are often followed by brief arousals to waking, greater predation pressure may lead to shorter sleep cycles, resulting in more opportunities to monitor the surrounding environment for predators (Lima et al., 2005; Van Twyver & Garrett, 1972; Voss, 2004). Applying a similar argument to phasing of sleep, the number of sleep bouts per day should be greater in species that face higher predation risk, because a polyphasic sleep pattern would avoid prolonged time in a vulnerable state of low consciousness (Ball, 1992; Capellini et al., 2008b; Stampi,1992; Tobler, 1989).

The hypothesis that increased perceived risk of predation leads to more frequent arousals per sleep bout has found support in studies at the individual level in wild birds (Gauthier-Clerc, Tamisier, & Cezilly, 1998, 2000, 2002; Lendrem, 1983, 1984) and laboratory rats (Broughton, 1973; Lesku et al., 2008; see below). In a comparative analysis in terrestrial mammals, however, both sleep-cycle length and the phasing of sleep were unrelated to surrogate measures of predation risk (sleep site exposure, social sleep behavior, and trophic level) that have been shown to impact sleep durations (Capellini et al., 2008b). This result may not be surprising in the light of studies on vigilance behavior in the wild, which show that a high scanning frequency seems to be employed to detect approaching predators (from a few seconds to a few minutes; Caro, 2005). These scanning rates would not be achieved even with the shortest sleep cycles that have been recorded in mammals, and thus shorter sleep cycles may not be an effective way to detect approaching predators. We argue that the species' trophic niche, energetics, and body mass may instead explain the evolution of the phasing of sleep (Capellini et al., 2008b) (see next section).

Other aspects of predation pressure may also influence the phasing of sleep. For example, species that are predated by generalist predators may be able to adjust the timing of their sleep period to minimize the risk of predation (Fenn & Macdonald, 1995; Lima et al., 2005). In this respect, a polyphasic sleep pattern is believed to be advantageous because it may be associated with a more flexible time budget (Lima et al., 2005; Tobler, 1989). Conversely, a species that is mostly predated by a specialist predator would benefit little from modifying its activity pattern, because the predator would adjust its own activity in accordance with that of the prey (Lima et al., 2005). Finally, drowsiness may represent a "state of vigilance with light sleep" that allows species under intense predation pressure to gain some of the benefits of sleep without the additional vulnerability associated with deeper sleep stages (Lima et al., 2005; Makeig, Jung, & Sejnowshi, 2000; Noser, Gygax, & Tobler, 2003).

Sleep, trophic niche, and energetics

Trophic niche might represent another important ecological factor that affects sleep architecture. We previously mentioned that "predators" sleep for longer periods than "prey"; this result was based on diet-based indices used as a proxy for trophic level (Allison & Cicchetti, 1976; Capellini et al., 2008a; Lesku et al., 2006). However, the finding that a more herbivorous diet is associated with shorter sleep times is also compatible with the hypothesis that trophic niche dictates how

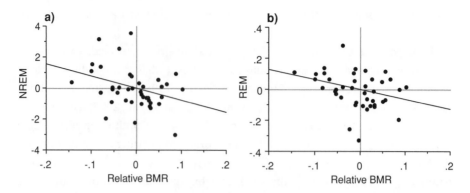

Figure 1.7. Sleep and energetics. Phylogenetically independent contrasts of NREM and REM sleep with basal metabolic rate (used as a proxy for total daily energy expenditure) after controlling for allometry (NREM: $t_{40} = -2.32$, $R^2 = 0.12$, $P = 0.026$; REM: $t_{40} = -2.08$, $R^2 = 0.10$, $P = 0.044$). (From Capellini et al., 2008a.)

much time is necessary to find, acquire, and process food; thus there might be trade-offs between foraging time (in this broad sense) and sleep time (Allison & Cicchetti, 1976; Capellini et al., 2008a; Elgar, Pagel, & Harvey, 1988). Although a direct comparative test of this hypothesis has not yet been carried out, primates with a more folivorous diet spend more time resting (which includes both quiet resting and sleep) relative to species with a frugivorous diet (Oates, 1987; see also Chapter 6 in this volume). This is probably because fruits are more dispersed in the environment and therefore more time is needed to find them.

Acerbi, McNamara, and Nunn (2008) argue that phasing of sleep and sleep durations are potentially influenced by how trophic resources are distributed in the environment relative to sleep sites. Using an agent-based model, the authors showed that when trophic resources are distributed in discrete patches and sleep sites are more distant from foraging sites, sleep time is reduced. Furthermore, sleep tends to be concentrated in one bout per day, so that travel time between foraging and sleep sites is minimized. This intriguing proposal has yet to be validated with field studies and comparative tests.

The energy requirement of an animal is an important biological trait that may link foraging effort and sleep time. Specifically, Allison and Cicchetti (1976) suggested that large-bodied species with high energy demands have less time available for sleep because, with their greater metabolic needs, these species must spend a greater proportion of the daily cycle foraging. Although sleep durations are unrelated to body mass after controlling for phylogeny (Capellini et al., 2008a; Lesku et al., 2006), comparative tests have shown that REM and NREM sleep time are inversely related to basal metabolic rate (a surrogate measure of total daily energy expenditure) after controlling for body mass (Capellini et al., 2008a) (see Figure 1.7).

In other words, species that have a higher metabolic rate than expected for their size sleep less. This result provides support for the hypothesis that a trade-off exists between time that is available for foraging and time that can be spent sleeping.

Finally, polyphasic sleep is associated with small body mass (Capellini et al., 2008b). We have suggested that this may be due to the limited fat reserves and high mass-specific metabolism of small mammals (Blackburn & Hawkins, 2004; Lindstedt & Boyce, 1984; Macdonald, 2006; Withers, 1992), which forces them to feed more frequently; hence they cannot spend long periods of time asleep and must instead adopt a polyphasic sleep pattern to meet their daily sleep and energy requirements. In agreement with this interpretation, shrews alternate short foraging and sleep (or rest) bouts, possibly because their small gut capacity limits ingestion rate (Saarikko, 1992; Saarikko & Hanski, 1990).

How plastic is sleep architecture in mammals?

Can individual mammals modify their sleep patterns in response to changes in environmental, ecological, and social factors? Or is sleep architecture relatively inflexible? We all have firsthand experience with pulling an "all nighter" when the need arises, and similar kinds of flexibility are likely to occur in wild animals. The majority of studies on plastic responses of sleep have been carried out in wild birds. These studies show that when birds perceive themselves to be under higher predation risk, they sleep less, arouse more frequently, and allocate more time to unihemispheric sleep at the expense of bihemispheric sleep (Gauthier-Clerc et al., 1998, 2000, 2002; Lendrem, 1983, 1984; Rattenborg, Lima, and Amlaner, 1999a, 1999b). Similarly, laboratory studies have shown that birds sleep less around the time of their seasonal migration, when they have to traverse large distances with little opportunity for sleep (Fuchs, Haney, Jechura, et al., 2006; Rattenborg et al., 2004).

In contrast to the growing literature on avian sleep flexibility, only two studies have assessed how mammals adjust their sleep patterns in response to increased predation risk (Broughton, 1973; Lesku et al., 2008). These revealed that sleep times are reduced and arousals to waking are more frequent in experimental rats after they are exposed to cats or humans mimicking predation in the laboratory (Broughton, 1973; Lesku et al., 2008). Sleep onset was delayed after the encounter with potential predators, and both NREM and REM sleep quotas were reduced. However, one study found that the mechanism by which this was achieved was different for each sleep state (Lesku et al., 2008). While NREM sleep time was reduced by shortening the duration of NREM sleep episodes but not their numbers (Figure 1.8), REM time was decreased by reducing the number but not the duration of REM sleep episodes, especially during early sleep bouts (Figure 1.9).

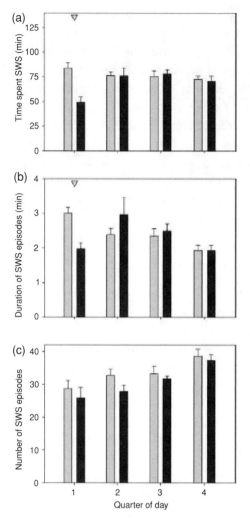

Figure 1.8. Plastic response of NREM sleep in rats after an encounter with humans mimicking predation. Relative to baseline condition (grey), rats that encountered a predator (black) showed reduced total time in deep SWS during NREM sleep (a), specifically by reducing NREM sleep episode length (b), but did not reduce the number of NREM sleep episodes (c). Significant differences between baseline and postencounter sleep are denoted by a triangle over the pairwise comparison. (From Lesku et al., 2008.)

Lesku and colleagues (2008) concluded that the onset of REM sleep is delayed because it is the most vulnerable sleep state – that is, because of higher arousal thresholds and the loss of muscle tone during REM. An alternative explanation, however, is that REM sleep is "less physiologically important" than NREM sleep (Horne, 1988). Thus, under selective pressure to reduce time asleep, REM sleep would be sacrificed to a greater extent than NREM sleep. Interestingly, time in

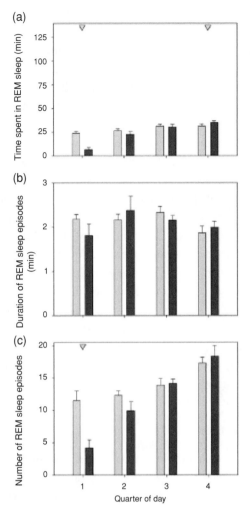

Figure 1.9. Plastic response of REM sleep in rats after an encounter with humans mimicking predation. Total time in REM sleep was reduced in rats exposed to predation risk (grey) relative to baseline (black) (a), specifically by decreasing the number of REM episodes (c) but not their length (b). Significant differences between baseline and postencounter sleep are denoted by a triangle over the pairwise comparison. (From Lesku et al., 2008).

light sleep stages – which are supposed to be less restorative than deep SWS sleep – appeared to be unaffected by predator encounters (Lesku et al., 2008). This might support our suggestion that transitional stages from waking into deep sleep cannot be compressed in time or skipped and therefore that monophasic sleep would be more efficient than polyphasic sleep (Capellini et al., 2008b) (also see above). Further studies should assess whether this plastic response in sleep architecture in laboratory rats represents a common response to predation across all mammals.

Finally, parasites represent an important ecological pressure, and various experimental studies have shown that sleep architecture is altered in response to an infection. In general, time spent in NREM sleep – specifically in SWS – increases with increasing body temperature in response to infection, while time in REM sleep is decreased (Bryant, Trinder, & Curtis, 2004; Majde, 2005). Further links between sleep and the immune system are suggested by the effect of sleep deprivation, which causes perturbations in immune function and effectiveness that may ultimately lead to death (Bryant et al., 2004; Majde, 2005). Thus, it may be that sleep serves an immune function and that flexibility in sleep architecture is required in order to meet the changing demands on the immune system. Comparative evidence supports this hypothesis, as longer REM and NREM sleep durations are associated with both greater numbers of immune cells and lower infection levels, indicating that species that have evolved longer sleep durations have been able to enhance their immune defenses (Preston, Capellini, McNamara, 2009). There is clearly a need to improve our understanding of how parasites have influenced the evolution of sleep architecture, and how facultative changes in sleep architecture might boost an animal's ability to withstand infection. Future studies should also explore how socioecological factors influence the likelihood of infection and how this in turn might affect the evolution of sleep architecture.

Conclusions and future directions

Recent comparative research has reevaluated the importance of ecology in the evolution of sleep. These studies have shown that predation pressure, trophic niche, and energy demands can, in part, explain patterns of interspecific variation in mammalian sleep architecture (Capellini et al., 2008a,b). Thus the ecological niche that animals inhabit can exert significant evolutionary pressure on sleep durations as well as on how sleep is organized across the daily cycle.

Further comparative and field research is needed to improve our understanding of sleep. In particular, it remains unclear to what extent socioecological factors and activity period affect mammalian sleep architecture (see Chapter 6 in this volume). The possibility that some mammals are able to sleep more efficiently by consolidating their sleep into a single uninterrupted time period has yet to be assessed and could represent a major advance in our understanding of mammalian sleep (Capellini et al., 2008b). Other gaps in our knowledge include the extent to which sleep varies in mammals that experience environmental seasonality (Barre & Petter-Rousseaux, 1988; Palchykova, Deboer, & Tobler, 2003) and how sleep might be constrained during the breeding season or during long-distance migration (as in birds, Fuchs et al., 2006; Rattenborg et al., 2004). Physiological measures of sleep intensity are needed to evaluate whether sleep efficiency coevolved with sleep durations and the phasing of sleep.

In closing, comparative studies of mammalian sleep have begun to reveal fundamental links between ecology and sleep architecture. With the development of new techniques to characterize sleep in the wild (Rattenborg et al., 2008), major advances in our understanding of how ecology has influenced sleep are likely to be just around the corner. Comparative analyses are certain to play an integral role in these advances, especially as data accumulate for more mammalian species.

Acknowledgments

We are grateful to Erica Harris and Nikita Patel for their help on the sleep database and logistic support; Lana Ruvinskaya for translating papers from Russian; Patrik Lindenfors, Sean O'Hara, Joann Chang, Meike Mohneke, and Timothy Morrison for help collecting the data on the ecological traits; Mark Pagel and Andy Purvis for advice on the comparative analysis, and John Lesku for comments on an earlier draft of this manuscript. This work was supported by NIMH (grant number 1R01MH070415-01A1) and the Max Planck Society (C. N. and B. P.). We thank Wiley-Blackwell and Elsevier for permission to reproduce the figures from Capellini et al. (2008a,b), and Lesku et al. (2008).

References

Acerbi, A., McNamara, P., & Nunn, C. L. (2008). To sleep or not to sleep: The ecology of sleep in artificial organisms. *BMC Ecology*, *8*, 10.

Affani, J. M., Cervino, C. O., & Marcos, H. J. A. (2001). Absence of penile erections during paradoxical sleep. Peculiar penile events during wakefulness and slow-wave sleep in the armadillo. *Journal of Sleep Research*, *10*, 219–228.

Allison, T., & Cicchetti, D. V. (1976). Sleep in mammals: Ecological and constitutional correlates. *Science*, *194*, 732–734.

Ambrosini, M. V., & Giuditta, A. (2001). Learning and sleep: The sequential hypothesis. *Sleep Medicine Reviews*, *5*, 477–490.

Ball, N. J. (1992). The phasing of sleep in animals. In C. Stampi (Ed.), *Why we nap. Evolution, chronobiology, and functions of polyphasic and ultrashort sleep* (pp. 31–49). Boston: Birkhäser.

Barre, V., & Petter-Rousseaux, A. (1988). Seasonal variation in sleep-wake cycle in *Microcebus murinus*. *Primates*, *29*, 53–64.

Benington, J. H., & Heller, H. C. (1994). Does the function of REM sleep concern non-REM sleep or waking? *Progress in Neurobiology*, *44*, 433–449.

Benington, J. H., & Heller, H. C. (1995). Restoration of brain energy metabolism as the function of sleep. *Progress in Neurobiology*, *45*, 347–360.

Berger, R. J. (1990). Relations between sleep duration, body weight and metabolic rate in mammals. *Animal Behaviour, 40*, 989–991.

Berger, R. J., & Walker, J. M. (1972). A polygraphic study of sleep in the tree shrew (*Tupaia glis*). *Brain Behavior and Evolution, 5*, 54–69.

Bert, J., Balzamo, E., Chase, M., & Pegram, V. (1975). Sleep of baboon, *Papio papio*, under natural conditions and in laboratory. *Electroencephalography and Clinical Neurophysiology, 39*, 657–662.

Bert, J., Pegram, V., Rhodes, J. M., Balzamo, E., & Naquet, R. (1970). A comparative study of two Cercopithecinae. *Electroencephalography and Clinical Neurophysiology, 28*, 32–40.

Blackburn, T. M., & Hawkins, B. A. (2004). Bergmann's rule and the mammal fauna of northern North America. *Ecography, 27*, 715–724.

Blomberg, S. P., & Garland, T. (2002). Tempo and mode in evolution: Phylogenetic inertia, adaptation and comparative methods. *Journal of Evolutionary Biology, 15*, 899–910.

Blomberg, S. P., Garland, T., & Ives, A. R. (2003). Testing for phylogenetic signal in comparative data: Behavioural traits are more labile. *Evolution, 57*, 717–745.

Broughton, R. J. (1973). Confusional sleep disorders: Interrelationship with memory consolidation and retrieval in sleep. In T. Boag & D. Campbell (Eds.), *A triune concept of the brain and behaviour* (pp. 115–127). Toronto: Toronto University Press.

Bryant, P. A., Trinder, J., & Curtis, N. (2004). Sick and tired: Does sleep have a vital role in the immune system? *Nature Reviews, Immunology, 4*, 457–467.

Campbell, S. S., & Tobler, I. (1984). Animal sleep: A review of sleep duration across phylogeny. *Neuroscience and Biobehavioral Reviews, 8*, 269–300.

Capellini, I., Barton, R. A., McNamara, P., Preston, B. T., & Nunn, C. L. (2008a). Phylogenetic analysis of ecology and evolution of mammalian sleep. *Evolution, 62*, 1764–1776.

Capellini, I., Nunn, C. L., McNamara, P., Preston, B. T., & Barton, R. A. (2008b). Energetic constraints, not predation, influence the evolution of sleep patterning in mammals. *Functional Ecology, 22*(5), 847–853.

Caro, T. (2005). *Antipredator defenses in birds and mammals*. Chicago: University of Chicago Press.

Elgar, M. A., Pagel, M. D., & Harvey, P. H. (1988). Sleep in mammals. *Animal Behaviour, 36*, 1407–1419.

Felsenstein, J. (1985). Phylogenies and the comparative method. *The American Naturalist, 125*, 1–15.

Fenn, M. G. P., & Macdonald, D. W. (1995). Use of middens by red foxes: Risk reverses rhythms of rats. *Journal of Mammalogy, 76*, 130–136.

Freckleton, R. P., Harvey, P. H., & Pagel, M. (2002). Phylogenetic analysis and comparative data: A test and review of evidence. *The American Naturalist, 160*, 712–726.

Fuchs, T., Haney, A., Jechura, T. J., Moore, F. R., & Bingman, V. P. (2006). Daytime naps in night-migrating birds: Behavioural adaptation to seasonal sleep deprivation in the Swainson's thrush, *Catharus ustulatus*. *Animal Behaviour, 72*, 951–958.

Galvão de Moura Filho, A. G., Huggins, S. E., & Lines, S. G. (1983). Sleep and waking in the three-toed sloth, *Bradypus tridactylus*. *Comparative Biochemistry and Physiology, A: Comparative Physiology, 76*, 345–355.

Garland, T., Bennett, A. F., & Rezende, E. L. (2005). Phylogenetic approaches in comparative physiology. *The Journal of Experimental Biology, 208*, 3015–3035.

Gauthier-Clerc, M., Tamisier, A., & Cezilly, F. (1998). Sleep-vigilance trade-off in green-winged teals (*Anas crecca crecca*). *Canadian Journal of Zoology, 76*, 2214–2218.

Gauthier-Clerc, M., Tamisier, A., & Cezilly, F. (2000). Sleep-vigilance trade-off in gadwall during the winter period. *Condor, 102*, 307–313.

Gauthier-Clerc, M., Tamisier, A., & Cezilly, F. (2002). Vigilance while sleeping in the breeding pochard *Aythya ferina* according to sex and age. *Bird Study, 49*, 300–303.

Harvey, P. A., & Pagel, M. (1991). *The comparative method in evolutionary biology*. Oxford: Oxford University Press.

Horne, J. A. (1988). *Why we sleep: The functions of sleep in humans and other animals*. Oxford: Oxford University Press.

Jouvet-Monier, D., & Astic, L. (1966). Study of sleep in the adult and newborn guinea pig. *Comptes Rendus des Séances de la Société de Biologie et de ses Filiales (Paris), 160*, 1453–1457.

Kushida, C. A. (2004). *Sleep deprivation: Basic science, physiology, and behavior. (Lung Biology in Health and Disease)*. New York: Marcel Dekker.

Lendrem, D. W. (1983). Sleeping and vigilance in birds. I. Field observations of the mallard (*Anas platyrhynchos*). *Animal Behaviour, 31*, 532–538.

Lendrem, D. W. (1984). Sleeping and vigilance in birds. II. An experimental study of the barbary dove (*Streptopelia risoria*). *Animal Behaviour, 32*, 243–248.

Lesku, J. A., Bark, R. J., Martinez-Gonzalez, D., Rattenborg, N. C., Amlaner, C. J., & Lima, S. L. (2008). Predator-induced plasticity in sleep architecture in wild-caught Norway rats (*Rattus norvegicus*). *Behavioural Brain Research, 189*, 298–305.

Lesku, J. A., Roth, T. C., Amlaner, C. J., & Lima, S. L. (2006). A phylogenetic analysis of sleep architecture in mammals: The integration of anatomy, physiology, and ecology. *The American Naturalist, 168*, 441–453.

Lima, S. L., Rattenborg, N. C., Lesku, J. A., & Amlaner, C. J. (2005). Sleeping under the risk of predation. *Animal Behaviour, 70*, 723–736.

Lindstedt, S. L., & Boyce, M. S. (1984). Seasonality, fasting endurance, and body size in mammals. *The American Naturalist, 125*, 873–878.

Macdonald, D. (2006). *The encyclopedia of mammals*. Oxford: Oxford University Press.

Majde, J. A. (2005). Links between the innate immune system and sleep. *The Journal of Allergy and Clinical Immunology, 116*, 1188–1198.

Makeig, S., Jung, T. P., & Sejnowski, T. J. (2000). Awareness during drowsiness: Dynamics and electrophysiological correlates. *Canadian Journal of Experimental Psychology, 54*, 266–273.

Martins, E., & Garland, T. (1991). Phylogenetic analyses of the correlated evolution of continuous characters: A simulation study. *Evolution, 45*, 534–557.

McNamara, P., Capellini, I., Harris, E., Nunn, C. L., Barton, R. A., & Preston, B. T. (2008). The Phylogeny of Sleep Database: A new resource for sleep scientists. *The Open Sleep Journal, 1*, 11–14.

Meddis, R. (1975). On the function of sleep. *Animal Behaviour, 23*, 676–691.

Mukhametov, L. M. (1984). Sleep in marine mammals. *Experimental Brain Research, 8*, 227–238.

Mukhametov, L. M. (1995). Paradoxical sleep peculiarities in aquatic mammals. *Sleep Research, 24A*, 202.

Noser, R., Gygax, L., & Tobler, I. (2003). Sleep and social status in captive gelada baboons (*Theropithecus gelada*). *Behavioral Brain Research, 147*, 9–15.

Nunn, C. L., & Barton, R. A. (2001). Comparative methods for studying primate adaptation and allometry. *Evolutionary Anthropology, 10*, 81–98.

Oates, J. F. (1987). Food distribution and foraging behavior. In B. B. Smuts, D. L. Cheney, R. M. Seyfarth, R. W. Wrangham, and T. T. Struhsaker (Eds.), *Primate societies* (pp. 197–209). Chicago: University of Chicago Press.

Palchykova, S., Deboer, T., & Tobler, I. (2003). Seasonal aspects of sleep in the Djungarian hamster. *BMC Neuroscience, 4*, 9–17.

Pillay, P., & Manger, P. R. (2004). Testing thermogenesis as the basis for the evolution of sleep phenomenology. *Journal of Sleep Research, 13*, 353–358.

Preston, B. T., Capellini, I., McNamara, P., Barton, R. A., & Nunn, C. L. (2009). Parasite resistance and the adaptive significance of sleep. *BMC Evolutionary Biology, 9*, 7.

Rattenborg, N. C., & Amlaner, C. J. (2002). Phylogeny of sleep. In T. L. Lee-Chiong, M. J. Sateia, & M. A. Carskadon (Eds.), *Sleep medicine* (pp. 7–22). Philadelphia: Hanley & Belfus, Inc.

Rattenborg, N. C., Amlaner, C. J., & Lima, S. L. (2000). Behavioral, neurophysiological, and evolutionary perspectives on unihemispheric sleep. *Neuroscience and Biobehavioral Reviews, 24*, 817–842.

Rattenborg, N. C., Lima, E. M., & Amlaner, C. J. (1999a). Half-awake to the risk of predation. *Nature, 397*, 397–398.

Rattenborg, N. C., Lima, S. L., & Amlaner, C. J. (1999b). Facultative control of avian unihemispheric sleep under the risk of predation. *Behavioural Brain Research, 105*, 163–172.

Rattenborg, N. C., Mandt, B. H., Obermeyer, W. H., Winsauer, P. J., Huber, R., Wikelski, M., et al. (2004). Migratory sleeplessness in the white-crowned sparrow (*Zonotrichia leucophrys gambelii*). *PLoS Biology, 2*, 0924–0936.

Rattenborg, N. C., Voirin, B., Vyssotski, A. L., Kays, R. W., Spoelstra, K., Kuemmeth, F., et al. (2008). Sleeping outside the box: Electroencephalographic measures of sleep in sloths inhabiting a rainforest. *Biology Letters, 4*(4), 402–405.

Rechtschaffen, A. (1998). Current perspectives on the function of sleep. *Perspectives in Biology and Medicine, 41*, 359–391.

Rechtschaffen, A., & Bergmann, B. M. (2002). Sleep deprivation in the rat: An update of the 1989 paper. *Sleep, 25*, 18–24.

Ruckebush, Y. (1963). Etude EEG et comportamentale des alternantes veille-sommeil chez lane [EEG and behavioral study of alternating waking and sleeping in the donkey]. *Comptes Rendus des Séances de la Société de Biologie et de ses Filiales, 157*, 840–844.

Saarikko, J. (1992). Risk of predation and foraging activity in shrews. *Annales Zoologici Fennici, 29*, 291–299.

Saarikko, J., & Hanski, I. (1990). Timing of rest and sleep in foraging shrews. *Animal Behaviour, 26*, 861–869.

Siegel, J. M. (2004). Sleep phylogeny: Clues to the evolution and function of sleep. In P. H. Luppi (Ed.), *Sleep: Circuits and functions* (pp. 163–176). Boca Raton, FL: CRC Press.

Siegel, J. M. (2005). Clues to the function of mammalian sleep. *Nature, 437*, 1264–1271.

Siegel, J. M. (2008). Do all animals sleep? *Trends in Neurosciences, 31*, 208–213.

Stampi, C. (1992). Evolution, chronobiology, and functions of polyphasic and ultrashort sleep: Main issues. In C. Stampi (Ed.), *Why we nap. Evolution, chronobiology, and functions of polyphasic and ultrashort sleep* (pp. 1–20). Boston: Birkhäser.

Steiger, A. (2003). Sleep and endocrinology. *Journal of Internal Medicine, 254*, 13–22.

Tobler, I. (1989). Napping and polyphasic sleep in mammals. In D. F. Dinges & R. J. Broughton (Eds.), *Sleep and alertness: Chronobiological, behavioral, and medical aspects of napping* (pp. 9–30). New York: Raven Press.

Tobler, I. (1995). Is sleep fundamentally different between mammalian species? *Behavioural Brain Research, 69*, 35–41.

Tobler, I. (2005). Phylogeny and sleep regulation. In M. H. Kryger, T. Roth, & W. C. Dement (Eds.), *Principles and practices of sleep medicine* (pp. 72–81). Philadelphia: W. B. Saunders.

Ursin, R. (1968). The two stages of slow-wave sleep in the cat and their relation to REM sleep. *Brain Research, 11*, 347–356.

Van Cauter, E., Plat, L., & Copinschi, G. (1998). Interrelations between sleep and the somatotropic axis. *Sleep, 21*, 553–566.

Van Twyver, H. (1969). Sleep patterns in five rodent species. *Physiology & Behavior, 4*, 901–905.

Van Twyver, H., & Garrett, W. (1972). Arousal threshold in the rat determined by "meaningful" stimuli. *Behavioral Biology, 7*, 205–215.

Voss, U. (2004). Functions of sleep architecture and the concept of protective fields. *Reviews in the Neurosciences, 15*, 33–46.

Wauquier, A., Verheyen, J. L., Van Den Broeck, W. A. E., & Janssen, P. A. J. (1979). *Electroencephalography and Clinical Neurophysiology, 46*, 33–48.

Withers, P. C. (1992). *Comparative animal physiology.* Orlando, FL: W. B. Saunders College Publishing.

Zepelin, H. (1989). Mammalian sleep. In M. H. Kryger, T. Roth, & W. C. Dement (Eds.), *Principles and practices of sleep medicine* (pp. 30–49). Philadelphia: W. B. Saunders.

Zepelin, H., Siegel, J. M., & Tobler, I. (2005). Mammalian sleep. In M. H. Kryger, T. Roth, & W. C. Dement (Eds.), *Principles and practices of sleep medicine* (pp. 91–100). Philadelphia: W. B. Saunders.

2

Sleep in insects

KRISTYNA M. HARTSE

Fond as the butterflies are of the light and sun, they dearly love their beds. Like most fashionable people who do nothing, they stay there very late. But their unwillingness to get up in the morning is equalled by their desire to leave the world and its pleasures early and be asleep in good time. They are the first of all our creatures to seek repose.

The Naturalist on the Thames, C. J. Cornish, 1902, p. 44

From these charming observations of insect quiescence made more than a century ago to current molecular and genetic studies in the fruit fly, the study of insect sleep during the last decade has evolved into a sophisticated field of inquiry for dissecting the potential cellular mechanisms controlling sleep in living organisms. The fundamental question of why we sleep continues to be unanswered, but it is likely that sleep in living organisms evolved from ancient origins (Allada & Siegel, 2008; Siegel, 2005). By examining insects, which have a long phylogenetic history, clues to the function and purpose of sleep may be discovered. Sleep in mammalian species such as humans, cats, and rodents has been well studied (Zeplin, Siegel, & Tobler, 2005). In contrast, there are relatively few systematic investigations of nonmammalian vertebrate sleep, and the literature is even sparser for invertebrate species. Insect sleep, with rare notable exceptions, is almost completely unstudied.

Why study sleep in insects? Although accurate estimates are difficult to obtain, the total number of insect species that have ever lived has been estimated at 100 million, and the number of living insect species is still very large, probably around 5 million (Grimaldi & Engel, 2005). In contrast, the number of living mammalian species is comparatively small, currently estimated at 4550 (Bininda-Emonds, Cardillo, Jones, et al., 2007). This diversity of insects suggests the potential for examining a broad spectrum of sleep behaviors and physiology. Furthermore, the study of insects holds significant potential for evaluating common, as well

34

as unique, molecular mechanisms controlling the expression of sleep in living organisms.

From an evolutionary standpoint, insects are ancient organisms with a longer evolutionary history than that of mammals. The earliest fossil evidence indicates that insects have been in existence for at least 400 million years, and there are modern insect orders that appeared about 250 million years ago (Engel & Grimaldi, 2004). Many living insect species have changed little from their fossil ancestors (Grimaldi & Engel, 2005). Living mammals, on the other hand, appeared in the fossil record relatively recently. Most placentals originated by 100 million to 85 million years ago, and all living placentals had appeared by 74 million years ago (Bininda-Emonds et al., 2007). One underlying assumption in studying organisms with a known fossil record is that the behavior and physiology of currently living organisms mirror, or at least plausibly reflect, the behavior and physiology of ancient ancestors. However, this assumption may result in erroneous conclusions about current functions of a behavior such as sleep. Of course, physiology and behavior are elements that are not typically preserved in the fossil record, although the presence of avian-like behavioral sleep has been suggested in a fossil dinosaur (Xing, 2004). Therefore, even in insects with a well-documented fossil record, the conclusions that can be drawn about the origins of sleep behavior are necessarily limited. This limitation does not, however, diminish the importance of discovering currently existing mechanisms supporting a behavior such as sleep.

Before we begin a review of insect sleep studies, let us consider first how sleep is defined. There are two major approaches to this definition. One set of definitions is based on well-known behavioral criteria (Hartse, 1994). These criteria include a species-specific posture, behavioral quiescence, and state reversibility with stimulation to distinguish sleep from torpor, coma, or death. The expression of sleep is influenced by both circadian and homeostatic factors. Circadian influences, or the timing of sleep within the 24-hour day, have been well studied from a behavioral and molecular perspective, particularly in insects (for reviews, see Denlinger, Giebultowicz, & Saunders, 2001; Hastings & Herzog, 2004; Zheng & Sehgal, 2008). The homeostatic factor or the expression of "sleep drive" provides an additional important behavioral tool for evaluating the presence of sleep, and this concept is integral to the definition of sleep. To return to a state of homeostasis, organisms that are deprived of sleep exhibit a rebound in sleep behaviors following the period of deprivation (Tobler, 2005).

In addition to behavior, electrophysiological criteria can define sleep. Mammals have been more intensively studied than any other group (Zeplin et al., 2005), and unique electrophysiology has been discovered to bear a close relationship with behavioral state. Two distinctive, cyclically alternating phases of electrophysiological sleep accompany behavioral sleep: NREM (non–rapid-eye-movement) and REM (rapid-eye-movement) sleep. During NREM sleep, high-amplitude cortical EEG (electroencephalograms) slow waves are present. REM sleep is characterized by a low-voltage, mixed-frequency EEG, rapid eye movements, and chin muscle atonia. In humans, dreaming is reported during this stage. As a result of the well-known relationship between behavior and electrophysiology, electrophysiology substitutes for the behavioral definition of sleep in most mammalian sleep studies.

Although the behavioral criteria for sleep have been applied to many different organisms, the electrophysiological definition of sleep and the implications for determining the "true" presence or absence of sleep in an organism based on electrophysiology have been swathed in controversy (for example, see Rattenborg, Lesku, Martinez-Gonzalez, et al., 2007; Rial, Nicolau, Gamundí, et al., 2007a,b). This debate is beyond the scope of the current review. In vertebrates it is well established that virtually all mammals as well as birds exhibit both NREM and REM sleep (Amlander & Ball, 1994; Zeplin et al., 2005). Fish, amphibians, and reptiles, however, have been variously reported to exhibit neither, both, or either NREM or REM sleep (for a review, see Hartse, 1994). There are few electrophysiological studies in invertebrates, but there is divergence in the electrophysiological correlates associated with invertebrate behavioral sleep (Brown, Piscopo, DeStefan, et al., 2006; Ramon, Hernandez-Falcon, Nguyen et al., 2004). As we shall see, almost no studies are available describing the electrophysiological correlates of behavioral sleep in insects.

With their sophisticated neuroanatomy, well-defined electrophysiology, and varied behavioral repertoires, it might reasonably be expected that sleep could be defined unambiguously in vertebrates. Yet significant issues arise in defining vertebrate sleep, and the same might also be expected to occur in defining insect sleep. Additional challenges in studying insects are presented from a number of perspectives. The diversity of living insect species is far greater than the diversity of living vertebrates. This suggests that the potential for variations in behavior and physiology among insects is exceedingly large, and the generalization of findings to other insect species or even to vertebrates may be difficult. With some exceptions, such as giant beetles (Williams, 2001), most insects are relatively small organisms with compact central nervous systems, making behavioral observations potentially difficult and the electrophysiological study of brain activity even more so. The environments occupied by many insects, such as burrows or arboreal habitats, do not lend themselves to naturalistic observations in the laboratory. Despite these challenges, there have been studies of cockroaches, scorpions, bees, and fruit flies that have added to our knowledge of sleep.

This review examines the evidence for sleep in insects, beginning with observational studies and continuing to the burgeoning literature on the molecular and genetic studies of sleep in the fruit fly, *Drosophila melanogaster*. Some of the issues under consideration include whether insects exhibit behavioral and electrophysi-ological signs of sleep, what the study of insects can tell us more generally about the molecular and genetic basis of sleep in other living organisms, and whether insect models can provide insights into the impact of sleep on issues such as human aging, longevity, and sleep disorders.

Early observational studies of insect sleep

An insect composes itself to sleep with its antennae folded. Some of the beetles adjust them to their breast; the butterfly seeks some particular aspect of a tree, and folds vertically its wings, throws back the antennae, and remains motionless and insensible to all external circumstances. When caterpillars, which are insatiable feeders, are observed resting immovable with their heads bent down, they are asleep.

(*Yearbook of Facts for 1864*, R. Hill, 1865, p. 282)

Many incidental descriptions of insect behavioral quiescence are scattered throughout the early scientific and entomological literature. The first observations of insect quiescence were performed in natural habitats, and the descriptions of quiescent behavior were not typically constrained by questions of whether these behaviors were merely "rest" or whether they were "sleep." Insects were most often judged to be quite unambiguously asleep in their natural environments. In these descriptions, the early naturalists not only documented circadian patterns of quiescent behavior but they also recognized, almost inadvertently, the now well-known behavioral criterion of species-specific postures, behavioral quiescence, and elevated arousal thresholds in describing insect sleep. An extensive early observa-tional field study documented behavioral sleep in wasps, bees, flies, dragonflies, grasshoppers, butterflies, and moths (Rau & Rau, 1916). In several different species of beetles, however, the presence of sleep was determined to be questionable. Also of interest is that the issues raised by Rau and Rau, as the result of their insect observations, included the role of sleep variability in determining longevity, the impact of sleep on ontogenetic and central nervous system development, and the evolutionary significance of sleep. These continue to be the very issues that lie at the core of present-day sleep research.

Systematic studies of insect sleep

With the exception of the literature on bees and the current literature on fruit flies, systematic studies on sleep in insects are few. However, recent stud-ies utilizing operationalized criteria to characterize sleep behavior suggest that sleep is present in insects. Two states of spontaneous behavioral quiescence, tonic immobilization and prostrate immobilization, were identified in the mosquito on the basis of posture and head position, and there was decreased responsiveness to mechanical stimulation during the latter state (Haufe, 1963). In the moth, five sleeping postures were described based on antenna position. In the fifth posture, arousal thresholds were markedly elevated in response to tactile stimulation, suggesting that this posture corresponded to sleep (Sogaard Andersen, 1968). Three states of vigilance, also based on behavior, have been identified in the scorpion: locomotor activity, alert immobility, and relaxed immobility (Tobler & Stalder, 1988). The scorpions were most responsive to a mechanical stimulus during activity and least responsive during relaxed immobility. Heart rate was highest during activity and lower during relaxed immobility. Following 12 hours of rest deprivation during the light period, there was a significant increase in relaxed immobility and a decrease of alert mobility, indicating a homeostatic response to deprivation. A similar response to rest deprivation has also been observed in the cockroach (Tobler, 1983).

These studies establish behavioral criteria for insect sleep as well as the homeostatic response to sleep loss in at least some insects, but they do not address the physiological consequences of sleep loss, which has been a topic of substantial interest in the mammalian literature. In the rat, pioneering work demonstrated that the consequence of prolonged sleep deprivation is an increase in energy metabolism followed by death (Rechtschaffen & Bergmann, 2002; Rechtschaffen, Gilliland, Bergmann, et al., 1983). If a similar physiological response to sleep deprivation could be established in insects, then the importance of sleep as a state necessary for the sustenance of life could be extended to organisms other than mammals and would suggest a more universal function for sleep across the phylogenetic scale.

The Pacific beetle cockroach, *Diploptera punctata* (Eschscholtz), does, in fact, display a physiological response to sleep deprivation similar to that of mammals (Stephenson, Chu, & Lee, 2007). Long-term deprivation of behavioral quiescence in these cockroaches was achieved under constant temperature and light conditions by combination of a CO_2 pulse and brief rotation of the experimental chamber. Oxygen consumption was measured during the deprivation procedure. The sleep-deprived group received one stimulus per minute; the control group received an equal number of stimuli per day but with periods free from stimulation lasting 3 hours. The effects on energy metabolism in the cockroach mirrored the effects in rats. Metabolic rate increased by 81.8% in the sleep-deprived group with respect to the control group. As in the rat studies, the most dramatic findings were on longevity. Beginning on day 17 of the deprivation procedure, there was an average of 0.57 death per day in the sleep-deprived group, such that half of the cockroaches had died after 30 days. The control group, on the other hand, averaged one death per 7.7 days. These findings suggest not only that sleep may serve a similar function in both mammals and insects in regulating energy metabolism but also that sleep is essential for life in widely varying branches of the phylogenetic tree.

Further detailed studies on the effects of energy metabolism and sleep deprivation are clearly needed before these broadly based conclusions can be made confidently. Even within vertebrate species, there is new evidence to suggest that sleep deprivation may not always have similar marked metabolic and lethal effects. For example, a recent study in pigeons, using sleep deprivation techniques similar to those employed in the rat studies, did not confirm the effects on mortality and energy expenditure (Newman, Paletz, Rattenborg, et al., 2008). However, the extent of sleep deprivation achieved in pigeons was not as extreme as that achieved in rats. In addition, only mild effects have been noted on telecephalic gene expression in the pigeon during the migratory season, when sleep is severely curtailed, as well as during enforced sleep deprivation – findings that are in contrast to the *Drosophila* findings following deprivation of quiescence (Jones, Pfister-Genskow, Cirelli, et al., 2008; J. A. Williams, Sathyanarayanan, Hendricks, et al., 2007). Thus, as in these studies in vertebrates, it would also be expected that there may be variable responses to sleep deprivation within insect species.

Sleep in bees

The literature on bee sleep is more abundant than for any other insect with the exception of *Drosophila*. It is estimated that bees originated about 120 million years ago, around the time of the diversification of flowering plants (Grimaldi & Engel, 2005). There are approximately 20,000 species of bees, but only a handful of studies have examined sleep in these insects, and most of these investigations have been performed in the honeybee, *Apis mellifera*.

In an elegant series of studies designed to answer the question of whether bees meet the criteria for sleep, Kaiser performed the first systematic studies of bee sleep (Kaiser, 1988). This work was prompted by the finding that optomotor interneurons in the optic lobes of forager honeybees displayed a circadian sensitivity to horizontal movement of a light stimulus. Sensitivity to this stimulus during the day was elevated when locomotor activity was high, and sensitivity decreased at night when locomotor activity was decreased during presumed sleep (Kaiser & Steiner-Kaiser, 1983). Through videotape analysis and behavioral observations, a clear pattern of quiescence during the dark portion of the light–dark cycle accompanied by specific body, head, and antenna postures was documented (Kaiser, 1988). In contrast to mammalian deep sleep, which occurs early in the night, the period of least antenna mobility in bees occurred late in the night, suggesting that the temporal organization of bee sleep may be different from that of mammals. Thoracic temperature and leg muscle activity were both measured. No independent endogenous temperature rhythm was revealed, and muscle activity was at its lowest when the bee was quiescent. The threshold for stimulation of grooming behavior by the application of an infrared light was highest toward the end of the night, consistent with the reduction in antenna mobility. These findings led the author to conclude that "forager honeybees experience a state of profound rest at night." However, also of note is that solitary bee species, both in the field and in the laboratory, exhibited quiescent behaviors different from those of honeybees in the laboratory, including a distinctive clamping of the mandibles onto a twig during behavioral quiescence, less antenna motility during sleep, and more prolonged periods of quiescence (Kaiser, 1995). These findings suggest that there may be multiple variations in sleep behavior between different bee species, just as there are among different mammalian species.

Although the honeybee's homeostatic response to sleep deprivation was not evaluated in this study, subsequent work revealed that 12 hours of sleep deprivation during the dark period resulted in an increase of antenna immobility per hour during the next dark period as well as an increase in the duration of bouts of antenna immobility (Sauer, Herrmann, & Kaiser, 2004). Application of a tilting device that enforced behavioral wakefulness in the deprivation experiments was applied during the day without effect on the animals' subsequent sleep. Thus the homeostatic response to sleep deprivation is confirmed in the honeybee.

Other physiological parameters have been investigated during sleep in bees, including ventilatory activity, heart rate, and neck muscle activity, all of which show specific relationships with the states of vigilance (Kaiser, 1988; Kaiser, Weber, & Otto, 1996). These variables also demonstrated distinct differences between coma and behavioral sleep in bees, supporting the conclusion that bee quiescence is a sleep-like state (Kaiser, 2002). Energy conservation has been proposed as one of the functions of sleep in homeothermic mammals (Zeplin & Rechtschaffen, 1974). However, ectothermic bees do not choose resting nocturnal ambient temperatures that maximize energy conservation. Presented with a temperature gradient ranging between 18 and 38°C, honeybees preferred a resting nocturnal range between 23 and 26°C (Kaiser, Faltin, & Bayer, 2002). These findings are similar to those of another study in which honeybees preferred a resting ambient temperature of 28°C in a gradient of 20 to 35°C (Schmolz, Hoffmeister, & Lamprecht, 2002). These findings suggest that bees do not choose to decrease their metabolic rates for maximum energy conservation during sleep and that sleep may serve additional functions besides energy conservation in insects.

Similar to changes in mammalian sleep across the life span, there is evidence for age-related changes in the sleep of honeybees. In a preliminary report, the well-known circadian periodicity of rest and activity in mature bees (foragers) was not observed in newly emerged bees (callows). The rest–activity cycle gradually developed over a period of about 3 weeks, even though behavioral sleep was observed in the arrhythmic callows. These findings suggest a possible age-related maturation of circadian rhythms (Sauer, Menna-Barreto, & Kaiser, 1998). However, social influences may also affect the development of circadian rhythms. Young bees housed with foragers experienced an acceleration in the development of circadian rest–activity cycles in comparison to young bees housed with other young bees (Meshi & Bloch, 2007).

In addition to the variable of age, more detailed recent observational studies of honeybee colonies suggest that bee sleep changes as a function of tasks performed within the colony (Eban-Rothschild & Bloch, 2008; Klein, Olzowy, Klein, et al., 2008). Three sleep stages were identified on the basis of body posture, antenna movements, and response to increasing light intensities in callows and foragers (Eban-Rothschild & Bloch, 2008). Both callows and foragers exhibited the same sleep stages, and they did not differ in their antenna movements during these stages. No differences were found between these two groups in the percentages of time asleep or in arousal thresholds during the same sleep stage. However, foragers showed a strong circadian pattern of activity during the day, with quiescence being concentrated during the night. Callows, on the other hand, tended to have sleep distributed throughout the 24 hours. In both groups brief awakenings interrupted sleep. A major significant difference between the two groups was that the callows typically made the transition from the second and third sleep stages to the first sleep stage, whereas foragers made the transition to waking states. Callows exhibited significantly more transitions between sleep stages than did foragers, indicating greater sleep fragmentation.

Another recent observational study has described bee behavioral sleep, in this case involving four honeybee castes: cell cleaners (young bees), nurse bees, food storers, and foragers (mature bees) (Klein et al., 2008). In contrast to the 24-hour sleep–wake periodicity in food storers and foragers, cell cleaners and nurse bees did not exhibit a strong circadian rhythmicity. Uninterrupted sleep bouts were longer in food storers and foragers than in cell cleaners and nurse bees, and foragers spent more time sleeping outside cells than did any of the other groups.

Although these studies in bee colonies are of great interest, there is a confound that has an impact on the interpretation of the findings. Do these changes in bee sleep reflect the process of aging, do they reflect the results of changes in tasks within the hive as the bee ages, or is there a combined effect of aging and task specificity on sleep? There is neuroanatomical evidence for plasticity in the honeybee brain in response to age as well as to the effects of experience. In the mushroom bodies, major sensory processing centers in the insect brain, there is an increase in the neurophil as well as in dendritic branching with increasing age (Farris, Robinson, & Fahrbach, 2001). Quite strikingly, additional neuroanatomical changes were associated with increasing foraging experience. In *Drosophila,* social experience affects both sleep amounts as well as dopaminergic and cyclic AMP signaling pathways (Ganguly-Fitzgerald, Donlea, & Shaw, 2006). Social enrichment is also associated with an increase in *Drosophila* synaptic terminals in ventral lateral neurons, which play a major role in the expression of circadian rhythms (Donlea, Ramanan, & Shaw, 2009). These findings suggest that not only age but also different tasks and social conditions are associated with neuroanatomical and neurochemical changes potentially affecting sleep in bees and possibly in other insects. Furthermore, the conclusions that can be drawn from isolated laboratory organisms may be different from those derived from organisms studied in natural environments.

In summary, there is a paucity of rigorous studies on sleep in insects. The number of species that have been studied is very small, but even in this small sample, there is significant variability in quiescent behaviors among ages, social groups, and individual species. With the exception of the electrophysiological recordings from bee optomotor neurons, the electrophysiological correlates of insect behavioral sleep and waking are virtually unknown. Even with these limitations, however, the evidence supports the position that sleep is present in insects: Behavioral quiescence is accompanied by species-specific postures, elevated arousal thresholds, state reversibility, and a homeostatic response to sleep deprivation. The metabolic response to prolonged sleep deprivation in the cockroach is similar to that in mammals. This last piece of evidence is particularly supportive of a similarity of sleep function between insects and mammals.

Drosophila: A model system for sleep

In 2000, the fruit fly, *Drosophila melanogaster*, was introduced as a model organism for studying the genetic and biochemical basis of sleep (Hendricks, Finn, Panckeri, et al., 2000a; Shaw, Cirelli, Greenspan, et al., 2000). Since those first studies, a new body of knowledge has emerged from *Drosophila* about the cellular mechanisms that control sleep and wakefulness. The serious health and social risks associated with human sleep disorders as well as their high prevalence in the population are even more compelling reasons to discover the fundamental causes for sleep disruption (National Sleep Foundation, 2008). Study of the molecular basis of sleep in *Drosophila* may enhance the potential for the development of effective treatments in humans.

At first glance, the fruit fly would appear to be an unlikely candidate for evaluating human sleep. However, there are at least four reasons why this organism presents distinct advantages as a possible model for elucidating sleep mechanisms (Hendricks, Sehgal, & Pack, 2000b). First, the genome of *Drosophila* has been well studied, and it comprises a large number of orthologs – or genes that are similar in genetic sequence and function, with a presumed common origin – to human disease genes (Rubin, Yandel, & Wortman, 2000). Second, with the exception of orexin/hypocretin, most of the major mammalian neurotransmitter systems affecting sleep have been identified in *Drosophila* (Agosto, Choi, & Parisky, et al., 2008; Andretic, van Swinderen, & Greenspan, 2005; Yuan, Joiner, & Sehgal, 2006). Third, *Drosophila* is a small, readily available organism easily housed and maintained in the laboratory. Fourth, its short life span and rapid reproductive cycle allow for outcomes of genetic and molecular manipulations to be assessed quickly. Besides *Drosophila*, two other organisms, the zebrafish (*Danio rerio*) and roundworm (*Caenorhabditis elegans*), have been proposed as candidates for model sleep systems because they also share these advantages as model organisms (Hendricks et al., 2000b; Raizen, Zimmerman, & Maycock, et al., 2008).

There are strong similarities between characteristics of *Drosophila* behavioral quiescence and mammalian sleep (Hendricks et al., 2000a; Huber, Hill, Holladay, et al., 2004; Shaw et al., 2000). Behavioral criteria for sleep established in mammals are also present in the fly. Flies exhibit a stereotypic posture during behavioral quiescence, and arousal thresholds are elevated during this quiescent state. There is a homeostatic response to rest deprivation, as well as age-related declines in amounts of quiescence, as in mammals. Administration of caffeine, a stimulant, results in a dose-dependent decrease in resting behavior, and antihistamines, which increase the amount of sleep in mammals, produce a similar response in flies. Modafinil and methamphetamine, pharmacological alerting agents, also have alerting effects in *Drosophila* (Andretic et al., 2005; Hendricks, Kirk, Panckeri, et al., 2003). Sleep, sleep-deprived states, and waking are accompanied by differential gene expression (Zimmerman, Rizzo, Shockley, et al., 2006). Finally, flies lacking the central clock gene, *period*, still exhibit a homeostatic response to rest deprivation, suggesting that sleep can be manipulated separately from the cycling of this circadian clock gene (Shaw et al., 2000). On the other hand, flies with a mutation of a second clock gene, *timeless*, did not respond to deprivation with a homeostatic increase in resting behavior (Hendricks et al., 2000a, b). These findings support a differential role for clock genes in the homeostatic response to sleep deprivation.

Besides confirming the behavioral evidence for sleep in *Drosophila*, new details of the pharmacological and molecular similarities to mammalian sleep have emerged. A question that then arises is whether these mechanisms in *Drosophila* are conserved across phylogenetic groups (for reviews, see Allada & Siegel, 2008; Zimmerman, Naidoo, Raizen, et al., 2008a). If they are unique to *Drosophila*, then their significance for understanding the regulation of sleep in the wider variety of living vertebrates and invertebrates is, of course, limited. On the other hand, if it can be demonstrated that these findings in *Drosophila* enhance the understanding of basic sleep mechanisms in many different organisms, including humans, then there is the potential that this research may be productively applied to discovering targeted treatments for human sleep disorders.

The behavioral similarities between fly sleep and mammalian sleep are, as we have seen, quite well established. Areas in which it would be desirable to establish additional similarities include the following.

Electrophysiology

Recording of surface EEG activity from various areas of the mammalian brain yields specific patterns of neural activity, which are used to identify not only the presence or absence of sleep but also different stages of sleep (Iber, Ancoli Israel, Chesson, et al., 2007). Subcortical recordings, performed primarily although not exclusively in mammals, yield additional detailed electrophysiological information from deep brain structures. In small organisms such as insects, the technical challenges of neural recordings in freely moving organisms are immense. In addition to this technical challenge, the homologies between various mammalian and insect brain structures that would allow for meaningful comparison have yet to be established.

Evidence exists for an electrophysiological correlate of behavioral sleep in *Drosophila* recorded from the mushroom bodies. These paired structures located in the protocerebrum of the *Drosophila* brain have been implicated in the expression of sleep (Joiner, Crocker, White, et al., 2006; Pitman, McGill, Keegan, et al., 2006). Chemical ablation and enhancement of cyclic-AMP–dependent protein kinase in the mushroom bodies results in reduced sleep amounts, and learning impairments induced by sleep deprivation are reversed by gene expression of the mushroom body dopamine D-1 receptors (Seugnet, Suzuki, Vine, et al., 2008). Electrophysiological recordings of local field potentials (LFPs) from the medial protocerebrum between the mushroom bodies have revealed bursting spike potentials that are prominent during behavioral waking and decrease during behavioral sleep (Nitz, van Swinderen, Tononi, et al., 2002). Further studies demonstrated that LFPs are not simply the product of movement but rather that variations in LFPs are associated with variations in arousal thresholds, suggesting that LFPs are a neural marker for sleep (van Swindern, Nitz, & Greenspan, 2004).

The absence of LFPs during *Drosophila* behavioral sleep is, as discussed by Nitz et al., not an electrophysiological equivalent to the presence of waveforms such as sleep spindles, slow waves, or K-complexes defining mammalian NREM sleep. Whether the state associated with the absence of LFPs is the *Drosophila* neurophysiological equivalent of mammalian NREM sleep is unknown. No electrophysiological evidence exists to suggest that REM sleep is present in *Drosophila*. Thus, although this electrophysiological pattern of LFP absence correlates with *Drosophila* behavioral sleep, no defining waveform that is present exclusively during sleep has been discovered.

Genetics

The genome of *Drosophila* is well studied, and over 60% of human genes have functional *Drosophila* orthologs (Nichols, 2006). This feature has particular importance for the discovery of mutant strains with known gene sequences as a means of revealing the underpinnings of human sleep disorders. For example, discovery of the mechanisms responsible for the expression of sleeplessness in flies might shed light on the factors responsible for insomnia in humans.

Short-sleeping fly strains have been identified, but it is clear that genetic short sleepers are rare. Ten lines of short sleepers were identified after approximately 5000 lines of mutant flies were screened (Cirelli, 2003). Most short sleepers responded to sleep deprivation with increased sleep duration and decreased sleep fragmentation. However, four lines that had no sleep rebound in response to deprivation were identified. Differences in sleep between male and female flies also emerged during this screening procedure, with females sleeping almost exclusively at night and males exhibiting an additional sleep period during the day. A second study also confirms the rarity of the genetic short sleeper (Wu, Koh, Yue, et al., 2008). After more than 5000 lines of mutant flies were screened, 7 short-sleeping mutant lines were identified. More detailed analysis of sleep patterns in the short sleepers revealed that the primary sleep defect was a significant reduction in the length of the nighttime sleep bout but not in the number of sleep bouts, suggesting that mechanisms controlling sleep maintenance rather than sleep initiation account for decreased sleep in these lines. Additionally, a DAT (dopamine active transporter) short-sleeping mutant has been discovered, reinforcing the body of work that implicates dopaminergic signaling in sleep mechanisms (Kume, Kume, Park, et al., 2005).

More detailed studies have been performed on mutant lines of short-sleeping flies – including *minisleep (mns)*, *hyperkinetic (Hk)*, and *sleepless (sss)* – to characterize the physiological changes, if any, associated with short sleep. The defect in these lines is linked to a mutation in the *Shaker* gene locus, which controls potassium ion channels responsible for cell membrane repolarization and neurotransmitter release (Bushey, Huber, Tononi, et al., 2007; Cirelli, Bushey, Hill, et al., 2005; Koh, Joiner, Wu, et al., 2008). In the *mns* mutation, which maps to the X chromosome, heterozygous female *mns* flies and wild flies slept for 9 to 15 hours per day, but male and homozygous female *mns* flies slept for only 4 to 5 hours per day. Like wild-type flies, *mns* flies showed a homeostatic response to sleep deprivation. No reproductive, behavioral, or performance deficits were observed in the short-sleeping flies. The *sss* mutants, both males and females, exhibited severe reductions of greater than 80% in their daily sleep. Of significance is that about 9% of flies showed no signs of sleep (Koh et al., 2008). Furthermore, this strain did not exhibit a homeostatic response to sleep deprivation. A recent report has also demonstrated an abnormal homeostatic response to sleep deprivation in the *amnesiac (amn)* mutant (Liu, Guo, Lu, et al., 2008).

Genetically shortened sleep may affect both life span and memory in *Drosophila*. The *mns* mutants had a decreased life span in comparison to controls, suggesting that lack of sleep contributed to decreased longevity. However, another short-sleeping fly mutant, *fumin (fum)*, did not exhibit decreased life spans (Kume, et al., 2005). In humans, the effects of sleep duration on life span are also variable. Increased human mortality associated with both decreased and increased sleep amounts has been reported (Dew, Hoch, Buysse, et al., 2003; Kripke, Garfinkel, Wingard, et al., 2002). In addition to possible effects on longevity, genetically shortened sleep may affect memory. Reductions in memory were reported only in *Hk* flies with alleles for short sleep, supporting a memory-enhancing function for sleep (Stickgold & Walker, 2005a,b). Debate continues to surround the role of sleep in memory consolidation (Vertes & Siegel, 2005). In addition to these findings of sleep alterations in mutant strains, genetic correlations have also been revealed between sleep and energy stores (Harbison & Sehgal, 2008).

Aging

The effects of aging on mammalian sleep have been well studied. In general, young organisms sleep longer and in a more consolidated fashion than do older organisms. A change in sleep patterns over the life span also occurs in the fly and provides another point of similarity with mammalian sleep, suggesting that *Drosophila* may also be used as a model to investigate the changes that occur with human aging (Shaw et al., 2000; Shaw, Ocorr, Bodmer, et al., 2008).

Like aging mammals, *Drosophila* exhibits a deterioration in sleep quality with age (Koh, Evans, Hendricks, et al., 2006). Monitoring across the life span reveals a decline in the rhythm strength of the sleep–wake cycle as well as an increase in sleep fragmentation. Premature aging of flies by exposure to elevated temperatures produces increased mortality as well as a more rapid disintegration of sleep–wake rhythms and sleep continuity. These findings suggest that physiological rather than chronological aging is most important in these age-related changes. Oxidative stress accompanying aging appears to be a critical mediating variable. Flies treated with paraquat, an oxidative stress–inducing agent, demonstrated a deterioration in sleep similar to that seen in prematurely aged flies. These findings have particular significance for aged organisms, because expression of the immune response is also affected by decreased sleep (Williams et al., 2007).

Signaling pathways and neurotransmitters

There is convincing evidence to suggest that the signaling pathways and neurotransmitters that modulate cellular activity and affect sleep in mammals and *Drosophila* are similar. One such modulating pathway in mammals is cyclic AMP (cAMP) signaling and cAMP response element binding (CREB) protein. In flies, increased cAMP signaling is associated with decreased quiescence and, conversely, increased cAMP signaling is associated with increased quiescence (Hendricks, Williams, Panckeri, et al., 2001). The blocking of CREB activity enhances the rebound to sleep deprivation, and an increase in CREB activity occurs during the period following rest deprivation, suggesting a restorative function for CREB. That these signaling effects are not the result of alterations in circadian clocks is suggested by the retention of a normal circadian distribution of rest and activity.

An additional signaling pathway that has been examined is the epidermal growth factor receptor (EGFR) (Foltenyi, Greenspan, & Newport, 2007). In mammals, the suprachiasmatic nucleus, which controls the circadian expression of sleep, is a site for transforming growth factor alpha (TGF-α), a ligand for EGFR. In *Drosophila*, activation of EGFR and extracellular signal–regulated kinase (ERK) is associated with excessive sleep. The elimination of Rho proteins, which stimulate EGFR signaling, is associated with increased sleep fragmentation. These pathways exert their effects via the *Drosophila* pars intercerebralis, which may be the equivalent of the mammalian hypothalamic–pituitary axis. Also of note is that EGFR activation within a single neuron of the worm *C. elegans* induces quiescence, suggesting that this signaling pathway has been conserved across species (Van Buskirk & Sternberg, 2007).

Several neurotransmitters – including serotonin, dopamine, and GABA – appear to have similar effects on the regulation of sleep and wakefulness in flies and mammals. One of the first neurotransmitters to be systematically studied in mammalian sleep was serotonin. Pharmacological inhibition of serotonin or surgical ablation of serotonin-containing raphe neurons in the mammalian brain is associated with marked reductions in sleep (Jouvet, 1969). In *Drosophila*, mutation of one of three serotonin receptors (d5-HT1A) also results in decreased and fragmented sleep, but with preservation of circadian rhythms. These changes were reversed by expression of this receptor. Administration of 5-HTP, the precursor of serotonin, to flies was associated with increased daytime sleep, similar to the increases in mammalian sleep following 5-HTP administration (Yuan et al., 2006).

Dopamine has been implicated in mediating arousal and the stimulating effects of modafinil and amphetamines in mammals (Winsor et al., 2001) Similarly, dopamine appears to have an important role in regulating arousal in *Drosophila*. Hyperactive *fumin (fmn)* mutants have a mutation of the dopamine transporter gene (DAT) and exhibit decreased sleep amounts, hyperresponsiveness to mild mechanical stimuli, and an attenuated sleep rebound following deprivation of quiescence (Kume et al., 2005). Also, pharmacological inhibition of dopamine synthesis in flies is associated with an increase in daytime sleep (Andretic et al., 2005). These findings provide support for dopamine as a transmitter modulating alertness in the fly, similar to the effects of dopamine in mammals.

Many benzodiazepine medications that induce sleep in humans are also GABA receptor agonists, suggesting that GABA plays an important role in the initiation and maintenance of sleep (Lancel, 1999). In *Drosophila*, recent work has elegantly demonstrated that sleep onset and sleep maintenance can be differentially affected by the manipulation of GABA receptors (Agosto et al., 2008).

In addition to these well-known neurotransmitters common to flies and mammals, octopamine, a biogenic amine widely present in the nervous systems of invertebrates, also has effects on sleep in *Drosophila* (Crocker & Sehgal, 2008). Mutations in the octopamine synthesis pathway resulted in increased sleep amounts, and feeding octopamine to flies reduced sleep. These findings suggest that there may be unique neurotransmitter systems regulating sleep in flies and other invertebrates. Similarly, there may also be neuroregulatory systems unique to mammals. For example, the role of orexin/hypocretin has been intensively studied in relation to the expression of narcolepsy in humans, and orexin/hypocretin has also been examined in zebrafish (Mignot, 2005; Yokogawa et al., 2007). However, it has not been discovered in *Drosophila* or other invertebrates.

This brief review is illustrative of the rapid advances being made in understanding the common effects of molecular signaling systems and neurotransmitters on sleep in *Drosophila* and mammals. There are also broader implications of this work for understanding the possible effects of, for example, pharmacological agents on human sleep. Clearly, forthcoming research will clarify these mechanisms further.

Methodological considerations in *Drosophila* studies

From many different perspectives, *Drosophila* is an attractive organism in which to study sleep. As we have seen, not only are the similarities to mammalian sleep striking but also, from a practical standpoint, complex molecular manipulations in several generations of flies can be performed over a relatively short period of time. Currently, it is difficult to imagine this same level of discovery in mammalian species other than, for example, rodents, which have a relatively short life span as compared to other mammals and can be practically housed in the laboratory. However, even though findings from the *Drosophila* studies can be described as nothing less than ground-breaking in elucidating sleep mechanisms, there must also be consideration given to the question of whether the methodology for measuring sleep in these studies is a valid one that will allow us to generalize findings to other organisms.

Prolonged behavioral monitoring of many flies is clearly impractical for large-scale sleep studies. It also seems unlikely, based on the anatomy of the insect brain and the absence of clear homologies to mammalian brain structures, that distinctive electrophysiological waveforms similar to those observed in sleeping mammals will soon be discovered in *Drosophila*. Unlike electrophysiological monitoring performed in the majority of mammalian studies, measurement of sleep in *Drosophila* has relied almost exclusively on automated ultrasound activity monitoring systems as well as the now widely used *Drosophila* Activity Monitoring System (DAMS) (Andretic & Shaw, 2005; Shaw et al., 2000).

DAMS measures fly activity and inactivity by breaks recorded in an infrared beam that crosses a small glass tube housing an individual fly. DAMS beam crossings operationally define sleep and wakefulness – that is, flies that interrupt the beam are considered to be awake, and an absence of beam breaks, typically for at least 5 minutes, is considered to reflect sleeping behavior. However, DAMS may be insensitive to small fly movements that do not break the beam as well as to large movements that occur out of the beam's range. These factors can result in inaccurate estimations of fly sleep (Zimmerman, Raizen, Maycock, et al., 2008b). Using detailed digital video subtraction analysis and comparing these measurements with DAMS, DAMS overestimated total fly sleep by 39% to 95% during the day and by 7% to 21% at night. Additional findings with digital video analysis included more accurate measurement of sleep bout length and an accentuated DAMS overestimation of both sleep time and bout duration in aged flies. However, of note is that the homeostatic response to sleep deprivation was confirmed with digital video analysis, indicating that the homeostatic response to sleep deprivation is not a function of inaccuracies in DAMS monitoring.

The identification of totally sleepless fly mutants or the description of subtle changes in sleep with aging, for example, would appear to hinge on accurate measurement systems that take into account subtle movements of the appendages and body. The study of Zimmerman et al. is of importance because it highlights the necessity for ongoing reassessment of assumptions concerning the measurement and definition of behavior prior to drawing conclusions about the effects of genetic, neuronal, or chemical manipulations, even in an organism as well studied as *Drosophila*. Innovative measurement tools and techniques that accurately reflect behavior will strengthen the relevance of *Drosophila* as a model organism for the study of sleep.

Discussion

The distinction between sleep and sleep-like or resting states is not always a clear one. This may particularly be the case for less well-known organisms, such as insects, which exhibit a wide variety of unique resting behaviors in unique environments. As we have seen, however, the convergence of behavioral, genetic, and cellular evidence convincingly suggests that sleep is present in at least some insects, including the fruit fly, honeybee, and a very small assortment of other insects.

If there is an underlying assumption that most if not all organisms sleep or at least rest, quiescent states, by default, must serve an important biological function in species survival. Otherwise the state of sleep, which precludes activities such as food gathering, reproduction, care of young, and protection of territory, is very unlikely to have been perpetuated in evolution over millions of years in thousands of species. New evidence in *Drosophila* for differential changes in pre- and postsynaptic proteins associated with sleep and waking suggests a specific, and perhaps universal, cellular function for sleep (Gilestro, Tononi, & Cirelli, 2009). However, whether the findings in insects studied to date can be generalized to other unstudied insect species is obviously unknown, and it is even less certain whether these findings can be generalized to other vertebrates. There are several examples of vertebrates in which sleep behavior does not conform to currently accepted definitions. Do "sleep swimming" fish, which exhibit fin movements throughout the night as they remain on the coral reef, truly sleep (Goldschmid, Holzman, Weihs, et al., 2004)? Postpartum killer whale and dolphin mothers as well as their offspring are continuously mobile after birth for several weeks. Do these cetaceans experience the neurophysiological and neurochemical benefits of sleep other than through behavioral quiescence (Lyamin, Pryaslova, Lace, et al., 2005)? How do birds physiologically sustain prolonged migrations in the face of greatly reduced sleep amounts (Rattenborg, 2006)? It is also likely that new investigations of more insect species will uncover unique behaviors and physiological expressions of sleep that do not fit our current definitions of sleep.

The *Drosophila* model, as well as *Danio rerio* and *C. elegans*, has enormous potential for new discoveries in improving treatment of human sleep disorders. There has been progress, for example, in identifying the genetic underpinnings of advanced sleep phase syndrome (Xu et al., 2005). Narcolepsy, a disabling disorder of excessive daytime sleepiness, is linked to a deficiency in the orexin/hypocretin system and has been described in zebrafish (Kaslin, Nystedt, Ostergard, et al., 2004; Yokogawa et al., 2007). Insomnia is a uniquely human sleep complaint affecting a large proportion of the population, and treatments for insomnia have been notoriously inadequate. The identification of sleepless fly mutants presents the possibility of developing targeted insomnia treatments based on the genetic and neuronal findings in sleepless organisms. *Drosophila* research has, in one example, already identified a biological marker, amylase, for sleep drive (Seugnet, Boero, Gottschalk, et al., 2006). Amylase is also easily assayed in human saliva, and similar to the findings in flies, increases in human saliva after extended sleep deprivation. It is not difficult to imagine that this assay could potentially be used to detect sleepiness in humans before overt sleep is obvious, thus avoiding such serious consequences as automobile or industrial accidents.

In summary, the preponderance of existing evidence supports the presence of sleep in living insects. The question of when sleep first appeared in evolution will probably never be adequately answered, but current studies of insects, in conjunction with evidence from the fossil record, indicate that sleep or at least a form of quiescence is an ancient behavior with ancient origins. Ongoing studies in *Drosophila* hold significant promise for further understanding sleep mechanisms and for potentially applying these findings to the treatment of disturbed human sleep.

References

Agosto, J., Choi, J. C., Parisky, K. M., Stilwell, G., Rosbash, M., & Griffith, L. C. (2008). Modulation of GABAA receptor desensitization uncouples sleep onset and maintenance in *Drosophila*. *Nature Neuroscience*, *11*, 354–359.

Allada, R., & Siegel, J. M. (2008). Unearthing the phylogenetic roots of sleep. *Current Biology*, *18*(15), R670–R679.

Amlander, C. J., & Ball, N. J. (1994). Avian sleep. In M. H. Kryger, T. Roth, & W. C. Dement (Eds.), *Principles and practice of sleep medicine* (2nd ed., pp. 81–94). Philadelphia: W. B. Saunders.

Andretic, R., & Shaw, P. J. (2005). Essentials of sleep recordings in *Drosophila*: Moving beyond sleep time. *Methods in Enzymology*, *393*, 759–772.

Andretic, R., van Swinderen, B., & Greenspan, R. J. (2005). Dopaminergic modulation of arousal in *Drosophila*. *Current Biology*, *15*, 1165–1175.

Bininda-Emonds, O. R. P., Cardillo, M., Jones, K. E., MacPhee, R. D. E., Beck, R. M. D., Grenyer, R., et al. (2007). The delayed rise of present-day mammals. *Nature*, *446*, 507–512.

Brown, R., Piscopo, S., DeStefan, R., & Giuditta, A. (2006). Brain and behavioral evidence for rest–activity cycles in *Octopus vulgaris*. *Behavioral Brain Research.*, *172*, 355–359.

Bushey, D., Huber, R., Tononi, G., & Cirelli, C. (2007). *Drosophila* hyperkinetic mutants have reduced sleep and impaired memory. *Journal of Neuroscience*, *27*, 5384–5393.

Cirelli, C. (2003). Searching for sleep mutants of *Drosophila melanogaster*. *BioEssays*, *25*, 940–949.

Cirelli, C., Bushey, D., Hill, S., Huber, R., Kreber, R., Ganetzky, B., et al. (2005). Reduced sleep in *Drosophila Shaker* mutants. *Nature*, *434*, 1087–1092.

Cornish, C. J. (1902). *The naturalist on the Thames*. London: Seeley and Col, Ltd.

Crocker, A., & Sehgal, A. (2008). Octopamine regulates sleep in *Drosophila* through protein kinase A–dependent mechanisms. *Journal of Neuroscience*, *28*, 9377–9385.

Denlinger, D. L., Giebultowicz, J. M., & Saunders, D. S. (Eds.). (2001). *Insect timing: Circadian rhythmicity to seasonality*. New York: Elsevier.

Dew, M. A., Hoch, C. C., Buysse, D. J., Monk, T. H., Begley, A. E., Houck, P. R., et al. (2003). Healthy older adults' sleep predicts all-cause mortality at 4 to 19 years of follow-up. *American Psychosomatic Society, 65*, 63–73.

Donlea, J. M., Ramanan, N., & Shaw, P. J. (2009) Use-dependent plasticity in clock neurons regulates sleep need in *Drosophila*. *Science, 324*, 105–108.

Eban-Rothschild, A. D., & Bloch, G. (2008). Differences in the sleep architecture of forager and young honeybees (*Apis mellifera*). *Journal of Experimental Biology, 211*(15), 2408–2416.

Engel, M., & Grimaldi, D. (2004). New light shed on the oldest insect. *Nature, 427*, 627–630.

Farris, S. M., Robinson, G. E., & Fahrbach, S. E. (2001). Experience- and age-related outgrowth of intrinsic neurons in the mushroom bodies of the adult worker honeybee. *Journal of Neuroscience, 27*, 6395–6404.

Foltenyi, K., Greenspan, R. J., & Newport, J. W. (2007). Activation of EGFR and ERK by rhomboid signaling regulates the consolidation and maintenance of sleep in *Drosophila*. *Nature Neuroscience, 10*, 1160–1167.

Ganguly-Fitzgerald, I., Donlea, J., & Shaw, P. J. (2006). Waking experience affects sleep need in *Drosophila*. *Science, 313*, 1775–1780.

Gilestro, G., Tononi, G. & Cirelli, C. (2009). Widespread changes in synaptic markers as a function of sleep and wakefulness in *Drosophila*. *Science, 324*, 109–112.

Goldschmid, R., Holzman, R., Weihs, D., & Genin, A. (2004). Aeration of corals by sleep swimming fish. *Limmol. Oceanography, 49*, 1832–1839.

Grimaldi, D., & Engel, M. S. (2005). *Evolution of the insects*. Cambridge: Cambridge University Press.

Harbison, S. T., & Sehgal, A. (2008). Quantitative genetic analysis of sleep in *Drosophila melanogaster*. *Genetics, 178*, 2341–2360.

Hartse, K. M. (1994). Sleep in insects and nonmammalian vertebrates. In M. H. Kryger, T. Roth, & W. C. Dement (Eds.), *Principles and practice of sleep medicine* (2nd ed., pp. 95–104). Philadelphia: W. B. Saunders.

Hastings, M. H., & Herzog. (2004). Clock genes, oscillators, and cellular networks in the suprachiasmatic nuclei. *Journal of Biological Rhythms, 19*, 400–413.

Haufe, W. O. (1963). Ethological and statistical aspects of a quantal response in mosquitoes to environmental stimuli. *Behaviour, 20*, 221–241.

Hendricks, J. C., Finn, S. M., Panckeri, K. A., Chavkin, J., Williams, J. A., Sehgal, A., et al. (2000a). Rest in *Drosophila* is a sleep-like state. *Neuron, 25*, 129–138.

Hendricks, J. C., Kirk, D., Panckeri, K. A., Miller, M. S., & Pack, A. I. (2003). Modafinil maintains waking in the fruit fly *Drosophila melanogaster*. *Sleep, 26*, 139–146.

Hendricks, J. C., Sehgal, A., & Pack, A. (2000b). The need for a simple animal model to understand sleep. *Progress in Neurobiology, 61*, 339–351.

Hendricks, J. C., Williams, J. A., Panckeri, K., Kirk, D., Tello, M., Yin, J. C.-P., et al. (2001). A non-circadian role for cAMP signaling and CREB activity in *Drosophila* rest homeostasis. *Nature Neuroscience, 4*, 1108–1115.

Hill, R. (1865). Zoology: The sleep of insects. In D. A. Wells (Ed.), *Annual of scientific discovery: Year book of facts and science and art for 1864* (pp. 278–282). Boston: Gould and Lincoln.

Huber, R., Hill, S., Holladay, C., Biesiadecki, M., Tononi, G., & Cirelli, C. (2004). Sleep homeostasis in *Drosophila melanogaster*. *Sleep, 27*(4), 628–639.

Iber, C., Ancoli Israel, S., Chesson, A., & Quan, S. F. (2007). *The AASM manual for the scoring of sleep and associated events: Rules, terminology, and technical specifications* (1st ed.). Westchester, IL: American Academy of Sleep Medicine.

Joiner, W. J., Crocker, A., White, B. H., & Sehgal, A. (2006). Sleep in *Drosophila* is regulated by adult mushroom bodies. *Nature, 441*, 757–760.

Jones, S., Pfister-Genskow, M., Cirelli, C., & Benca, R. M. (2008). Changes in brain gene expression during migration in the white-crowned sparrow. *Brain Research Bulletin, 76*, 536–544.

Jouvet, M. (1969). Biogenic amines and the states of sleep. *Science, 163*, 32–41.

Kaiser, W. (1988). Busy bees need rest, too. *Journal of Comparative Physiology, A: Sensory, Neural, and Behavioral Physiology, 163*, 565–584.

Kaiser, W. (1995). Rest at night in some solitary bees – A comparison with the sleep-like state of honey bees. *Apidologie, 26*, 213–230.

Kaiser, W. (2002). Honey bee sleep is different from chill coma – Behavioural and electrophysiological recordings in forager honey bees. *Journal of Sleep Research (Suppl.), 11*, 115.

Kaiser, W., Faltin, T., & Bayer, G. (2002). Sleep in a temperature gradient: Behavioural recordings from forager honey bees. *Journal of Sleep Research (Suppl.), 11*, 115–116.

Kaiser, W., & Steiner-Kaiser, J. (1983). Neuronal correlates of sleep, wakefulness, and arousal in a diurnal insect. *Nature, 301*, 707–709.

Kaiser, W., Weber, T., & Otto, D. (1996). Vegetative physiology at night in honey bees. In N. Elsner & H.-U. Schnitzler (Eds.), *Proceedings of the 24th Gottingen Neurobiology Conference 1996* (Vol. 2). New York: Verlag Stuttgart.

Kaslin, J., Nystedt, J. M., Ostergard, M., Peitsaro, N., & Panula, P. (2004). The orexin/hypocretin system in zebrafish is connected to the aminergic and cholinergic systems. *Journal of Neuroscience, 24*, 2678–2689.

Klein, B. A., Olzowy, K. M., Klein, A., Saunders, K. M., & Seeley, T. D. (2008). Caste-dependent sleep of worker honey bees. *Journal of Experimental Biology, 211*, 3028–3040.

Koh, K., Evans, J. M., Hendricks, J. C., & Sehgal, A. (2006). A *Drosophila* model for age-associated changes in sleep–wake cycles. *Proceedings of the National Academy of Sciences of the United States of America, 103*, 13843–13847.

Koh, K., Joiner, W. J., Wu, M. N., Yue, Z., Smith, C. J., & Sehgal, A. (2008). Identification of SLEEPLESS, A sleep-promoting factor. *Science, 321*, 372–376.

Kripke, D., Garfinkel, L., Wingard, D. L., Klauber, M. R., & Marler, M. R. (2002). Mortality associated with sleep duration and insomnia. *Archives of General Psychiatry, 59*, 131–136.

Kume, K., Kume, S., Park, S. K., Hirsh, J., & Jackson, F. R. (2005). Dopamine is a regulator of arousal in the fruit fly. *Journal of Neuroscience, 25*, 7377–7384.

Lancel, M. (1999). Role of GABAA receptors in the regulation of sleep: Initial sleep responses to peripherally administered modulators and agonists. *Sleep, 22*, 33–42.

Liu, W., Guo, F., Lu, B., & Guo, A. (2008). *Amnesiac* regulates sleep onset and maintenance in *Drosophila melanogaster*. *Biochemical and Biophysical Research Communications, 372*, 798–803.

Lyamin, O. I., Pryaslova, J., Lace, V., & Siegel, J. M. (2005). Continuous activity in cetaceans after birth. *Nature, 435*, 1177.

Meshi, A., & Bloch, G. (2007). Monitoring circadian rhythms of individual honey bees in a social environment reveals social influences on postembryonic ontogeny of activity rhythms. *Journal of Biological Rhythms*, *22*, 343–355.

Mignot, E. (2005). Narcolepsy: Pharmacology, pathophysiology, and genetics. In M. H. Kryger, T. Roth, & W. C. Dement (Eds.), *Principles and practice of sleep medicine* (4th ed., pp. 761–770). Philadelphia: W. B. Saunders.

National Sleep Foundation. (2008). *Sleep in America poll*. Retrieved December 9, 2008, from www.sleepfoundation.org.

Newman, S., Paletz, E., Rattenborg, N. C., Obermeyer, W., & Benca, R. M. (2008). Sleep deprivation in the pigeon using the disk-over-water method. *Physiology and Behavior*, *93*, 50–58.

Nichols, C. D. (2006). *Drosophila melanogaster* neurobiology, neuropharmacology, and how the fly can inform central nervous system drug discovery. *Pharmacology and Therapeutics*, *112*, 677–700.

Nitz, D. A., van Swinderen, B., Tononi, G., & Greenspan, R. J. (2002). Electrophysiological correlates of rest and activity in *Drosophila melanogaster*. *Current Biology*, *12*, 1934–1940.

Pitman, J. L., McGill, J. J., Keegan, K. P., & Allada, R. (2006). A dynamic role for the mushroom bodies in promoting sleep in *Drosophila*. *Nature*, *441*, 753–756.

Raizen, D. M., Zimmerman, J. E., Maycock, M. H., Ta, U. D., You, Y., Sundaram, M. V., et al. (2008). Lethargus is a *Caenorhabditis elegans* sleep-like state. *Nature*, *451*, 569–572.

Ramon, F., Hernandez-Falcon, J., Nguyen, B., & Bullock, T. H. (2004). Slow wave sleep in crayfish. *Proceedings of the National Academy of Sciences of the United States of America*, *101*, 11857–11861.

Rattenborg, N. C. (2006). Do birds sleep in flight? *Naturwissenschaften*, *93*, 413–425.

Rattenborg, N. C., Lesku, J. A., Martinez-Gonzalez, D., & Lima, S. L. (2007). The nontrivial functions of sleep. *Sleep Medicine Reviews*, *11*, 405–409.

Rau, P., & Rau, N. (1916). The sleep of insects: An ecological study. *Annals of the Entomological Society of America*, *9*, 227–274.

Rechtschaffen, A., & Bergmann, B. M. (2002). Sleep deprivation in the rat: An update of the 1989 paper. *Sleep*, *25*, 18–24.

Rechtschaffen, A., Gilliland, M. A., Bergmann, B. M., & Winter, J. B. (1983). Physiological correlates of prolonged sleep deprivation in rats. *Science*, *221*, 182–184.

Rial, R. V., Nicolau, M. C., Gamundí, A., Akaârir, M., Aparicio, S., Garau, C., et al. (2007a). Sleep and wakefulness, trivial and nontrivial: Which is which? *Sleep Medicine Reviews*, *11*, 411–417.

Rial, R. V., Nicolau, M. C., Gamundí, A., Akaârir, M., Aparicio, S., Garau, C., et al. (2007b). The trivial function of sleep. *Sleep Medicine Reviews*, *11*, 311–325.

Rubin, G. M., Yandel, M. D., & Wortman, J. R. (2000). Comparative genomics of the eukaryotes. *Science*, *287*, 2204–2215.

Sauer, S., Herrmann, E., & Kaiser, W. (2004). Sleep deprivation in honey bees. *Journal of Sleep Research*, *13*, 145–152.

Sauer, S., Menna-Barreto, L., & Kaiser, W. (1998). The temporal organization of rest and activity in newly emerged honeybees kept in isolation – initial results. *Apidologie*, *29*, 445–447.

Schmolz, E., Hoffmeister, D., & Lamprecht, I. (2002). Calorimetric investigations on metabolic rates and thermoregulation of sleeping honeybees (*Apis mellifera carnica*). *Thermochimica Acta*, *382*, 221–227.

Seugnet, L., Boero, J., Gottschalk, L., Duntley, S., & Shaw, P. (2006). Identification of a biomarker for sleep drive in flies and humans. *Proceedings of the National Academy of Sciences of the United States of America*, *103*, 19913–19918.

Seugnet, L., Suzuki, Y., Vine, L., Gottschalk, L., & Shaw, P. J. (2008). D1 receptor activation in the mushroom bodies rescues sleep-loss–induced learning impairments in *Drosophila*. *Current Biology*, *18*, 1110–1117.

Shaw, P., Ocorr, K., Bodmer, R., & Oldham, S. (2008). *Drosophila* aging 2006/2007. *Experimental Gerontology*, *43*(1), 5–10.

Shaw, P. J., Cirelli, C., Greenspan, R. J., & Tononi, G. (2000). Correlates of sleep and waking in *Drosophila melanogaster*. *Science*, *287*, 1834–1837.

Siegel, J. M. (2005). Clues to the functions of mammalian sleep. *Nature*, *437*, 1264–1271.

Sogaard Andersen, F. (1968). Sleep in moths and its dependence on the frequency of stimulation in *Anagasta kuehniella*. *Opuscula Entomologica*, *33*, 15–24.

Stephenson, R., Chu, K. M., & Lee, J. (2007). Prolonged deprivation of sleep-like rest raises metabolic rate in the Pacific beetle cockroach, *Diploptera punctata* (Eschscholtz). *Journal of Experimental Biology*, *210*, 2540–2547.

Stickgold, R., & Walker, M. P. (2005a). Memory consolidation and reconsolidation: What is the role of sleep? *Trends in Neurosciences*, *28*, 408–415.

Stickgold, R., & Walker, M. P. (2005b). Sleep and memory: The ongoing debate. *Sleep*, *28*, 1225–1227.

Tobler, I. (1983). Effect of forced locomotion on the rest–activity cycle of the cockroach. *Behavioral Brain Research*, *8*(3), 351–360.

Tobler, I. (2005). Phylogeny of sleep regulation. In M. H. Kryger, T. Roth, & W. C. Dement (Eds.), *Principles and practice of sleep medicine* (4th ed., pp. 77–90). Philadelphia: W. B. Saunders.

Tobler, I., & Stalder, J. (1988). Rest in the scorpion – A sleep-like state? *Journal of Comparative Physiology, A: Sensory, Neural, and Behavioral Physiology*, *163*, 227–235.

Van Buskirk, C., & Sternberg, P. W. (2007). Epidermal growth factor signaling induces behavioral quiescence in *Caenorhabditis elegans*. *Nature Neuroscience*, *10*, 1300–1307.

van Swindern, B., Nitz, D. A., & Greenspan, R. J. (2004). Uncoupling of brain activity from movement defines arousal states in *Drosophila*. *Current Biology*, *14*, 81–87.

Vertes, R. P., & Siegel, J. M. (2005). Time for the sleep community to take a look at the purported role of sleep in memory processing. *Sleep*, *28*, 1228–1229.

Williams, D. M. (2001). *Largest*. In T. J. Walker (Ed.), *University of Florida book of insect records* (http://recbk.ifas.ufl.edu/recbk.htm).

Williams, J. A., Sathyanarayanan, S., Hendricks, J. C., & Sehgal, A. (2007). Interaction between sleep and the immune response in *Drosophila*: A role for the NFkB Relish. *Sleep*, *30*, 389–400.

Winsor, J., Nishino, S., Sora, I., Uhl, G. H., Mignot, E., & Edgar, D. M. (2001). Dopaminergic role in stimulant-induced wakefulness. *Journal of Neuroscience*, *21*, 1787–1794.

Wu, M. N., Koh, K., Yue, Z., Joiner, W. J., & Sehgal, A. (2008). A genetic screen for sleep and circadian mutants reveals mechanisms underlying regulation of sleep in *Drosophila*. *Sleep*, *31*, 465–472.

Xing, X. (2004). A new troodontid dinosaur from China with avian-like sleeping posture. *Nature*, *431*, 838–841.

Xu, Y., Padiath, Q., Shapiro, R., Jones, C. R., Wu, S. C., Saigoh, N., et al. (2005). Functional consequences of a CKI mutation causing familial advanced sleep phase syndrome. *Nature*, *434*, 640–644.

Yokogawa, T., Marin, W., Faraco, J., Pezeron, G., Appelbaum, L., Zhang, J., et al. (2007). Characterization of sleep in zebrafish and insomnia in hypocretin receptor mutants. *PLoS Biology*, *5*, 2379–2397.

Yuan, Q., Joiner, W. J., & Sehgal, A. (2006). A sleep-promoting role for the *Drosophila* serotonin receptor 1A. *Current Biology*, *16*, 1051–1062.

Zeplin, H., & Rechtschaffen, A. (1974). Mammalian sleep, longevity, and energy metabolism. *Brain, Behavior, and Evolution*, *10*, 425–470.

Zeplin, H., Siegel, J. M., & Tobler, I. (2005). Mammalian sleep. In M. H. Kryger, T. Roth, & W. C. Dement (Eds.), *Principles and practice of sleep medicine* (4th ed., pp. 91–100). Philadelphia: W. B. Saunders.

Zheng, X., & Sehgal, A. (2008). Probing the relative importance of molecular oscillations in the circadian clock. *Genetics*, *178*, 1147–1155.

Zimmerman, J. E., Rizzo, W., Shockley, K., Raizen, D., Naidoo, N., Mackiewicz, M., et al. (2006). Multiple mechanisms limit the duration of wakefulness in *Drosophila* brain. *Physiological Genomics*, *27*(3), 337.

Zimmerman, J. E., Naidoo, N., Raizen, D. M., & Pack, A. I. (2008a). Conservation of sleep: Insights from nonmammalian model systems. *Trends in Neurosciences*, *31*, 371–376.

Zimmerman, J. E., Raizen, D. M., Maycock, M., Maislin, G., & Pack, A. I. (2008b). A video method to study *Drosophila* sleep. *Sleep*, *31*, 1557–1598.

3

Schooling by continuously active fishes: Clues to sleep's ultimate function

J. LEE KAVANAU

Introduction

Aquatic habitats were the cradle of sleep many million years before sleep evolved in terrestrial animals. Yet these habitats were the last to be explored in seeking sleep's ultimate function, which I have suggested to be the enabling of highly efficient brain operation at all times. Although the sleep of most fishes is essentially indistinguishable from that of terrestrial vertebrates, by exploiting the rich variety and greater permissiveness of aquatic habitats, some fishes have bypassed a need for sleep. In three continuously active states, they purportedly achieve comparable, and even greater, benefits than is provided by sleep, yet remain perpetually vigilant.

I propose that "schooling" (swimming synchronously in polarized groups) by these fishes plays a major role in the lack of a need for sleep. Thus, by schooling, they are able to achieve sleep's benefits without closing or occluding their eyes, namely, a great reduction in the average school member's reception and processing of external sensory input. Because the evident benefits of schooling by some fishes substitute for the obscure benefits of sleep by closely related fishes, the evident benefits give clues to the obscure ones. These clues support views on the ultimate function of sleep.

After reviewing circumstances relating to the evolution of sleep in terrestrial animals, relevant topics in the lives of fishes are treated, emphasizing their sleep and its awake, almost equivalent functions with eyes open.

Origin and ultimate function of sleep in terrestrials

Recent approaches to the problem of the function of sleep from the perspectives of brain functions and Darwinian natural selection are unraveling sleep's long-standing mysteries. Merely from sleep's continued existence, an evolutionist would anticipate that it maintains an overall high level of efficiency of brain operation. That high level very likely was achieved by the brain's postponing to the new state, sleep, all nonurgent activities that conflicted with the crucial activities of wakefulness.

Which brain activities are nonurgent during waking? And which brain activities, in the interests of efficient operation, can safely be postponed until sleep? Here, neuroscience provides potential answers. The major, safely delayed, brain activity carried out during sleep is memory processing (reviewed by Maquet, 2001). This processing consists largely of consolidating recently acquired short-term memories (i.e., converting them to long-term memories) and reinforcing (maintaining and/or strengthening) the long-term memories already stored (Kavanau, 2005).

Why cannot this memory processing occur safely and efficiently during a continuous state of wakefulness, as it does in many invertebrates? In other words, why is much memory processing usually postponed until sleep? To answer this question, one must turn back evolutionary time to the early appearance of brains and simpler collections of nerve cells in nonsleeping invertebrates. In those ancient times, natural selection would have favored animals with versatile nerve cells, that is, nerve cells that had evolved the capacity to carry out more than one function. We recognize these circumstances as a "fundamental dogma of neuroscience" (Rauschecker, 1995). According to this dogma, memories are stored in the same collections of nerve cells involved in processing, analyzing, and controlling responses to the circumstances to be remembered.

Such circuits with multiple functional capacities were highly adaptive for invertebrates with the simple, stationary lifestyles in which the circuits evolved. This versatility, however, led to conflicts in later, more complex, highly mobile lifestyles that many invertebrates and almost all vertebrates achieved. For example, consider circuits in highly mobile vertebrates that were occupied with a crucial function during waking, say escape from a newly encountered predator. The same circuits would not have been able to carry out, simultaneously and highly efficiently, the much lower priority functions of consolidating recently acquired memories, including the memory of how the escape occurred, and reinforcing already-stored long-term memories.

These consolidations and reinforcements would have had to wait until a subsequent less critical period, namely, "restful waking" or primitive sleep. Such delays in the consolidation of memories of immediately preceding actions are accommodated by the inherently slow rates at which synaptic efficacies decay (Kavanau, 1997a). These slow decay rates provide the "bridging" conditions that underlie the existence of short-term memory.

Just such potential conflicts in highly mobile animals in complex environments – an inability of brain circuitry with multiple functional capacities to cope efficiently with all its functions simultaneously – is suggested to have been a selective pressure for earliest sleep.

Detailed focal vision's role in the origin of primitive sleep

The evolutionary progression toward primitive sleep is thought to have begun when animals were evolving increasingly complex, highly mobile lifestyles and detailed focal vision (vision that recreates a complex scene). Such vision requires enormous amounts of neural processing (Llinás & Paré, 1991). Almost half the neocortical circuitry of the primate, for example, is devoted to representing the pictorial world. Visual processing is vastly more complex than that for any other sense, probably even more than that for all other senses combined.

In those ancient times, the lifestyles of animals acquiring great mobility and detailed focal vision would have become markedly altered. With sharper discriminations and engagement in multifarious new activities, including fast, wide-ranging movements and rapid actions and reactions, it would have become crucial to retain greatly increased stores of memories for the long term.

In these circumstances, the parallel processing capacity of some brain regions would have become severely taxed. It would have become increasingly difficult for circuitry in these regions to meet waking demands associated with crucial, largely unpredictable hazards and routine but essential activities, and at the same time meet needs to consolidate and reinforce large stores of memories. Interference between multiple brain activities, even at a less taxing level, has long been known (Kavanau, 1997a).

In essence, an adaptation that initially conferred great efficiency of brain operation, before the evolution of great mobility and detailed focal vision, would have become increasingly less efficient as these highly mobile, more complex visual lifestyles evolved had not compensating adaptations evolved in parallel – first restful waking and ultimately primitive sleep.

Accordingly, the selective pressure for primitive sleep would appear to have been the need to resolve the developing potentialities for conflict. This could have been achieved most readily through the provision of a period with a greater degree of brain unresponsiveness to outside occurrences ("drastic and global isolation of the cerebral cortex"; Buzsáki, 2006) than exists during restful waking, namely, primitive sleep.

By providing a portion of the 24-hour cycle when enormously increased needs for memory processing could be accommodated, primitive sleep obviated possible conflicts of memory processing with crucial waking brain activities. The latter activities, of course, are largely processing and responding to increasingly complex and varied visual inputs. It follows from this proposed mode of origin that entirely sessile and very slow moving invertebrates – such as mussels, sea anemones, and some worms – would have no need for sleep. It also follows that sleep would be engaged in only during that part of an animal's existence when danger was at a relative minimum and rapid movements usually were unnecessary – for example, during the night for day-active animals.

The long-sought ultimate function of sleep thus appears to be an "enabling" one. Sleep enables the brain to operate with high efficiency at all times. It does this by providing an "offline" state to which low-priority brain activities, particularly the enormously increased need for processing memories, can be deferred. Little wonder that identifying this ultimate function has been elusive, because the principal avenues of investigation of sleep in the past have been, and mostly remain, along medically oriented rather than evolutionarily oriented lines.

This scenario for primitive sleep's origin does not preclude subsequent or accompanying evolution of other proximate functions than memory processing that may have become essential. Indeed, several other such functions now play important roles. These include rest and rejuvenation and many deep-seated rhythmical changes that influence physiological processes (Kavanau, 1997a).

Two conspicuous phenomena that appeared, superficially, to contradict this paradigm long obscured sleep's ultimate function. One of these was the continued need for sleep by blind mammals, despite a total absence of visual processing. This apparent conflict is resolved by findings that once visual neocortical regions have evolved highly specialized roles, the regions do not simply "lie unused" in blind mammals. Instead, they adapt and assume active new roles serving other, often enhanced sensory modalities, with resulting continuing potentials for conflict with memory processing (see Kavanau, 1997b, 1998).

The other conspicuous, superficially contradictory phenomenon is encountered in this chapter – namely, the absence of sleep in some fishes despite their possession of excellent vision. Concerning this phenomenon, any methods by which active, awake animals could, with a high degree of safety, reduce or avoid reception of, or the need to process, complex visual inputs also would be candidates for enabling highly efficient brain operation. Just such methods, involving both behavioral and environmental influences, have evolved in some fishes and reconcile the second apparently contradictory phenomenon (Kavanau, 1997a,b, 1998).

Examination of the biology, ecology, and behavior of fishes reveals the existence of deep-seated relationships between various forms of waking activity, sleep, and efficiency of brain operation. Included are three waking vigilance states that essentially provide the benefits of – usually even greater benefits than – sleep. These states are "stationary schooling," "active schooling," and "pelagic cruising."

"Stationary schooling" describes schools that remain at the same location. "Active schooling" describes schools that also travel. In "pelagic cruising," individuals or groups, whether schooling or not, travel long distances in pelagic environments.

These topics are treated in the remainder of this chapter, as exemplified by the habits and habitats of teleosts. The findings reinforce the previous proposals regarding adaptations that achieve continuous high efficiency of brain operation. In the waking vigilance states of schooling and pelagic cruising, reception and processing of sensory information are greatly reduced. On the one hand, this enables highly efficient memory processing. On the other (except in stationary schoolers), the accumulation of ontogenetic (experiential) memories and the need for them are greatly reduced.

Space does not permit a comparable treatment of sharks, some of which also are active continuously. Their morphology, behavior, and ecology, however, greatly resemble those of large scombrids (Kavanau, 1998), a prominent expression of the relationship that "convergent evolution of form and function dominates the epipelagic [upper pelagic] ocean" (Hamner, 1995).

Biology, ecology, and behavior of teleosts

Origin, history, and distinguishing features

Teleosts constitute the most numerous higher bony fishes of the subclass Actinopterygii (ray-finned fishes), including more than 99.9% of living representatives. All Devonian (410 to 360 Mya [million years ago]) ray fins were marine, probably pelagic. A massive extinction occurred about 5 million years before the end of the Devonian. An estimated 75% of piscine families became extinct. Few new groups appeared after this event, and piscine diversity was slow to recover. Ray fins underwent a great burst of diversification in the late Paleozoic and early Mesozoic, and have since increased continually in importance.

The main line of evolutionary progression was through a series of generalized carnivores with increasing improvements in basic feeding mechanisms, more powerful swimming, greater agility, and greater potentials for adaptive radiations. By Cretaceous times (129 to 65 Mya), all major phyletic lines had become established, occupying both marine and freshwater environments. All major groups weathered the Cretaceous extinctions comparatively unscathed, prefacing a very marked increase in teleost diversity and abundance. The teleost "explosion" of the early Eocene (about 55 Mya) was the most dramatic evolutionary radiation in vertebrate history (Janvier, 1996).

The key feature defining teleosts is the presence of small, paired uroneural bones in the tail. These help stiffen and support the tail fin, giving both greater swimming power and potential for evolving diverse body shapes. Additional distinguishing features include (1) highly modified jaws; (2) large ventral gill-arch bones; (3) many specializations of head musculature; (4) development of a symmetrical homocercal caudal fin and narrow-based, highly mobile paired fins; and (5) a more rigid vertebral column.

The versatile designs of the teleostean buccal cavity enhance suction feeding and food handling, making possible rapid adjustments in the gape, biting force, and protrusion of the jaws. By accommodating changes in nature and behavior of prey, these abilities have adapted teleosts to a multitude of different feeding procedures. They can utilize virtually every food resource encountered (Gerking, 1994; Long, 1995).

Vision, the eye, the retina, olfaction, and audition

Vision

The senses, largely vision, play key roles in piscine schooling. Vision is excellent in most teleosts. Its sensitivity generally equals that of humans. The visual system discriminates a wide range of differences in shape and color, making vision the most important sense for prey detection, chase and capture, and communication (Guthrie, 1981).

Piscivorous attacks on individuals of a school of prey fishes that are "confused," swim erratically, or stray from the main body well illustrate the importance of visual cues in predation and provide suggestive evidence of schooling's antipredator effectiveness (Keenleyside, 1979). Notwithstanding vision's importance, there is considerable evidence that olfaction (separate from the sense of taste) generally mediates chemical signals. These are involved in various behaviors, including food detection, habitat selection, homing, migration, and every aspect of reproduction (Noakes & Godin, 1988).

The eye

Most optical structures follow the basic vertebrate plan. The lens is spherical, optically symmetrical, and usable across its entire extent. Resolving power is determined by retinal properties, partly cone concentration and number of supporting ganglia (Tamura, 1968). Underwater images are generally of poor quality. Because of backscatter, objects of high contrast rapidly fade from sight at 40 to 60 meters, even in clearest water (Lythgoe, 1979).

In the aquatic environment, where color-contrast detection and hue discrimination assume great importance, visual acuity beyond 4 to 5 feet may serve no useful purpose (Guthrie, 1986). Diurnal species tend to have smaller eyes than nocturnals. Among the exceptions, primarily nocturnal moray eels and brotulids (deep-water, primarily marine forms) have relatively small eyes. Their predation probably relies primarily on olfaction and tactile senses (Breder, 1967).

From the present perspectives, the most significant deviation from the basic vertebrate plan is the absence in all but a very few species of eyelids or other eye-occluding structures. In some species this leads to novel methods of reducing visual input (often by physically occluding one or both eyes by partial or total burial in sand). Although the eye enlarges throughout life, the field of view remains unchanged. The rate of eye growth relative to body growth depends on the importance of vision and may be under social control in some species (Fernald, 1991).

The retina

Retinal structure is more complex in fishes than in mammals. Cone cells are of composite morphology. They may be single, double (with differing visual pigments), twins (usually with the same pigment), triple, and even quadruple. Cones of diurnal species are more numerous and often smaller than in rod-dominated nocturnals. Not all types of cones migrate in all species (the "retinomotor response"; see Wagner, 1990). Teleostean ultraviolet photoreceptors provide perhaps the greatest vertebrate spectral sensitivity range, extending from below 3500 to above 8000 angstroms (Douglas & Hawryshyn, 1990).

Spectral sensitivities of cones optimize contrast in the aquatic environment (Fernald, 1993). Visual acuity generally increases with age and decreasing angle between adjacent cones. Proportions of scotopic pigments (rhodopsin and porphyropsin), which become seasonally altered in some forms, also tend to correlate with properties of the photic environment. Teleosts with forward-located eyes have restricted but useful binocular vision (36 degrees of overlap in the African cichlid, *Haplochromis bartoni*). Adaptations for binocular vision are found only in deep-sea forms (e.g., stalked and tubular eyes). (Douglas & Hawryshyn, 1990; Fernald, 1988; Munz & McFarland, 1977; Wagner, 1990).

The acousticolateralis system and inner ear

The lateral line system (both mechano- and electrosensitive) enables a fish to perceive not only water currents and turbulence directly around its body but also surface waves. It also appears to be involved in monitoring speed and direction of travel of nearest neighbors in schools, functions that overlap partially with those of vision (Partridge & Pitcher, 1980). For example, as a station-keeping device, the lateral line of sprats, which "hardly ever collide," apparently senses earliest changes in a neighbor's acceleration as well as changes of position and distance (through monitoring of the nearest neighbor's tail) (Denton & Gray, 1988).

Indirectly, the lateral line detects low-frequency sound waves (20 to 500 hertz) (Breder, 1967). Besides being implicated in schooling behavior, it functions in predator avoidance, the detection and localization of prey in the dark (Bleckmann, 1993), feeding, and social communication.

The inner ear (labyrinth) is the major organ for detection of underwater sounds, linear acceleration, and gravity. Teleosts are acutely sensitive to sounds, with sound communication apparently widespread. Social interactions often involve intricate signals, operating sequentially and originating from different modalities, often including sound (Myrberg, 1997).

Teleosts commonly produce low-frequency sounds when disturbed by predators, during reproductive activity, and when subjected to noxious stimuli. Sensitivity, however, is rather limited (less than 2 to 3 kilohertz) compared to that of "higher" vertebrates. Many species are acutely sensitive to sound pressure. They can locate particular sound sources in three dimensions. Both ears are essential for an acoustical sense of space. Presumably fishes are able to resolve pressure and particle displacement, postulated to be carried out in the central nervous system (Popper, Rogers, Saidel, et al., 1988). Together, the inner ear and acousticolateralis system of adults play several important roles (Hawkins, 1993; Noakes & Godin, 1988).

Sleep, sleep-swimming, restful waking, activity, and schooling

Sleep and restful waking: Free-living fishes

Among some teleosts sleep is as pronounced as among terrestrials, and sleepers are as "withdrawn" or more so; moreover, the need for sleep correlates with varied and complex daily activities. One cannot always distinguish between sleep and restful waking during inactivity. Positions adopted vary, not only in different groups but in closely related congeners (Norman, 1931). Wrasses (family Labridae, mostly diurnal) lie on the substrate on their sides. Plaice (*Pleuromecies*) and dabs (*Limanda*) are found just above the substrate. Many sleeping fishes occupy species-specific positions, often partially or fully buried in sand (Weber, 1961). Those most lively during the day, the multicolored coral species, are least active at night, sleeping hidden between coral blocks and rocks. In most species, nocturnal respiration slows and response thresholds markedly increase. Some pelagic species float in open water. The multitude of tropical species living in shallow coastal areas and coral reefs regularly sleep repeatedly at the same sites. Some embed themselves in the substrate or occupy caves or other retreats.

A horizontal inactive position is most frequent, with the head higher than the tail. Some nearly stand on their heads or lie on one side. Others even float on their backs. Primitive, predatory bichirs (*Polypterus,* freshwater, air-breathing) spent their nights at the same sites "packed together like sardines in a box" (Weber, 1961). In some cases, one member apparently acted as a lookout. Likewise, while a shoal of about 60 Mediterranean wrasses *(Symphodus ocellatus)* slept buried in the sand for five consecutive nights, one member remained visible.

When stirred from deep slumber, triggerfishes *(Pseudobalistes)* took up to 5 minutes to awaken completely. Individuals not uncommonly slept under overhanging ledges, pressed against the underside. During inactivity, they generally held their fins smoothly against their bodies. In the transition to inactivity, gill beats per minute declined 78% in goldfish *(Carassius auratus)*. The highly diurnal cichlids (family Cichlidae) and the yellow perch *(Perca flavescens)* remained immobile at night (Muntz, 1990).

Sleep and restful waking: Captive fishes

Laboratory studies of a reef-dwelling wrasse *(Irideo bivittata)* revealed its behavioral inactivity to resemble that of sleeping mammals (Tauber, Weitzman, & Korey, 1939). There was gross overall inactivity, decreased responsiveness to alerting stimuli, and diminished and irregular respiratory rate. During the night numerous species were observed draped over or propped against rocks or lying in the sand.

Others buried themselves partially or completely under the sand or lay against or on top of one another on the sandy bottom. In this state they could be touched lightly for several minutes without evoking alerting responses. Sometimes they could be lifted almost to the water surface before becoming fully alert and fleeing (Tauber et al., 1939). Diurnal parrotfishes (family Scaridae) were also essentially immobile in darkness or dim light, with diminished respiratory rate (Munz & McFarland, 1977).

Diurnal cichlids *(Tilapia mossambica)* also had increased response thresholds and lowered respiratory rates (Reebs, 1992; Shapiro & Hepburn, 1976). Postures adopted by restful waking or sleeping fishes were also studied intensively (Siegmund, 1969). Perch *(Perca fluviatilis)* lay on the substrate at night for 5 to 10 hours with most fins closed. Tench *(Tinca tinca)* lay on the substrate during the day (for 4 to 15 hours) with all fins closed. For both species, gill-beat rates decreased as these fish became inactive (to 65% of normal in tench; Reebs, 1992).

Sleep, sleep-swimming, restful waking, and activity: Coral-reef communities

Coral reefs have the most complex fauna of any biotype. Much of their diversity consists of various adaptations to the rich reef habitats. Short periods of sleep during twilight are crucial to the survival of many inhabitants because predation by large piscivores is concentrated heavily then. Although the small forms are still in their diurnal stationary schools as sunset approaches, they become increasingly vulnerable as they disperse to feed. Solitary diurnal fishes often hug the reef surface, staying close to refuges too small for predator entry. The few diurnal forms that venture far from shelter are usually in large schools (Levine, 1993).

Most adults are strongly site-attached (Lowe-McConnell, 1977). Only the young of some species gather in schools. Elders hold feeding and breeding territories (Helfman, 1993). Smaller diurnal species spend the night quiescently in crevices among rocks. Some lie on their sides, usually at least partially covered with sand. In aquaria, they enter shells or bury themselves in sand. Others settle on the bottom, usually under at least partial cover, remaining continually quiet. Still others lie quietly on their sides in protective holes or nonconfining locations (Gerking, 1994). Some species display a strong affinity for specific sites several days running.

Most nocturnal species form daytime resting aggregations. Some are seen only occasionally by day. They remain deep in the shadows of overhanging rocks or ledges or just inside small caves. Others remain under cover among the rocks, motionless and flush against sheltered faces. Still other nocturnals hide among algae in shallow water. Sweepers (Pempheris adusta) feed all night in darkness in open water, far from the reef. Each returns daily to rest or sleep in the same cave (Levine, 1993).

About 50% of fishes in some habitats are piscivores; roughly the remainder are herbivores; this holds true the world over (Gerking, 1994). Bottom feeders of great diversity dominate coral reefs. Many nocturnal predators feed on various nocturnal crustaceans. The most successful feeding by generalized predators during the day is by ambushing and stalking. Diurnal feeders also include sessile-invertebrate specialists. Some predators seek out prey in their confined daytime and nighttime resting sites.

At dusk, diurnal species take reef cover, in an order corresponding roughly to increasing size, in a regular sequence requiring about 20 minutes. Few fishes are left by 10 to 15 minutes after sunset, when the vast majority drop suddenly to the reef below. The nocturnal species do not leave their resting sites until about 30 minutes after sunset, during which time the light level decreases 1000-fold (Munz & McFarland, 1977). Then they emerge rather suddenly. As a result, there is a precipitous decrease in the level of activity at dusk, culminating in a "quiet period" lasting about 20 minutes (twilight lasts 70 to 85 minutes; Helfman, 1993). During this time few fishes except stragglers and crepuscular predators are out. All small fishes keep close to the sheltering substrate.

Then, very abruptly, the water once more seems full of fishes. But the new arrivals are large-eyed nocturnal forms, moving away from the reef to feed, either into the water column or to distant sandy flats or sea-grass beds, more so on dark nights (Lowe-McConnell, 1987). The reef may appear deserted. Seeking cover often replays in reverse order to leaving it, generally beginning about 40 to 50 minutes before sunrise (but 10 to 15 minutes before by medium to large herbivores).

Hobson (1972) suggests that this well-ordered series of events is shaped by the threat of crepuscular predators. Although predators' eyes function poorly relative to prey capabilities during daylight and darkness, predators have a great visual advantage during the 30 minutes after sunset and before sunrise. Because the retinomotor response in different prey species requires 20 to 70 minutes (Blaxter, 1988), these species would be at great risk if they left resting sites at such times (Hobson, 1972, 1974; Munz & McFarland, 1977). Natural selection has also maximized the spectral sensitivity of the prey's scotopic twilight vision. Both their acute vision and behavior reduce twilight predation on them.

Severe hypoxia sometimes develops at night in the inner sections of several species of branching hard corals (e.g., *Stylophora pistillata*) in shallow coral reefs in the Gulf of Aqaba, Red Sea. This results from normal blocking of impinging currents by coral structures. It is overcome by three species of sheltered, resident, site-attached, day-foraging zooplanktivorous fishes (e.g., *Dascyllus marginatus*) in a unique mutualistic relationship (Goldshmid, Holzman, Weihs, et al., 2004; Nilsson, Hobbs, & Ostlund-Nilsson, 2007). The fishes overcome hypoxia by engaging the entire night in an unusual "sleep-swimming" mode. In this mode, while holding positions or inhabiting quasi-separate swimming zones, they enhance water replenishment through energetic, high-frequency fin motions. Stroke frequencies are about twice the rate seen during normal swimming.

Schooling

It is a very widespread practice for fishes to swim together synchronously in polarized groups, known as "schools," which may contain many thousands of members (Cushing & Jones, 1968). If the "schooling" occurs at a relatively fixed location, it is termed "inactive" or "stationary" schooling; whereas if the schools travel, it is termed "active" schooling. Fully developed schooling is the habit of roughly 2000 marine and 2000 freshwater piscine species (Shaw, 1970).

The ability to school depends largely but not exclusively on vision. Thus, schools usually disperse in dim light (at 1/1000 to 1 lux in different species) or darkness (Blaxter, 1965; Partridge & Pitcher, 1980). But nonpolarized or poorly polarized aggregations may persist (Blaxter, 1965). Mackerels (*Scomberomorus* spp.) may disband and reform dense schools with variations in daylight. Their nighttime schooling is thought to depend on their bioluminescence (Sette, 1950). On the other hand, the schooling of tunas in the eastern tropical Pacific occurs in the "dark of the moon" and absence of bioluminescence (Tiews, 1963).

The participation of other sensory modalities in effecting schooling is shown by findings with temporarily blinded individual saithe (*Pollachius virens*), though not with other tested species. The saithe were able to join active schools and maintain their position in them indefinitely, but their reaction times were slower. Schooling was not possible, however, if the lateral line was sectioned. Lateral line section alone led to more accurate side-by-side orientation of neighbors, with eyes more closely in apposition (in normal formation, the heads of fishes are at the midpoints of their neighbors' bodies). Lateral line field-flow monitoring apparently is sufficient for "normal" schooling, whereas visual information is required to elicit normally timed evasive behavior (Partridge & Pitcher, 1980).

In most species, though, one gains the impression that vision is essential; for example, for the young catfish (genus *Corydoras*). "Neither blinded fishes nor normal fishes in the dark ever aggregate. . . . " (Bowen, 1931). Very little light suffices for some species. Thus, the tetra (genus *Aphyocharax*) will school at a light level as low as 1/1000 that on a clear moonless night (John, 1964).

School formation is based on an inherent mutual attraction. Concerning its basis, in addition to protecting small species from predation (see below), schooling is believed to provide the ultimate benefit of sleep – enabling highly efficient brain operation at all times (Kavanau, 1998). Unlike during sleep, however, vigilance is not largely sacrificed. Rather, the school, collectively, continues to receive, process, and respond to sensory inputs from the environment, with the momentary extent of participation by individual school members dependent on their positions within the school.

Processing all incoming sensory information and earliest responding is effected primarily by individuals in the van and secondarily by those in other peripheral positions. Individuals in a school's interior need merely orient with respect to neighbors and follow their movements. Because of this "division of labor," a schooling individual's average needs for processing complex visual and other sensory inputs are greatly reduced.

Thus although periods of sleep provide more time for memory processing by virtue of sensory input being essentially excluded, periods of schooling accomplish the same result by merely reducing the individual schooler's average sensory input. Both circumstances enable highly efficient brain operation. Whereas the benefits that derive from schooling are evident (as discussed above and below), sleep's benefits are obscure.

For small fishes, schooling can play a large role in protection from predation through synchronized movements, increased vigilance, and information transfer between members, ensuring that all become aware of impending danger. Tightly knit schools of damselfishes (family Pomacentridae) are said to be almost impervious to predation during the day, though they are more vulnerable at dusk, as their vision fails. However, circumstances also exist in which predation is facilitated.

Schooling is found in anatomically primitive as well as advanced piscine groups. The strength and rigidity of the habit is attested to by observations that departures from schooling are clearly associated with special circumstances (Breder, 1967). Some teleosts school from the beginning of independent locomotion until death, including periods of breeding. Others school only during certain stages. Thus only the fry or newborn of many species school (Keenleyside, 1979). These immatures are going through a period of intense learning, when needs for memory processing are at a peak. Their schooling during these phases doubtless is analogous to the greater periods of sleep required in neonate and infant mammals as compared to adults.

School members continually reappraise the costs and benefits of schooling, as reflected in decisions to join, stay with, or leave the school (Pitcher & Parrish, 1993). For example, in mixed schools of two species of juvenile parrotfishes, one species continued to school while the other sought shelter when attacked by predator models (Helfman, 1986). Schooling may optimize foraging benefits, being quite prevalent among species whose principal food is plankton. It may confer hydrodynamic advantages and facilitate reproduction and juvenile growth.

That schooling under threat of predation generally is adaptive is shown by the facts that (1) many group-living fishes react to an approaching predator by clumping more closely and swimming away in a polarized school; (2) the schooling tendency is well developed in guppies (Poecilia reticulata) that are heavily preyed on by characins (family Characidae) and cichlids, but this tendency is absent in guppies that experience very little predation; and (3) when water bodies become greatly restricted during low-water periods, schooling by Amazonian characins is their main adaptation for escaping predators. Most of these advantages of schooling also apply for shoals, usually the more so the larger the shoal (Breder, 1967, 1976; Pitcher & Parrish, 1993; Shaw, 1970; 1978).

Some fishes school stationarily, close to rocky bottoms offshore; others hover in large, stationary schools (often of mixed species) over sand adjacent to rocky bottoms. The smallest juveniles may feed on plankton in the water column in "feeding clouds." When threatened, they descend immediately to the reef (McFarland & Hillis, 1982). Daytime-schooling fishes include many carangids (a family of marine fishes, including jacks, pompanos, jack mackerels, and scads), lutjanids (snapper-like fishes), pomadasyids (grunts) certain mullids (goatfishes), and sciaenids (drums and jackknifes).

Active (traveling) schools usually involve conspecifics of similar age and size engaged in the same activities at a given time, with a high degree of synchrony and polarized swimming (except during feeding and courtship) (Magnuson, 1978). Thus the participants act in concert, moving forward simultaneously, keeping equal distances apart, and changing direction at apparently the same moment (within

0.15 to 0.25 seconds). This applies even when they are executing complicated evasive maneuvers (Bleckmann, 1993).

Fishes in some schools follow the diurnal vertical planktonic migrations, moving down and swimming in dense schools by day and moving up and dispersing to feed on the plankton at night (Lowe-McConnell, 1977). Some fishes spend the day in dense schools close to shore, forming milling masses or "tight balls" when under attack.

School members are often equipotential and schools usually are leaderless, although some position preferences do exist (Pitcher & Parrish, 1993). All the saithe in a still, nonpolarized ball follow the first to move out, all polarized on parallel paths (Wardle, 1985). In pods of young catfishes, leadership always falls to the fish in the van or the individual farthest in the direction of turning (Bowen, 1931).

At nightfall, many stationary schools break up, with individuals scattering to feed on small nocturnal invertebrates. For some small fishes that feed in stationary schools, individuals home not only to a particular reef site but often also to a specific location therein (McFarland & Hillis, 1982). Younger fishes of some species form shoals (nonpolarized, unsynchronized aggregations). Older ones hold territories (Helfman, 1993).

Many fishes that are loosely aggregated or even scattered when they are not feeding form tighter aggregates or schools as they begin to feed. Planktivores often feed in tight schools. Schooling also is characteristic of many shallow-water species while feeding in the water column. Some fishes swimming close to the bottom often travel in shoals or schools (Keenleyside, 1979).

Scombrids

Further attention concerning piscine biology, ecology, and behavior is directed largely to members of the family Scombridae (namely tunas, mackerels, bonitos, albacores). These are typically pelagic carnivores and mostly continuously swimming, active schoolers. There are two categories: (1) those inhabiting open waters far from land, such as large tunas, and (2) those frequenting coastal waters or sites a few 100 kilometers from land. The latter generally are small tunas, bonitos, and various mackerel. Continuous swimming is a way of life for most scombrids – "astounding bundles of adaptations for efficient and rapid swimming." It has coevolved with extremes of adaptations for reduced drag, efficient swimming, and high levels of energy utilization (Magnuson, 1978).

Scombrids: Adaptations for sustained high-speed cruising

Scombrids are swift, epipelagic predators, characteristically adapted for sustained, high-speed cruising (Moyle & Cech, 1996). The bodies of swimming tunas and bonitos scarcely flex. Significant lateral movement occurs only in the peduncle and caudal fin. The high, stiff, lunate or semilunate, widely forked, tapered, and enlarged caudal fin provides essentially all of the thrust. Its rays extend over the last vertebra, giving the relatively inflexible tail more driving force than if it were hinged, as in many other teleosts (Herald, 1961).

The pectoral and caudal fins have high aspect ratios, with resulting drag reduction (Magnuson, 1978). In mackerels, however, there is a moderate amount of flexibility to the body, and the tail swings through a relatively large arc in propulsion (Collette, 1977). The maneuverable pectoral fins are spread out at low or cruising speeds. At increasing speeds they are retracted progressively until, at high speeds, they are folded back into a groove, with the body alone providing lift.

Several tuna species, the eastern Pacific bonito (*Sarda chiliensis*) and the Atlantic mackerel (*Scomber scombrus*), have no gas bladder. They are obligate "ram gill ventilators" that sink if they do not gain lift continuously by swimming (Collette, 1977). Magnuson (1978) suggests that lack of a gas bladder increases vertical mobility in prey capture and predator avoidance, particularly near the surface. Minimal speeds of tunas to maintain hydrodynamic equilibrium range from 1.04 to 3.59 km/h (Magnuson & Weininger, 1977).

To meet the great oxygen consumption needs for such exertions, the gills have a much greater surface area. Maximal rates of oxygen consumption in active, free-swimming skipjacks are more than twice as great as in nonscombrids (Hazel, 1993). Generally, the rate of oxygen consumption may increase more than 100% for an increase in swimming speed of one body length per second. Anaerobic metabolism also increases (Jobling, 1993).

Adult scombrids probably are the world's fastest-swimming fishes, possessing the energy and structure for immense migrations at great speeds (referred to here as active schooling or pelagic cruising, depending on whether schools are formed). Yellowfin (*Thunnus albacares*) and wahoo (*Acanthocybium solandri*) tunas attain speeds up to 70 km/h during sprints of 10 to 20 seconds. Many cruise at 50 km/h (Imamura, 1951; Beamish, 1978; Smith & Heemstra, 1986).

Scombrids: Distribution and migrations

Scombrids range around the world in tropical, temperate, and even cold seas, often over water where bottom contact is lost (Herald, 1961). The extremely well-adapted tunas travel long distances at high efficiencies in the open ocean, where food resources are minimal (Magnuson & Weininger, 1977). Their important habitats are the complex of ocean currents composing the equatorial circulation and the great current gyres poleward of the equator as well as seas at suitable latitudes (Brock, 1965).

These voracious, slashing carnivores are notorious for attacking fishes, octopuses, and squid alike as they negotiate the open oceans (Gerking, 1994). Some cross the great oceans. In such crossings, they are lean when leaving the Straits of Florida in June and arriving in Norway in August, apparently without substantial food during journeys that can exceed 3 months (Rivas, 1977).

Tunas are basically warm-water species at temperatures up to 26 to 29°C, exceptionally as low as 5°C (Nakamura, 1954). Distribution and movement mirror oceanic regions of favorable environment. Many species move poleward during the summer months (Brock, 1965). Schools feed and swim much of the time at depths well below the surface (Sharp & Vlymen, 1977). The epipelagic habitat of tunas, mostly in tropical seas, is one of the most unproductive, nutritively dilute, and patchily distributed marine environments.

Notwithstanding these limitations, physiological properties imposed by adaptations for predatory efficiency – rather than food availability – are growth-limiting for adult skipjacks (*Katsuwonus pelamis*) and yellowfins (Kitchell, Neill, Dizon, et al., 1977). Most prespawning tunas are nomadic. Of the others, only spawners can be considered to be "directed" migrants (e.g., Atlantic bluefins, *Thunnus thynnus*). Nomadic behavior provides for rapid extension of ranges into "new" habitats and for utilization of sporadic or patchy blooms in areas peripheral to the "average" habitats.

Skipjacks are migratory, living in the open sea and following warm currents. Adults (maximum weight, 22 kilograms) inhabit the upper thermocline (Dizon, Brill, & Yuen, 1977; Yuen, 1970). The highly migratory bluefin group penetrates far into much cooler waters. They have very efficient heat exchangers (peripheral or cutaneous) and much greater weight (some exceeding 200 kilograms) and thermal inertia (even tending to maintain a fairly constant body temperature), For the Atlantic bluefin, these waters include the south polar seas, Iceland, and the Arctic Circle (Sella, 1952).

The bigeye tuna (*Thunnus obesus*) is usually found in still deeper and colder waters and tolerates the greatest temperature range. In the tropical western and central Pacific, the bigeye tuna is associated with the permanent thermocline (Brock, 1965; Collette, 1977). It can vary whole-body conductivity by a factor of 100 by disengaging and re-engaging its heat exchangers (Hazel, 1993). Yellowfins tolerate highest temperatures (Brock, 1965).

Many mackerels, lacking heat exchangers, tend to stay in the warm oceanic surface layer, although they have been found in abundance at temperatures as low as 7° to 8°C. Adults are distinctly open-sea inhabitants but rarely beyond continental shelf waters. They apparently spend the winter in the upper part of the continental shelf at depths rather greater than 200 to 400 meters (Sette, 1950).

The very slender, fusiform wahoos are migratory and oceanic; these are found around the world in tropical and subtropical waters.

Scombrids: Active schooling

Scombrids generally form large, highly polarized, active schools in their hunting (mostly on much smaller, densely schooling fishes or plankton) and spawning migrations (Shimada & Van Campen, 1951). Schools tend to be mixed (comprising 100 to 5000 individuals) with a strongly developed tendency to assort by size). The larger fishes travel in schools of fewer individuals, or at the bottoms of schools containing members of different sizes.

Schooling may begin as early as the postlarval and juvenile stages (Collette & Nauen, 1983; Pitcher & Parrish, 1993). As many as 50% of all juvenile teleosts school (Shaw, 1978). Atlantic bonito schools tend to be very large. Swimming depth varies in different species. Skipjacks descend to at least 70 meters and yellowfins to 150 meters. Although they may feed at swimming depth, tunas characteristically strike upward from below to take prey near the surface (Muntz, 1977). Many small mackerels feed principally on plankton (*Calanus finmarchicus*) in the Gulf of Maine (Sette, 1950).

For skipjack and yellowfin tunas, at least, mixed schools form when single-species schools are attracted together by an external stimulus, usually food (Yuen, 1962). Large-mouthed, piscivorous larvae, exemplified by those of the Pacific mackerel (*Scomber japonicus*), typically feed from a rigid position with a straight body. They capture other piscine larvae as well as their siblings with a forward lunge and engulfment. Prey too large to engulf are manipulated (Gerking, 1994).

Scombrids: Continuous swimming

"The tuna are, without doubt, the most highly specialized fishes in regard to sustained, high levels of locomotor activity.... they swim continually, never stopping to rest" (Magnuson, 1977); ".... all life's activities...are done on the move.... time of day, temperature, dissolved oxygen, and food deprivation have little influence on sustained swimming" (Magnuson, 1978), ".... the extremely streamlined tuna is an open-ocean fish that moves constantly, indulges in long migrations, and pursues fast-swimming schools of smaller fishes" (Migdalski & Fichter, 1976).

Other evidence of continuous activity by scombrids derives from fishery practices. Thus, shallow-water netting of tunas must be done in the evening or at night, when tuna are unable to discern the net meshes; otherwise they would take evasive action. On the other hand, setting of lines for hook-and-line fishing ordinarily begins before dawn (Nakamura, 1954). In other words, tunas are continuously active, and fishing must be tailored to ambient light conditions and visual capabilities.

Scombrids: "Partial warm-bloodedness"

Although many continuously active teleosts have partial warm-bloodedness (PWB), especially the larger scombrids, this is not essential for continuous activity. Thus, many continuously swimming fishes, such as mackerels and carcharhinoid sharks, are pure ectotherms. PWB occurs in only 13 species of the tuna tribe. These are unique in having the red oxidative muscles, placed internally along the vertebral column, that power their sustained swimming. Groups possessing only regional endothermy in the brain and eyes are the largest teleosts in the oceans. They travel 8000 kilometers or more in crossing the ocean (termed "pelagic cruising" if not in schools) on yearly migrations (Block, 1991; Collette, 1977; Smith & Heemstra, 1986).

Cranial endothermy is achieved via heat generation in either the lateral or superior rectus eye muscle (a "furnace beneath the brain"). Tunas achieve PWB by having the major arteries and veins for blood transport between the heart, gills, and rete located close to the skin. This enables them to transport cool blood to and from the rete without absorbing much swimming-muscle heat. Countercurrent heat exchangers also occur in tunas' head regions and viscera (Block, 1991; Hazel, 1993).

PWB extends the horizon of the tolerable thermal habitat, increases prowess in prey capture, shortens digestion times, and increases metabolic rates, all of which generally facilitate sustained activity, predatory success, and predator evasion. As a consequence, PWB tends to ensure a successful, largely routine existence, the implications of which are discussed below.

Overview: Why some fishes require no rest or sleep

This topic is introduced by recalling the ultimate function of sleep as deduced from the proposed circumstances surrounding its origin in terrestrial animals. Sleep enables the brain to operate with high efficiency at all times. It achieves this by eliminating the possibility of interference of memory processing with crucial waking activities (chiefly reception, processing, and responding to complex visual input) and vice versa. Sleep provides the "offline" periods to which the greatly increased needs for memory processing are largely confined.

If some fishes could eliminate interference between memory processing and crucial waking activities by some other means, they too, like many invertebrates, would have no need for sleep. Emphasizing "by some other means," this is precisely the manner by which it is proposed that certain fishes bypass a need for sleep. They accomplish the crucial elimination of interference by two other principal means.

The first means is direct and is facilitated simply by the possession of greatly reduced memory loads – the fewer the number of memories needing to be processed, the more readily processing is accomplished and the more efficient the operation of the brain. The second principal means is by schooling rather than sleeping. Schooling greatly reduces the average need to receive and process complex visual input by school members. In this way, it eliminates interference with memory processing, thereby achieving efficient brain operation at all times. Both principal means ultimately depend on the greater permissiveness of aquatic environments.

Pelagic cruisers, individually or in small groups, receive no schooling benefit as such, nor is any needed. Because their ancestors led a monotonous existence in a monotonous environment for many millions of years, much of it in dark or nearly dark habitats, their stores of phylogenetic (inherited) memories are greatly reduced. With memory processing being facilitated both directly by virtue of lesser numbers of memories needing to be processed and indirectly by lesser needs to process visual input, continuous, highly efficient brain operation is enabled.

For these animals, there would be little need or selection for lengthily persisting use-dependent alterations in synaptic efficacy – that is, for many long-term memories. Altered synaptic efficacies could be maintained merely through short-term activity dependence – that is, through frequent use.

The lifestyles of continuously active, stationarily schooling fishes lie at the other end of the spectrum. During their foraging periods they lead a complex existence in often complex environments. Their highly efficient memory processing during schooling depends on a lack of interference by the greatly reduced need to process complex visual inputs. The survival of these fishes depends heavily on the ability to retain memories lengthily. Their nervous systems must preserve activity-dependent alterations of synaptic efficacy and possess a large repertoire of long-term memories.

Actively schooling fishes receive memory-processing benefits both directly from their possession of lesser stores of memories and indirectly from lesser average needs to process sensory inputs by virtue of schooling. The degree of their direct memory reduction and processing benefits, as for pelagic cruisers, would depend on their degree of monotony of lifestyle and exposure to environmental monotony.

Influences of a comparatively routine pelagic existence

Among continuously swimming pelagic fishes, it is suggested that most memories are reinforced almost entirely through frequent use. These fishes are exposed repeatedly to essentially the same topographically featureless, largely pelagic environments that their ancestors encountered over spans of many millions of years.

It is unlikely that they experience much in the way of new behavioral needs during predation and escape from predators. Thus, in essence, they never encounter selective pressures for sleep – a need to eliminate interference between the waking brain's urgent orchestration of crucial activities and its nonurgent need to process memories. The lengthy existence of these fishes in unchanged or little changed environments is reflected in the little outward change seen in them, in many cases over millions of years. Recall in this connection, ".... convergent evolution of form and function dominates the epipelagic ocean...." (Hamner, 1995).

These observations are not meant to imply that different circumstances requiring the possession of long-term memories are not encountered occasionally. With sensory input and processing being greatly reduced in pelagic environments, particularly at night, consolidation and reinforcement of memories for the long term would not interfere with the brain's management of almost entirely routine, repetitive daily activities. The much greater interference that can occur in more complex lifestyles is suggested by the finding that forgetting of learned foraging sequences by sticklebacks on brine shrimp begins after only 2 days of withdrawal (Huntingford, 1993).

Most of the "well-developed patterns of escape, hunting, prey capture, and procurement of passive food" by pelagic fishes draw on phylogenetic (inherited) memories (Kavanau, 1997a). Except for very large fishes, predator avoidance and predation would usually be in use on a more or less daily basis and would be reinforced through use. This mode of existence deviates significantly from that of foraging piscine species with diel cycles of activity and sleep and/or restful waking. In them, learning and flexibility of behavior play important, not minor, roles. They must respond to predator pressures and mating opportunities, modify feeding and foraging strategies, and cope with environmental variability, particularly seasonal.

For example, as positions of patches and types of prey change, such a predator must repeatedly modify its behavior (Hart, 1993). Similarly, seasonal or even yearly switches between preferred food items may be necessary if a more efficient competitor alters the profitability balance (Welcomme, 1985). Reef dwellers must be capable of complex, visually related tasks; they must recognize and discriminate between the many coral species, identify and remember predators in and around their territories, and be able to recall the locations of resting sites.

Additionally, reef-dwelling fishes can acquire knowledge about social traditions via learning (Helfman, 1993), whereas intertidal teleosts learn and remember details of their environment, limit their movements to restricted areas, and home to these areas if displaced (Gibson, 1993). Only pelagic environments are sufficiently unvarying to allow a predator to survive employing almost completely stereotypic behavior.

A high proportion of piscine bodies are composed of muscles used in forward propulsion (Moyle & Cech, 1996). Probably not a day goes by without most muscles being in frequent use. Even reproductive systems and reproductive behaviors of scombrids probably receive more continuous use and the latter occur much less competitively than in terrestrial vertebrates. Many teleosts apparently mate indiscriminately, and gonadal sex reversal occurs in some during development; some are self-fertilizing, others are simultaneous hermaphrodites. Among still others, males do not exist (Turner, 1993).

It would follow that, of all vertebrates, the large, continuously swimming fishes probably give the most purely instinctive responses. We may be seeing a reflection of this in the assessment of Philippe Cousteau (Cousteau & Cousteau, 1970) that "...the shark is the most mechanical animal I know and his attacks are totally senseless...."

As in many invertebrates, some complex piscine behaviors show no evidence of adaptive change: they are performed precisely the first time, developing with remarkable stability despite environmental perturbations (Huntingford, 1993). Numerous examples could be given of feeding and habitat preferences, twilight migratory behavior, and diel foraging patterns that develop, not through learning but as a consequence of maturational changes. For example, swimming and feeding modes of Pacific mackerel develop in three stages, with the successive development of the caudal fin, jaws, and pharyngeal teeth (Noakes & Godin, 1988); that is, purely phylogenetic memories are involved.

Regarding the matter of brain complexity, the brains of teleosts are highly variable, small, and compact, with the exception that the optic lobes (tectum) often are very large. Direct correlations are evident between brain size and the complexity of teleost behavior (Guthrie, 1990). I would suggest, in the light of these circumstances, that the scombrid brain need store very few ontogenetic (learned or experiential) memories for the long term.

This is not to imply that scombrids are incapable of learning and storing memories of learned responses. Indeed, learning is believed to occur in all teleosts under natural conditions (Shaw, 1970). It implies only that, by leading an existence that rarely involves novel experiences, scombrids are rarely called on to store new ontogenetic memories. Of course, following the theses advanced here, teleosts that do not swim continuously but engage in restful waking or sleep (or school stationarily throughout the day – see below), constituting the vast majority – would be expected to lead an existence in which daily events were much less predictable, where a response learned (or a site visited) one day might need to be recalled many days later.

Obtaining the essential benefits of sleep by schooling

The evident protective benefits of schooling for small fishes would not be needed by large teleosts. Yet many large pelagic species spend most or all of their lives in schools (Breder, 1967; Wootton, 1992). The key consideration pointing to a probable ultimate function of schooling relates to brain operations facilitated by lesser needs, on the average, for sensory processing. The fishes at inner positions of schools need not exercise the full range of their sensory capabilities. They have no need to "listen," "smell," or process complex visual information. They need only maintain awareness of their position with respect to nearest neighbors.

Even school members at most of the peripheral positions would not require processing of complex visual input of more than one eye (internal "churning" of schools exposes one member after another to peripheral locations; Breder, 1967). On the average, then, the amount of sensory processing carried out in the brain of a school member is greatly reduced compared to the amount in alert, solitary swimmers.

In effect, the burden of sensory processing is shifted from individuals to the entire school collectively (Grunbaum, 1997; Norris & Schilt, 1988; Warburton, 1997). One of many phenomena consistent with these interpretations is the finding that stationarily schooling bluegill sunfishes (*Lepomis macrochirus*) disperse and seek cover in dense vegetation when available (Helfman, 1986; Magurran, 1993). At such sites, even less sensory processing would be required than during schooling, and operation of the piscine brain would approach more closely to, or achieve, maximal efficiency.

One could have suspected that the benefits of schooling and sleep overlap from the knowledge that stationary schooling can be even more cyclically stereotyped than sleep and often is temporally coincident with it. Thus juvenile grunts (family Haemulidae) in large daytime stationary schools stream off of their patchreefs following highly predictable historical routes to their feeding grassbeds at predictable times (within 30 seconds to 1 minute after sunset). They return along the same routes to the same patchreefs just before sunrise. The precise timing of these events avoids exposure during the "quiet periods."

Both migrations begin at the same light level, when the migrating fishes have a relative advantage over their predators. Such twilight migrations by stationarily schooling fishes to and from reefs are a common phenomenon that occurs at the same times that many other fishes are terminating and initiating activity (Helfman, 1986; McFarland, Ogden & Lythgoe, 1979). Moreover, as noted above, many fishes school only as fry (Pitcher & Parrish, 1993), indicating a much greater need for memory processing at that stage.

Also suggestive of analogy, descriptions of behavior in stationary schools also would apply to resting and sleeping in groups. Thus, there is no apparent behavioral differentiation, no persistent tie or pair bond, no leader or dominance, and no overt aggression. And contact schools (pods or "tightly packed masses") of various species (Bowen, 1931; Keenleyside, 1979) recall Weber's (1961) description of sleeping bichirs packed together like sardines in a box.

It will be evident from the foregoing that the reductions in the average amount of sensory processing by individuals in stationary schools of small fishes also would accrue to larger, actively schooling fishes. These reductions might be of greater or lesser significance, dependent on the amount of attention-requiring sensory inputs from neighboring environments. Reductions also will accrue to varying degrees to solitary or small numbers of pelagic cruisers. For them, though, the greatly reduced need for processing sensory input would follow from the monotony and barren nature of pelagic environments and from a great deal of time spent virtually in total darkness.

Accordingly, all three activities, stationary and active schooling and pelagic cruising, fall into the category of vigilance states that confer comparable or even greater benefits than those of sleep. As regards efficiency of brain operation, these benefits, though of varied origin, consist of elimination of interference between the brain's need to processing sensory information for crucial waking activities and its need to process memories.

The correlation that generally exists between metabolic rate (often seen, as noted earlier, as a decrease in respiratory rate) and degree of activity also points to piscine schooling as a relatively restful state. Thus, lowering the metabolic rate of mature sunfishes induces school formation. Normally these fishes school only in the winter and as juveniles (Morrow, 1948). Failure to school for some period of time in a member of a schooling species probably is the equivalent of loss of some rest or sleep in nonschooling species.

Similar considerations may apply to varying degrees to individuals in other aquatic animal aggregates, including schools of sighted invertebrates, such as squids, cuttlefishes, and crustaceans. Recent studies suggest close parallels between aggregates of crustaceans and fish schools as regards internal structure and possibly also of function (Ritz, 1997). Krill, for example, have prominent, mobile, compound eyes. When not feeding, they often form large, tightly packed, highly organized, fast-swimming schools (Hamner, 1984).

Flexibility in piscine activity phasing

Within the framework of the paradigm presented to account for sleep in some fishes, and continuous activity of others, it is evident that piscine sleep might be acquired and dispensed with in response to relatively minor alterations of behavior and ecology. This, in fact, is a common occurrence among free-living fishes, where otherwise "nocturnal" or "diurnal" fishes become continuously active during periods of parental care (Helfman, 1993; Magguran, 1993). For example, some otherwise cyclically active species fan the eggs day and night (Reebs, 1992).

Summary

Many continuously active fishes without need for rest or sleep are of a highly derived nature, dependent on specializations of morphology, physiology, and lifestyles. At one extreme are the very large fishes that swim continuously in schools, small groups, or solitarily. They occupy essentially featureless pelagic habitats, lead a monotonous, essentially routine existence, and consequently have a greatly reduced need for sensory processing. They also require relatively few onto-genetic memories, most of which are short term and reinforced by frequent use. Their lesser needs for processing visual inputs and memories enable highly effici-ent operation of their brains at all times – the proposed ultimate function of sleep.

Continuously active fishes, at the other extreme, are exemplified by small reef dwellers that lead complex lives in rich habitats, a lifestyle that requires many long-term memories. They school stationarily during the day and disperse at night to feed. The complexity of their lives during foraging is no less than that of their close relatives that sleep. On the average, the brains of members of stationary schools need process greatly reduced amounts of sensory inputs. This reduction enables highly efficient reinforcement of their large stores of memories, just as does sleep in their close relatives.

Between these extremes there exist many other species of continuously active fishes of greatly varying intermediate lifestyles.

Thus, fishes provide a complete spectrum of sleeping to sleepless, in closely related members of a single class, a situation not to be found in any terrestrial ani-mal. The ultimate function of sleep in terrestrials – enabling highly efficient brain operation at all times – has been deduced on the basis of existing knowledge and principles of Darwinian natural selection. In the realm of fishes, a similar ultimate function is deduced, but on a more solid foundation. Both sleep and stationary schooling enable highly efficient brain operation at all times. The deduction of this ultimate function from stationary schooling of some fishes is based on observ-able influences that substitute for the sleep of other related fishes. Thus, it greatly reinforces the comparable but more remote deduction of sleep's ultimate function for terrestrials based on Darwinian evolutionary principles.

References

Beamish, F. H. A. (1978). Swimming capacity. In W. S. Hoar & D. J. Randall (Eds.), *Fish physiology* (Vol. 7, pp. 101–187). London: Academic Press.

Blaxter, J. H. S. (1965). Effect of change of light intensity on fish. *International Commission Northwest Atlantic Fisheries Special Publications, 6*, 647–661.

Blaxter, J. H. S. (1988). Sensory performance, behavior, and ecology of fish. In J. Atema, R. R. Fay, A. N. Popper, & W. N. Tavolga (Eds.), *Sensory biology of aquatic animals* (pp. 203–232). New York: Springer-Verlag.

Bleckmann, H. (1993). Role of the lateral line in fish behaviour. In T. J. Pitcher (Ed.), *Behaviour of teleost fishes* (pp. 201–246). London: Chapman & Hall.

Block, B. A. (1991). Endothermy in fish: Thermogenesis, ecology, and evolution. In P. W. Hochachka & T. P. Mommsen (Eds.), *Biochemistry and molecular biology of fishes* (Vol. 1, pp. 269–311). Oxford: Elsevier.

Bowen, E. S. (1931). The role of the sense organs in aggregations of *Ameiurus melas*. *Ecological Monographs*, *1*, 1–35.

Breder, C. M., Jr. (1967). On the survival of fish schools. *Zoologica*, *52*, 25–40.

Breder, C. M., Jr. (1976). Fish schools as operational structures. *Fisheries Bulletin*, *47*, 471–502.

Brock, V. E. (1965). A review of the effects of the environment on the tuna. *International Commission Northwest Atlantic Fisheries Special Publications*, *6*, 75–92.

Buzsáki, G. (2006). *Rhythms of the brain*. New York: Oxford University Press.

Collette, B. B. (1977). Adaptations and systematics of the mackerels and tunas. In G. D. Sharp & A. E. Dizon (Eds.), *The physiological ecology of tunas* (pp. 7–39). New York: Academic Press.

Collette, B. B., & Nauen, C. E. (1983). Scombrids of the world. *Rome: FAO Fisheries Synopsis*, *125*(2), 1–137.

Cousteau, J. Y., & Cousteau, P. (1970). *The shark: Splendid savage of the sea*. New York: Doubleday & Co, Inc.

Cushing, D. H., & Jones, F. R. H. (1968). Why do fish school? *Nature*, *218*, 918–920.

Denton, E. J., & Gray, J. A. B. (1988). Mechanical factors in the excitation of the lateral lines of fishes. In J. Atema, R. R. Fay, A. N. Popper, & W. N. Tavolga (Eds.), *Sensory biology of aquatic animals* (pp. 596–617). New York: Springer-Verlag.

Dizon, A. E., Brill, R. W., & Yuen, H. S. H. (1977). Correlations between environment, physiology, and activity and the effects on thermoregulation in skipjack tuna. In G. D. Sharp & A. E. Dizon (Eds.), *The physiological ecology of tunas* (pp. 233–239). New York: Academic Press.

Douglas, R. H., & Hawryshyn, C. W. (1990). Behavioural studies of fish vision: An analysis of visual capabilities. In R. Douglas & M. Djamgoz (Eds.), *The visual system of fish* (pp. 373–418). London: Chapman & Hall.

Fernald, R. D. (1988). Aquatic adaptations in fish eyes. In J. Atema, R. R. Fay, A. N. Popper, & W. N. Tavolga (Eds.), *Sensory biology of aquatic animals* (pp. 435–466). Berlin: Springer-Verlag.

Fernald, R. D. (1991). Teleost vision: Seeing while growing. *Journal of Experimental Zoology*, *5*(Suppl.), 167–180.

Fernald, R. D. (1993). Vision. In D. H. Evans (Ed.), *The physiology of fishes* (pp. 161–189). London: CRC Press.

Gerking, S. D. (1994). *Feeding ecology of fish*. New York: Academic Press.

Gibson, R. N. (1993). Intertidal teleosts: Life in a fluctuating environment. In T. J. Pitcher (Ed.), *The behaviour of teleost fishes* (pp. 513–536). London: Chapman & Hall.

Goldshmid, G., Holzman, R., Weihs, D., & Genin, A. (2004). Aeration of corals by sleep-swimming fish. *Limnology Oceanography*, *49*, 1832–1839.

Grunbaum, D. (1997). Schooling as a strategy for taxis in a noisy environment. In J. K. Parrish & W. M. Hamner (Eds.), *Animal groups in three dimensions* (pp. 257–281). Cambridge: Cambridge University Press.

Guthrie, D. M. (1981). Visual central processes in fish behavior. In J. P. Ewert, R. R. Capranica, & D. J. Ingle (Eds.), *Advances in vertebrate neuroethology* (pp. 381–412). London: Plenum Press.

Guthrie, D. M. (1986). Role of vision in fish behavior. In T. J. Pitcher (Ed.), *The behavior of teleost fishes* (pp. 75–113). Baltimore: Johns Hopkins University Press.

Guthrie, D. M. (1990). The physiology of the teleostean optic tectum. In R. Douglas & M. Djamgoz (Eds.), *The visual system of fish* (pp. 279–343). London: Chapman & Hall.

Hamner, W. M. (1984). Aspects of schooling in *Euphausia superba*. *Journal of Crustacean Biology*, *4*(Spec. No. 1), 67–74.

Hamner, W. M. (1995). Predation, cover, and convergent evolution in epipelagic oceans. *Marine and Freshwater Behaviour and Physiology*, *26*, 71–89.

Hart, P. J. B. (1993). Teleost foraging: Facts and theories. In T. J. Pitcher (Ed.), *The behaviour of teleost fishes* (pp. 254–284). London: Chapman & Hall.

Hawkins, A. D. (1993). Underwater sound and fish behaviour. In T. J. Pitcher (Ed.), *The behaviour of teleost fishes* (pp. 129–169). London: Chapman & Hall.

Hazel, J. R. (1993). Thermal biology. In D. H. Evans (Ed.), *The physiology of fishes* (pp. 427–467). London: CRC Press.

Helfman, G. S. (1986). Behavioral responses of prey fishes during predator-prey interactions. In M. E. Feder & G. V. Lauder (Eds.), *Predator-prey relationships* (pp. 135–156). Chicago: University of Chicago Press.

Helfman, G. S. (1993). Fish behavior by day, night, and twilight. In T. J. Pitcher (Ed.), *The behaviour of teleost fishes* (pp. 479–512). London: Chapman & Hall.

Herald, E. S. (1961). *Living fishes of the world*. New York: Doubleday & Co.

Hobson, E. S. (1972). Activity of Hawaiian reef fishes during the evening and morning transitions between daylight and darkness. *Fisheries Bulletin of the National Marine Fisheries Service*, *70*, 715–740.

Hobson, E. S. (1974). Feeding relationships of teleostean fishes on coral reefs in Kona, Hawaii. *Fisheries Bulletin of the National Marine Fisheries Service*, *72*, 915–1031.

Huntingford, F. A. (1993). Development of behavior in fish. In T. J. Pitcher (Ed.), *The behaviour of teleost fishes* (pp. 59–83). London: Chapman & Hall.

Imamura, Y. (1951). The Japanese skipjack fishery (W. G. Van Campen, trans.). *Washington, D.C.: U. S. Fish and Wildlife Service Special Scientific Report, Fisheries*, *49*, 1–67.

Janvier, P. (1996). *Early vertebrates*. Oxford: Clarendon Press.

Jobling, M. (1993). Bioenergetics: Feed intake and energy partitioning. In J. C. Rankin & F. B. Jensen (Eds.), *Fish ecophysiology* (pp. 1–44). London: Chapman & Hall.

John, K. R. (1964). Illumination, vision, and schooling of *Astyanax mexicanus* (Fillipi). *Journal of the Fisheries Research Board of Canada*, *23*, 547–562.

Kavanau, J. L. (1997a). Memory, sleep, and the evolution of mechanisms of synaptic efficacy maintenance. *Neuroscience*, *79*, 7–44.

Kavanau, J. L. (1997b). Origin and evolution of sleep: Roles of vision and endothermy. *Brain Research Bulletin*, *42*, 245–264.

Kavanau, J. L. (1998). Vertebrates that never sleep: Implications for sleep's basic function. *Brain Research Bulletin*, *46*, 269–279.

Kavanau, J. K. (2005). Evolutionary approaches to understanding sleep. *Sleep Medicine Reviews*, *9*, 141–152.

Keenleyside, M. H. A. (1979). *Diversity and adaptation in fish behaviour*. Berlin: Springer-Verlag.

Kitchell, J. F., Neill, W. H., Dizon, A. E., & Magnusen, J. J. (1977). Bioenergetic spectra of skipjack and yellowfin tunas. In G. D. Sharp & A. E. Dizon (Eds.), *The physiological ecology of tunas* (pp. 357–368). New York: Academic Press.

Levine, J. S. (1993). *The coral reef at night*. New York: Harry N. Abrams.

Llinás, R. R., & Paré, D. (1991). Of dreaming and wakefulness. *Neuroscience*, *44*, 521–535.

Long, J. A. (1995). *The rise of fishes*. Baltimore: The Johns Hopkins University Press.

Lowe-McConnell, R. H. (1977). *Ecology of fishes in tropical waters*. London: Edward Arnold.

Lowe-McConnell, R. H. (1987). *Ecological studies in tropical fish communities*. Cambridge: Cambridge University Press.

Lythgoe, J. N. (1979). *The ecology of vision*. Oxford: Clarendon Press.

Magnuson, J. L. (1977). *Foreword*. In G. D. Sharp & A. E. Dizon (Eds.), *The physiological ecology of tunas* (pp. xi–xiii). New York: Academic Press.

Magnuson, J. L. (1978). Locomotion by scombrid fishes; hydromechanics, morphology, and behaviour. In W. S. Hoar & D. J. Randall (Eds.), *Fish physiology* (Vol. 7, pp. 239–313). New York: Academic Press.

Magnuson, J. L. & Weininger, D. (1977). Estimation of minimum sustained speed and associated body drag of scombrids. In G. D. Sharp & A. E. Dizon (Eds.), *The physiological ecology of tunas* (pp. 287–311). New York: Academic Press.

Magurran, A. E. (1993). Individual differences and alternative behaviours. In T. J. Pitcher (Ed.), *The behaviour of teleost fishes* (pp. 441–477). London: Chapman & Hall.

Maquet, P. (2001). The role of sleep in learning and memory. *Science, 294*, 1048–1052.

McFarland, W. N., Ogden, J. C., & Lythgoe, J. N. (1979). The influence of light on the twilight migration of grunts. *Environmental Biology of Fishes, 4*, 9–22.

McFarland, W. N., & Hillis, Z. M. (1982). Observations on agonistic behavior between members of juvenile French and white grunts. *Bulletin of Marine Science, 32*, 255–268.

Migdalski, E. C., & Fichter, G. S. (1976). *The freshwater and saltwater fishes of the world*. New York: A. A. Knopf.

Morrow, J. E., Jr. (1948). Schooling behavior in fishes. *Quarterly Review of Biology, 23*, 27–38.

Moyle, P. B., & Cech, J. J., Jr. (1996). *The fishes: An introduction to ichthyology*. Upper Saddle River, NJ: Prentice Hall.

Muntz, W. R. A. (1977). The visual world of the amphibia. *Handbook of Sensory Physiology, VII*(5), 275–307.

Muntz, W. R. A. (1990). Stimulus, environment, and vision in fishes. In R. H. Douglas & M. B. A. Djamgoz (Eds.), *The visual system of fish* (pp. 491–511). New York: Chapman & Hall.

Munz., F. W., & McFarland, W. N. (1977). Evolutionary adaptations of fishes to the photic environment. *Handbook of Sensory Physiology, VII*(5), 193–274.

Myrberg, A. A. (1997). Underwater sound: Its relevance to behavioral functions among fishes and mammals. *Marine and Freshwater Behaviour and Physiology, 29*, 3–21.

Nakamura, H. (1954). Tuna longline fishery and fishing grounds (Van Campen, W. G., trans.). *Washington, D.C.: U.S. Fish and Wildlife Service Special Scientific Report, Fisheries, No. 112.*

Nilsson, G. E., Hobbs, J. P. A., & Ostlund-Nilsson, S. (2007). Tribute to P. L. Lutz: Respiratory physiology of coral-reef teleosts. *Journal of Experimental Biology, 210*, 1673–1686.

Noakes, D. L. G., & Godin, J. G. J. (1988). Ontogeny of behavior and concurrent developmental changes in sensory systems in teleosts. In W. S. Hoar & D. J. Randall (Eds.), *Fish physiology, XIB* (pp. 345–395). New York: Academic Press.

Norman, J. R. (1931). *A history of fishes*. New York: F. H. Stokes.

Norris, K. S., & Schilt, C. R. (1988). Cooperative societies in three-dimensional space: On the origin of aggregations, flocks, and schools, with special reference to dolphins and fish. *Ethology and Sociobiology, 9*, 146–179.

Partridge, B. L., & Pitcher, T. J. (1980). The sensory basis of fish schools: Relative roles of lateral line and vision. *Journal of Comparative Physiology, 135*, 315–325.

Pitcher, T. J., & Parrish, J. K. (1993). Functions of shoaling behavior in teleosts. In J. G. Pitcher (Ed.), *The behaviour of teleost fishes* (pp. 363–439). London: Chapman & Hall.

Popper, A. N., Rogers, P. H., Saidel, W. M., & Cox, M. (1988). Role of the fish ear in sound processing. In J. Atema, R. R. Fay, A. N. Popper, & W. Tavolga (Eds.), *Sensory biology of aquatic animals* (pp. 687–710). New York: Springer-Verlag.

Rauschecker, J. (1995). Developmental plasticity and memory. *Behavioural Brain Research, 66*, 7–12.

Reebs, S. (1992). Sleep, inactivity, and circadian rhythms in fish. In M. A. Ali (Ed.), *Rhythms in fishes* (pp. 127–135). New York: Plenum Press.

Ritz, D. A. (1997). Costs and benefits as a function of group size: Experiments on a swarming mysid *Paramesopodopsis rufa* Fenton. In J. K. Parrish & W. M. Hamner (Eds.), *Animal groups in three dimensions* (pp. 194–224). Cambridge: Cambridge University Press.

Rivas, L. R. (1977). Preliminary models of annual life history cycles of the North Atlantic bluefin tuna. In G. D. Sharp & A. E. Dizon (Eds.), *The physiological ecology of tunas* (pp. 369–393). New York: Academic Press.

Sella, M. (1952). Migrations and habitat of the tuna (*Thunnus thynnus*) (Van Campen, W. G., trans.). *Washington, D.C.: U.S. Fish and Wildlife Service Special Scientific Report, Fisheries, 76*, 1–20.

Sette, O. E. (1950). Biology of the Atlantic mackerel (Scomber scombrus) of North America, Part II. Migrations and habits. *U.S. Fisheries & Wildlife Service Fisheries Bulletin, 51*(49), 251–358.

Shapiro, C. M., & Hepburn, H. R. (1976). Sleep in a schooling fish, *Tilapia mossambica. Physiology and Behavior, 16*, 613–615.

Sharp, G. D., & Vlymen, W. J., III. (1977). The relation between heat generation, conservation, and the swimming energetics of tuna. In G. D. Sharp & A. E. Dizon (Eds.), *The physiological ecology of tunas* (pp. 213–232). New York: Academic Press.

Shaw, E. (1970). Schooling in fishes: Critique and review. In L. R. Aronson, E. Tobach, D. S. Lehrman, & J. S. Rosenblatt (Eds.), *Development and evolution of behavior* (pp. 452–480). San Francisco: W. H. Freeman.

Shaw, E. (1978). Schooling fishes. *American Scientist, 66*, 166–175.

Shimada, B. M., & Van Campen, W. G. (1951). Tuna fishing in Palau waters. *Washington, D.C.: U.S. Fish and Wildlife Service Special Scientific Report, Fisheries, No. 42*, 1–26.

Siegmund, R. Lokomotorische (1969). Aktivitat und Ruheverhalten bei einheimischen Susswasserfischen (Pisces, Percidae, Cyprinidae). Biologisches Zentralblatt, *88*, 295–312.

Smith, M. M., & Heemstra, P. C. (1986). *Smiths' sea fishes.* New York: Springer-Verlag.

Tamura, T. (1968). Fundamental studies on the visual sense in fish. In H. Kristjonsson (Ed.), *Modern fishing gear of the world* (pp. 543–547). London: Fishing News (Books) Ltd.

Tauber, E. S., Weitzman, E. D., & Korey, S. R. (1939). Eye movements during behavioral inactivity of certain Bermuda reef fish. *Communications in Behavioral Biology, 3*, 131–135.

Tiews, K. F. W. (1963). Behavior and physiology. *FAO (FAU UN) Fisheries Report, 6*(1), 44–46.

Turner, G. (1993). Teleost mating behaviour. In T. J. Pitcher (Ed.), *The behaviour of teleost fishes* (pp. 307–331). London: Chapman & Hall.

Wagner, H. J. (1990). Retinal structure of fishes. In R. Douglas & M. Djamgoz (Eds.), *The visual system of fish* (pp. 109–157). London: Chapman & Hall.

Warburton, K. (1997). Social forces in animal congregations: Interactive, motivational, and sensory aspects. In J. K. Parrish & W. M. Hamner (Eds.), *Animal groups in three dimensions* (pp. 313–336). Cambridge: Cambridge University Press.

Wardle, C. S. (1985). Swimming activity in marine fish. In M. S. Laverack (Ed.), *Physiological adaptations of marine animals* (pp. 521–540). Scarborough: The Co. of Biologists Ltd.

Weber, E. (1961). Über Ruhelagen von Fischen. *Zeitschrift für Tierpsychologie*, *18*, 517–533.

Welcomme, R. L. (1985). River fisheries. *FAO Fisheries Technical Paper (Rome)*, *262*, 1–330.

Wootton, R. J. (1992). *Fish ecology*. New York: Chapman & Hall.

Yuen, H. S. H. (1962). Schooling behavior within aggregations composed of yellowfin and skipjack tuna. *FAO (FAO UN) Fisheries Report*, *6*(3), 1419–1429.

Yuen, H. S. H. (1970). Behavior of skipjack tuna, *Katsuwonus pelamis*, as determined by tracking with ultrasonic devices. *Canadian Fisheries Research Board*, *27*, 2071–2079.

4

What exactly is it that sleeps? The evolution, regulation, and organization of an emergent network property

JAMES M. KRUEGER

Abstract

It is posited that sleep is a network-emergent property of any viable group of interconnected neurons. Animals ranging from jellyfish to all homeotherms sleep. Biochemical sleep-regulatory events, including cytokines and nuclear factor kappa B (NF-kB), are shared by insects and mammals. It seems likely that these sleep-regulatory events evolved from metabolic-regulatory events and that sleep is a local use-dependent process. Relationships between sleep and tumor necrosis factor (TNF) are used to examine the local use-dependent sleep hypothesis. ATP released during neurotransmission is posited to drive the production and release of cytokines, such as TNF, that, in turn, act within a biochemical sleep homeostat in the short term – via adenosine, nitric oxide, and prostaglandins – to enhance non–rapid-eye-movement (NREM) sleep. In the long term, TNF and other sleep-regulatory substances, via NF-kB activation, enhance expression of receptors such as adenosine A1 and glutamate amino-3-hydroxy-5-methylisoxazoleproprionic (AMPA) receptors. Changes in the expression of these receptors will change the sensitivity of neurons and thereby change synaptic efficacy. Such actions suggest that sleep mechanisms cannot be separated from a connectivity function of sleep at the local network level. The need for sleep is derived from the experience-driven changes in neuronal microcircuitry that necessitate the stabilization of synaptic networks to maintain physiological regulatory networks and instinctual and acquired memories. The need for unconsciousness is derived from the local

This work was supported by grants from the National Institutes of Health (Grant Nos. NS25378, NS31453, NS27250, and HD36520) and the Keck Foundation.

use-dependent sleep mechanisms. Thus sleep regulatory substances such as TNF alter input–output relationships of the neuronal assemblies within which they are made, thereby divorcing real-time adaptive outputs from environmental inputs.

Introduction

Most if not all animals possessing a brain sleep. For example, thus far every animal displaying the behavioral sleep characteristics of species-typical postures and daily timing of inactivity, periods of increased arousal thresholds and rebound after loss (the latter often called sleep homeostasis), that has a complex network of neurons, ranging from jellyfish and insects to all mammals and birds, sleeps (e.g., Hendricks, Finn, Panckeri, et al., 2000; Kavanau, 2005; Rattenborg, Amlaner, & Lima, 2001; Sauer, Herrmann, & Kaiser, 2004; Sauer, Kinkelin, Herrmann et al., 2003; Siegel, 2005; Tobler, 2005). Further, although there is consensus among sleep researchers that sleep is a whole-organism property, the whole brain is not needed for sleep to occur. For instance, despite millions of cases of stroke, there is not one case in the literature of a poststroke long-term survivor (5 or more days) who fails to sleep. These and other data suggest that sleep is very robust, that it is self-organizing, that no specific area in the brain is necessary for sleep, and that sleep is an intrinsic property of any surviving viable group of neurons. Such conclusions lead to very different ways of thinking about what exactly it is that sleeps. They also lead to fundamental questions of logic concerning sleep regulation; how can one sensibly discuss or meaningfully research the regulation of something if one does not know what or where that something is? These conclusions reshape the way one views sleep pathologies and consequently have the potential to manifest in different treatment paradigms. These conclusions have bearing on brain organization of sleep and sleep functions. Herein these issues are addressed from this unique perspective.

The evolution of sleep

It is not possible to rigorously recreate evolutionary history, but one logical scenario is presented here. The light and temperature rhythms associated with the earth's rotation were probably used by early single cells – or even primordial conjuncts of metabolites – to time their use, activation, or production of metabolites and metabolic enzymes to the availability and energetic ease of processing nutrients (Figure 4.1).

This association would have established early in evolution, perhaps even before the inception of life and circadian rhythms of metabolic rest–activity cycles. As single cells evolved in complexity, so would chemical signaling, with the development of ligands and effector molecules (receptors) within the cells to time and efficiently

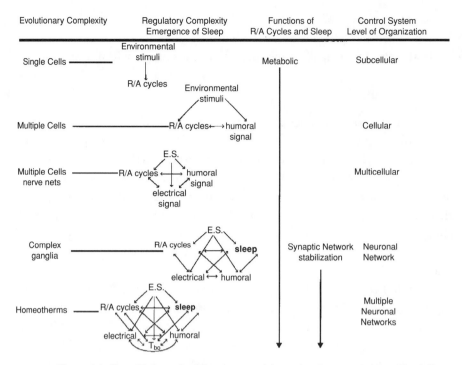

Figure 4.1. Sleep likely evolved from rest–activity cycles that were initiated by daily light–dark temperature-driven changes in metabolism. This diagram suggests that all organisms lacking complex ganglia can be considered in the wake state because, until ganglia evolved and the associated experience-dependent connectivity, there was no need for sleep. It also suggests that sleep evolved as an emergent network property rather than as a property existent in single neurons. Abbreviations: R/A, rest/activity; E.S., environmental stimuli; T_{bo}, body temperature.

regulate metabolism. It seems likely that such intracellular chemical signals might escape the cell from time to time and interact with effector molecules of nearby cells to signal the metabolic state of the first cells and perhaps coordinate the metabolic state of the recipient cell to the first cell for some adaptive purpose. As multicellular organisms evolved, such humoral signaling became even more sophisticated and was used to coordinate various physiological and simple behavioral functions of the whole animal. Further, the humoral signals themselves were influenced by the emergent rest–activity cycles and external environmental cues. Some of the ligand–receptor interactions were associated with ion movements across cell membranes. Although single cells such as bacteria have such electrical events, their use in higher-order information processing was maximized with the development of neurons, ganglia, and brains. As the degree of complexity evolved, the levels of organization at which control systems operated also expanded (Figure 4.1). New, higher-level control systems were integrated into and layered over previously existing control systems. With this complexity the ability to learn was greatly enhanced and became associated with specialized cells (neurons). Organisms became much more adaptive to environmental challenges in the sense that mobility was greatly enhanced and was directed in part by the integration of memories into real-time responsiveness to the environment. The evolutionary advantage of such complex information processing is obvious. This strategy, however, came at the cost of the development of epigenetic neuronal plasticity (Edelman, 1987). The flexible neuronal connectivity is experience-dependent and is metabolically expensive. The use-driven constant change in the brain's microcircuitry required a functional mode – that is, sleep – to ensure the stability of synaptic networks that encode innate and learned memories (Kavanau, 1994; Krueger & Obál, 1993).

It is likely that sleep developed from behavioral rest because during rest, niche-appropriate inactivity was already in hand. Similarly, sleep biochemical regulatory events also were probably derived from the regulatory events already regulating rest. Indeed, recent evidence suggests that clock genes involved in regulating circadian rhythms also partake in sleep regulation (Franken, Thomason, & Heller, et al., 2007) (Figure 4.2). Further, because the inactivity reduced metabolic demand and the circadian timing was fine-tuned to link inactivity to availability of nutrients and enzymatic efficacy, it seems likely that metabolic signals were used in sleep-control systems to target sleep to networks that had been metabolically active. This hypothesis is developed further later where we invoke the idea that ATP released with neurotransmission is a key signal that allows neural circuits to keep track of prior wake activity.

From this discussion and the comparative sleep literature (reviewed in Kavanau, 2006; Sauer, Kinkelin, Herrmann, et al., 2003; Tobler, 2005), it is concluded that sleep clearly occurs in insects and other higher-order animals. This conclusion has far-reaching implications. For instance, because the anatomical structures of brains are different the involvement of a particular circuit in a specific species in sleep regulation, although important to that species, may provide little information germane to the issues of what sleeps and the purpose of sleep. Indeed, in *Drosophila,* the circuits involved in sleep bear little relation to mammalian sleep-regulatory circuits, although the humoral sleep-regulatory molecular networks seem to be shared (Chen, Gardi, Kushikata, et al., 1999; Foltenyi, Greenspan, & Newport, 2007; Kushikata, Fang, Chen, et al., 1998; Williams, Sathyanarayanan, Hendricks, et al., 2007). This issue is returned to later, where we posit that sleep-regulatory circuits play a role in coordinating sleep of neuronal assemblies.

Before concluding this section on the evolution of sleep, two issues are raised. First, do individual neurons sleep? There are definitional difficulties in answering this question. Thus, animal behavior is used to define sleep; we are not aware

The Sleep Homeostat

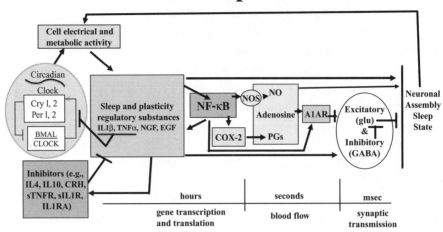

Figure 4.2. The sleep homeostat is composed of multiple sleep regulatory substances that act over different time periods. These substances are also involved in neural plasticity, and this may have bearing on sleep function. Collectively they form parallel and series pathways and feedback loops. Our current knowledge of the biochemical (humoral) regulation of sleep is more extensive than that shown. This figure was derived from work done in mammals. However, subsequently some of the elements were also implicated in sleep regulation in *Drosophila melanogaster,* including NF-kB (Williams et al., 2007), adenosine (Hendricks et al., 2000), and the EGF signaling network (Foltenyi et al., 2007). The cell types involved are not shown but likely involve every cell type in the brain (e.g., see Figure 4.3). The ultracomplexity of the sleep homeostat likely reflects its long evolutionary history. The longer times associated with the transcription and translation events and protein half-life on the left side of the figure offer a mechanism by which the brain can keep track of past sleep/wake history. The direct sleep-promoting activities of these proteins involve labile substances with shorter half-lives – e.g., NO, adenosine. We propose that this sleep homeostat operates within local neuronal networks such as cortical columns and thereby affects local state (see text). Abbreviations: IL-1, interleukin-1 beta; TNF, tumor necrosis factor; NGF, nerve growth factor; EGF, epidermal growth factor; NF-kB, nuclear factor kappa B; COX, cyclo-oxygenase; NOS, nitric oxide synthase; NO, nitric oxide; PG, prostaglandins; A1AR, adenosine A1 receptor; glu, glutamate; GABA, gamma aminobutyric acid; IL-4, interleukin-4; IL-10, interleukin 10; CRH, corticotrophin releasing hormone; sTNFR, soluble TNF receptor; sIL1R, soluble IL-1 receptor; IL-1RA, IL-1 receptor antagonist.

of any instance where the electrical or metabolic activities of an *individual* cell have been shown to be causative of any complex animal behavior such as sleep. Cortical columns, an example of an experimentally accessible neuronal network, are considered the fundamental unit of information processing in the waking animal (Kock, 2004). We can show that cortical column state properties possess all the definitional characteristics of whole-organism sleep; this cannot be done for individual cells. Further, neuronal inactivity, defined by the number of action potentials, cannot be used to define sleep because such inactivity might be part of a larger process. For example, receptor field surround inhibition is used in sensory systems to enhance signal contrast; a single neuron in such a field may not fire, but such silence cannot be considered sleep because it is part of a much bigger process within which the silence is used to enhance a signal. Similarly, reduction of metabolic activity cannot be used to define sleep at the cell level, because if we did so, we would conclude that bacteria – and even certain nonliving chemical systems that are subjected to daily changes in temperature – sleep. Further, it does not seem parsimonious to propose that individual neurons with their relatively short evolutionary history can sleep if we accept that bacteria with billions of years of evolution under their belt do not sleep. It seems more likely that sleep is a network-emergent process (Figure 4.1).

Second, all animals coevolved in the presence of microbes that preceded them by billions of years. There are multiple examples of endosymbiotic relationships between microbes and humans – for example, mitochondria are likely of bacterial origin. Indeed, there are proteins in mammalian brain that bind bacterial cell wall components and signal their presence. The brain expression of at least one of these proteins, peptidoglycan recognition protein, is enhanced by sleep loss (Rehman, Taishi, Fang, et al., 2001). Further, microbial products alter sleep via such pathogen-associated molecular pattern-recognition receptors (Majde & Krueger, 2005). Neurobiologists often avoid such information because it reduces the primacy of the neuron. Nevertheless, serious discussion of the evolution of sleep or any other higher-order function of the nervous system must be inclusive of broader historical issues.

Sleep regulation

This section focuses on the biochemical regulation of sleep and its dependency on metabolic activity because there is already evidence that these mechanisms are shared by fruit flies and mammals – for example, epidermal growth factor (Kushikata et al., 1998) and NF-kB (Chen et al., 1999). This evidence is used to illustrate the use-dependency of sleep. Then, in the next section, it is used in combination with additional data to draw conclusions about what it is that sleeps.

The neuronal circuits involved in sleep regulation are mentioned only in passing because, as noted, although important for individual species, they differ substantially between insects and mammals and are thus unlikely to provide information on what sleeps and on sleep function. There are many good reviews dealing with mammalian sleep-regulatory circuitry (e.g., Jones, 2003; McGinty & Szymusiak, 2003; Saper, Scammell, & Lu, 2005; Steriade, 2003). However, a fundamental premise of this literature is that such circuits impose sleep on the brain; hence sleep is viewed as being initiated by these circuits. Our theory, presented later, posits, in contrast, that sleep is initiated within any neuronal assembly depending on past use. The sleep-regulatory circuit paradigm also fails to explain many well-known sleep phenomena. For instance, it does not offer any explanation of performance detriments associated with prolonged wakefulness, sleep function, sleep homeostasis, many sleep parasomnias such as sleepwalking, recurrence and reorganization of sleep after lesions, and so on. In contrast, the local use-dependent sleep theory addresses all these issues.

We have known for many years that mammalian cerebrospinal fluid contains sleep-promoting substances that accumulate during wakefulness (Miller, Goodrich, & Pappenheimer, 1967; reviewed in Obál & Krueger, 2003). Many of these substances have been identified (Figure 4.2). It is not possible, however, to isolate sleep as the independent variable because most if not all physiological parameters change with sleep. As a consequence, investigators have developed lists of criteria that need to be met before a substance can be considered a sleep-regulatory substance (Borbély & Tobler, 1989; Inoué, 1989; Jouvet, 1984; Krueger & Obál, 1994). Criteria common to these lists are (1) The substance if injected should enhance sleep; (2) if inhibited, sleep should be reduced; (3) the level of the substance should vary in brain with sleep propensity; (4) the substance should act on sleep-regulatory circuits (but see next, where we posit that sleep is a fundamental property of any neuronal assembly – by extension the substance should induce the sleep-like state in such assemblies); and (5) the substance should be altered in pathological states associated with enhanced sleepiness. All of these criteria have been met by interleukin-1 beta (IL-1β), tumor necrosis factor alpha (TNF-α), growth hormone–releasing hormone, adenosine, and prostaglandin D_2 for NREM sleep regulation. For REM sleep regulation, prolactin, nitric oxide (NO), and vasoactive intestinal polypeptide also meet these requirements (reviewed in Obál & Krueger, 2003). By way of example, the evidence dealing with TNF is expanded on here.

Cytokines such as TNF have long evolutionary histories; they date to the early jawless vertebrates 500 million years ago, and related cytokines date to invertebrates occurring at least 850 million years ago (Opp, 2005). Cytokines are best known for their roles in the immune system, but they may initially have evolved for different purposes (Opp, 2005). TNF was first implicated in sleep regulation about 20 years ago by the finding that it has the capacity to enhance NREM sleep (Shoham, Davenne, Cady, et al., 1987). In subsequent years, a plethora of additional evidence suggests that TNF is a sleep-regulatory substance (reviewed in Krueger, Rector, & Churchill, 2007; Obál & Krueger, 2003).

Central or systemic injection of TNF-α enhances the duration of NREM sleep and electroencephalographic (EEG) delta (1/2 to 4 hertz) power (an index of sleep intensity) and/or sleepiness in every species thus far tested, including mice, rats, rabbits, sheep, and humans (e.g., Dickstein, Moldofsky, Lue, et al., 1999). In contrast, inhibition of TNF using anti-TNF antibodies, the soluble TNF receptor, or a TNF siRNA reduces spontaneous NREM sleep and/or EEG delta power (Taishi, Churchill, Wang, et al., 2007). IL-4, IL-10, and IL-13 inhibit either the actions or production of TNF and sleep (reviewed in Obál & Krueger, 2003). Mice lacking the TNF 55-kDa receptor sleep less than corresponding control mice; this effect is manifest primarily during a 12-hour period surrounding the transition between night and day (Fang, Wang, & Krueger, 1997). Hypothalamic and cerebral cortical levels of TNF mRNA (Bredow, Taishi, Guha-Thakurta, et al., 1997) and TNF protein (Floyd & Krueger, 1997) vary about 2- and 10-fold respectively across the day, with higher levels associated with greater sleep propensity. Sleep deprivation is associated with enhanced brain TNF levels (Taishi, Gardi, Chen, et al., 1999); if TNF is blocked, sleep rebound after sleep deprivation is greatly attenuated (reviewed in Obál & Krueger, 2003). Further, inhibition of TNF also blocks the increases in NREM sleep induced by an acute mild increase in ambient temperature (Takahashi & Krueger, 1997). If TNF is injected into or near the anterior hypothalamic sleep-regulatory circuits, NREM sleep is enhanced (Kubota, Li, Guan, et al., 2002; Terao, Matsumura, Yoneda, et al., 1998). In contrast, injection of the soluble TNF receptor into this area reduces spontaneous sleep (Kubota et al., 2002). In humans, plasma levels of TNF vary with EEG delta power (Darko, Miller, Gallen, et al., 1995). In those pathologies associated with fatigue, sleepiness, or excess sleep that have been examined thus far, enhanced circulating TNF levels occur. The list includes postmyocardial infarction, preeclampsia, postdialysis fatigue, sleep apnea, insomnia, rheumatoid arthritis, postviral fatigue syndrome, Acquired Immune Deficiency Syndrome (AIDS), and alcoholism (reviewed in Krueger, Rector, & Churchill, 2007; Majde & Krueger, 2005). Treatment of rheumatoid arthritis (Franklin, 1999) and sleep apnea patients (Vgontzas, Zoumakis, Lin, et al., 2004) with a soluble TNF receptor reduces their fatigue or sleep. If sleep apnea patients are treated surgically, TNF plasma levels are restored to normal and sleepiness is reduced (Kataoka, Enomoto, Kim, et al., 2004). Further, if subjects are given low doses of endotoxin, plasma levels of TNF and sleep are enhanced (Mullington, Korth, Hermann et al., 2000). Circulating levels of TNF likely affect sleep via the vagus nerve, because vagotomy in rats blocks intraperitoneal TNF-enhanced NREM sleep (Kubota, Fang, Guan, et al., 2001). A TNF polymorphic variant, G-308A, is associated with sleep apnea (Riha, Brander, Vennelle, et al., 2005) and metabolic syndrome (Sookoian, Gonzalez, & Pirola, 2005), a condition exacerbated by sleep loss (Spiegel, Knutson, Leproult, et al., 2005). TNF is associated with the development of insulin resistance; thus sleep loss–enhanced TNF may be a causal factor in sleep loss–associated insulin resistance. Such data strongly implicate TNF in sleep regulation and in pathologies associated with sleep loss. Similar data, but not quite as extensive, exists for the other substances shown in Figure 4.2.

The downstream mechanisms of TNF's role in sleep regulation indicate its involvement in a biochemical network involving many of the other substances implicated in sleep regulation. TNF signaling is complex. Mature, soluble TNF, a 17-kDa protein, is cleaved from a 26-kDa membrane-associated protein. Both forms can have biological activity. Further, the transmembrane form can act as a receptor as well as a ligand (Eissner, Kolch, & Scheurich, 2004). The intracellular domains of the two TNF receptors lack intrinsic enzymatic activity; rather, they recruit a variety of cytosolic adaptor proteins, and this helps to account for the pleiotropic actions of TNF. The extracellular domains can be cleaved to form soluble TNF receptors and the 55-kDa soluble TNF receptor is a normal constituent of cerebrospinal fluid (Puccioni-Sohler, Rieckmann, Kitze, et al., 1995). Its role in sleep regulation is unknown, although – as mentioned – it can inhibit sleep. One major TNF-activated signaling pathway activates new gene transcription while another leads to cell death. The cell death pathway, via caspases 3 and 8, appears to be a rare physiological event, because TNF-responsive gene products function to prevent cell death (reviewed Ledgerwood, Pober, & Bradley, 1999) and indeed can be neuroprotective (Fontaine, Mohand-Said, Hanoteau, et al., 2002; Yang, Lindholm, Konishi, et al., 2002).

TNF promotes NF-kB activation; in turn, TNF production is enhanced by NF-kB activation (Figure 4.2) (reviewed in Obál & Krueger, 2003). NF-kB is a transcription factor usually acting as an enhancer element for a wide array of genes, including other cytokines such as IL-1, nerve growth factor (NGF), epidermal growth factor (EGF), and other substances involved in sleep regulation, such as the adenosine A1 receptor, the gluR1 component of AMPA receptors, cyclo-oxygenase, and nitric oxide synthase (Figure 4.2). There may be some degree of specificity for NF-kB activation to sleep. Sleep loss enhances hypothalamic and cortical NF-kB activation (Brandt, Churchill, Rehman, et al., 2004; Chen et al., 1999). Adenosine also elicits NF-kB nuclear translocation in basal forebrain via the adenosine A1 receptor (Basheer, Rainnie, Porkka-Heiskanen, et al., 2001). Finally, an inhibitor of NF-kB shortens the duration of NREM sleep (Kubota, Kushikata, Fang, et al., 2000). Thus, via the actions of TNF on NF-kB and the NF-kB–enhanced enzymes and receptors, shorter-lived molecules known to be involved in sleep regulation are recruited into the sleep regulatory biochemical cascade, including adenosine, NO, and prostaglandins. That these mechanisms are also likely involved in the regulation of local cerebral blood flow highlights the relationships between cellular metabolism, sleep, and blood flow; it might be fruitful to know if these relationships extend beyond mammals.

TNF may also provide a bridge between the circadian rhythm and the sleep homeostat regulatory influences on sleep. Thus, there are daily rhythms in brain

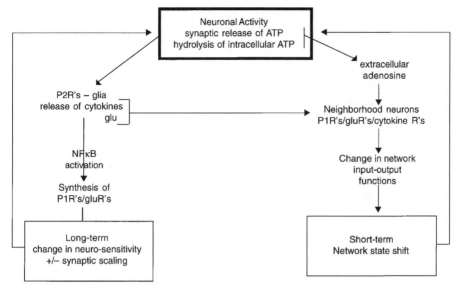

Figure 4.3. Synaptic release of ATP conjointly with neurotransmission provides a mechanism by which the brain can keep track of past activity via cytokine release and the subsequent effects on gene expression. Via conversion of ATP to adenosine rapid (seconds to minutes) (*right side*) changes in network state near the site of release via adenosine receptors are possible. The activity–ATP-induced cytokine release from glia (*left side*) in turn activates NF-kB, and this alters expression of receptors such as the adenosine A1 receptor, a P1 receptor, and the gluR1 subunit of the AMPA receptor. Both of these receptors are involved in sleep (Basheer, Strecker, Thakkar, et al. 2004; Bazhenov, Timofeev, Steriade, et al. 2002). Their expression will alter the cell's sensitivity to adenosine and glutamate. This mechanism is thought to be a long-term synaptic scaling mechanism (see text). How the individual cells determine whether upscaling (AMPA) or downscaling (adenosine) is needed and the relationship of that mechanism to sleep remains unknown. Abbreviations: P2R, purine type 2 receptors; P1R, purine type 1 receptors; NF-kB, nuclear factor kappa B; glu, glutamic acid; R, receptors.

TNF protein levels of about 10-fold in the hypothalamus and cortex (Floyd & Krueger, 1997). Removal of the TNF 55-kDa receptor results in sleep deficits that are limited to a period of about 8 hours at the transition between night and day (Fang et al., 1997). Finally, TNF inhibits expression of some clock genes via interfering with CLOCK-BMAL1−induced activation of E-box regulatory elements (Cavadini, Petrzilka, Kohler, et al., 2007) (Figure 4.2).

Upstream events involved in TNF synthesis and release suggest a close relationship to cell activity and metabolism in the brain. Activity in neurons is translated into pre- and postsynaptic events that manifest in both the short run and long run. Thus ATP is coreleased with neurotransmitters (Figure 4.3) (reviewed in Farber & Kettenmann, 2006); in turn, some of that ATP is hydrolyzed to adenosine. Adenosine acts on neurons via the adenosine A1 receptor to hyperpolarize cells via

K$^+$ channels (Basheer et al., 2004). ATP also acts via P2X7 receptors on microglia to induce the release of IL-1 and TNF (Bianco, Pravettoni, Colombo, et al., 2005; Gabel, 2007; Hide, Tanaka, Inoué, et al., 2000; Suzuki, Hide, Ido, et al., 2004). These substances, in turn, act on cells bearing their receptors to activate NF-kB. TNF also acts on astrocytes in conjunction with ATP via P2Y1 receptors to release glutamate (Domercq, Brambilla, Pilati, et al., 2006). Glutamate itself can induce TNF production and release (De, Krueger, Simasko, et al., 2005). A second method used to demonstrate activity-dependence of cytokines uses the whisker stimulation–somatosensory cortex rat model. Thus, after 2 hours of whisker twitching, TNF immunoreactivity in neurons is enhanced in the cortical column that receives afferent input from the whisker, while enhanced TNF immunoreactivity is not as evident in adjacent cortical columns (Fix, Churchill, Hall, et al., 2006). It is not currently known if this increase of neuronal TNF expression results from TNF uptake from the microglia-released TNF or if it is synthesized in the neuron of its expression.

The upstream and downstream events involved in TNF central nervous system actions begin to illustrate the molecular complexity of sleep regulation and its connection to present and past activity within the brain. An additional action of TNF suggests that it has a role in synaptic efficacy. Thus, TNF enhances cytosolic Ca^{2+} levels (De, Krueger, Simasko, et al., 2003) and AMPA receptor expression (Yu, Cheng, Wen, et al., 2002). If TNF is inhibited, AMPA-induced postsynaptic potentials and AMPA-induced changes in Ca^{2+} are reduced, suggesting that this action is physiological. Application of a TNF siRNA to the cortex reduces gluR1 mRNA (gluR1 is a subunit of the AMPA receptor) (Taishi et al., 2007). AMPA receptors are involved in EEG delta wave power and in synaptic plasticity (Bazhenov et al., 2002; Beattie, Stellwagen, Morishita, et al., 2002). Direct evidence indicates that TNF is involved in synaptic scaling (Stellwagen & Malenka, 2006). Such data suggest a TNF-dependent mechanism for the reconfiguration of synaptic weights (Albensi & Mattson, 2000; Malinow & Malenka, 2002) and that this is inseparable from sleep.

TNF and other sleep regulatory substances (SRSs) have the capacity to act locally within the cortex to alter a sleep phenotype, EEG delta power, and cortical column state. Thus unilateral application of TNF to the surface of the cortex enhances EEG delta power during NREM sleep – but not during REM sleep or waking – on the ipsilateral side but not on the contralateral side, suggesting that sleep is more intense on the side receiving TNF (Yoshida, Peterfi, Garcia-Garcia, et al., 2004). Further, if the soluble TNF receptor is applied unilaterally to the cortex, the enhanced EEG delta power induced by prior sleep deprivation is reduced on the ipsilateral side only. Further, application of a TNF siRNA reduces spontaneous EEG delta power during NREM sleep on the side receiving it but not on the opposite side

(Taishi, et al., 2007). These unilateral changes in EEG delta power are associated with changes in Fos and IL-1 protein immunoreactivity in the corresponding cortical areas and reticular thalamus (Churchill, Yasuda K., Yasuda T., et al., 2005), suggesting the involvement of a biochemical sleep-regulatory cascade (Figure 4.2) and known thalamocortical sleep regulatory circuitry (Steriade, 2003). These data coupled with what is known about use-dependent production of TNF strongly support the notion of sleep being targeted to active circuits and being initiated at a local network level. Further, local application of TNF to cortical columns is associated with changes in the state of the cortical column (Churchill, Rector, Yasuda, et al., 2006). Such changes suggest that sleep is a fundamental property of neuronal networks; this is the subject of the next section.

Organization of sleep

The evidence discussed concerning TNF's involvement in sleep regulation as well as other biochemical sleep-related evidence brought us to the idea that the whole brain was not required to participate in sleep and that sleep was initiated as a local event (Krueger & Obál, 1993). There are extensive other data in support of this hypothesis. Sleep intensity, a sleep phenotype determined from EEG delta power, is dependent on prior use and is targeted and localized to areas disproportionately used during prior wakefulness. EEG delta power is enhanced in the left somatosensory cortex compared to the right during NREM sleep after prolonged right-hand stimulation prior to sleep onset (Kattler, Dijk, & Borbely, 1994). Other evidence is consistent with the idea that sleep is a regional property of brain dependent on prior activity. In mice, rats, chickens, pigeons, humans, and cats, if a localized area is disproportionately stimulated during waking, EEG delta power in that area is enhanced during subsequent NREM sleep (Cottone, Adamo, & Squires, 2004; Ferrara, De Gennaro, Curcio, et al., 2002; Huber, Ghilardi, Massimini, et al., 2004; Iwasaki, Karashima, Tamakawa, et al., 2004; Miyamoto, Katagiri, & Hensch, 2003; Yasuda T., Yasuda K., Brown, et al., 2005; Vyazovskiy, Borbély, & Tobler, 2000). There are also several findings showing that cerebral blood flow during sleep is enhanced in those areas disproportionately stimulated during prior waking (Drummond, Brown, Stricker, et al., 1999; Maquet, 2001). Finally, the developmental plasticity literature (Frank, Issa, & Stryker, 2001; Marks, Shaffery, Oksenberg, et al., 1995; Mascetti Rugger, Vallortigara, et al., 2007) and the learning literature demonstrating replay of neuronal electrical patterns associated with waking learning tasks (Ji & Wilson, 2007) indicate changes in the EEG during sleep are targeted to areas activated during prior waking.

As mentioned in the introduction to this chapter, if a subject survives a brain lesion, whether experimental or pathological, for a few days or more, it sleeps.

There apparently are no reported cases of subjects with complete lack of sleep, including those with fatal familial insomnia (Montagna, 2005). This is an important metafinding for sleep research because it indicates that sleep is a property of any surviving group of neurons. Additional evidence suggests that parts of the brain can be awake while other parts are asleep. From comparative studies, it is clear that many species of birds and marine mammals exhibit unihemispheric sleep (Mukhametov, 1984; Rattenborg et al., 2001). Perhaps the best example of this is the sleep of the bottle-nose dolphin; high-amplitude delta sleep never occurs simultaneously in both of this creature's cerebral hemispheres (Mukhametov, 1984). A defining characteristic of NREM sleep, EEG delta waves, has a local cortical origin (Steriade, 2003). Further, isolated cortical islands that retain their blood flow wax and wane through periods of high-amplitude delta waves (Kristiansen & Courtois, 1949). Clinical evidence also indicates that the brain can be awake and asleep simultaneously (e.g., parasomnias such as sleepwalking) (Mahowald & Schenck, 2005).

Cerebral cortical columns are experimentally accessible well-defined examples of neuronal assemblies and are posited to be a basic processing unit of the awake brain (Koch, 2004). They as well as other neuronal assemblies are also posited to be the minimal component of brain manifesting a sleep state (Krueger & Obál, 1993, 2003). The input–output properties of individual cortical columns can be characterized using sensory stimulation as the input and evoked response potentials (ERPs) as the output. Rector, Topchiy, and Rojas (2005a) have shown that somatosensory and auditory cortical columns oscillate between at least two states; one of those states has sleep-like characteristics. Thus the amplitudes of ERPs are greater during sleep than during awake. If the rat is asleep when ERPs are determined, ERP amplitude is greater in most of the columns measured; conversely, if the rat is awake, most of the columns exhibit lower-amplitude ERPs. Because some columns are found in a sleep-like state during whole-animal wake episodes and some columns are in the wake-like state during sleep, it suggests that sleep is a fundamental property of the column. The probability of finding a column in the sleep-like state is increased if the column is excessively stimulated before determination of state. Further, cortical column sleep is homeostatic in the sense that the probability of occurrence of the sleep-like state is dependent on the amount of time it was previously in the waking state. The longer the column is in a wake-like state, the higher the likelihood that it will make the transition to the sleep-like state (Rector, Topchiy, Carter, et al., 2005b). Finally, using the ERP to define state and a conditioned learning paradigm, Rector and colleagues demonstrated that there are behavioral consequences to cortical column state. Rats trained to lick in response to stimulation of a single facial whisker have a greater incidence of behavioral errors if the somatosensory cortical column receiving afferent input from the stimulated whisker is in the sleep-like state (Walker, Topchiy, Kouptsov, et al., 2005).

The biochemical sleep mechanisms involving TNF and other related SRSs (Figure 4.2) are consistent with the idea that sleep is a local activity-dependent network property. Regional changes in sleep intensity (EEG delta power), blood flow and replay of electrical patterns during sleep also indicate a relationship between waking activity and sleep and that sleep is a regional/local process. The comparative and clinical literatures also clearly indicate that sleep is a property of something less than the whole brain. Finally, direct evidence suggests that neural assemblies oscillate between states.

Mechanistically our hypothesis is summarized as follows (Krueger et al., 2007). Neurotransmission is associated with synaptic corelease of ATP with the neuro-transmitter. The consequent increase in extracellular ATP thus provides a measure of prior local neuronal activity. The ATP is detected by nearby purine type 2 receptors on glia causing the release of sleep-regulatory cytokines such as TNF and IL-1, and this provides for the translation of prior neuronal activity into local levels of sleep-regulatory substances. These substances in turn, by a slow process (gene transcription/translation), alter electrical properties of nearby neurons by altering their own production and that of receptor populations, such as AMPA and adenosine receptors, on the neurons. The sleep-regulatory substances also, by a fast process (diffusion for short distances), directly interact with their receptors on neurons and alter electrical properties. Further, ATP itself breaks down, releasing extracellular adenosine, which, in turn, acts on adenosine receptors, again altering electrical potentials on the nearby neurons. These events are happening locally, and the collective electrical changes result in a shift in input–output relationships within the local neuronal assemblies that originally exhibited the increase in activity. In a mathematical model, the local neuronal assemblies synchronize (also called phase locking) with each other because they are loosely connected to each other via neurons and humorally (Roy, personal communication). Further, we think that the sleep-regulatory circuits and the associated activation networks also serve to ensure the synchronization of neuronal assembly state for niche-adaptation purposes. This view, that sleep is a local use-dependent process, has profound implications for sleep function; this is discussed in the next section.

Sleep function

The evolutionary costs of sleep are high; one does not normally eat, drink, socialize, or reproduce during sleep, and sleep subjects one to predation. The evolution of sleep in the face of these high costs suggests an important adaptive value to sleep. At one level, important for evolutionary fitness, we know a function of sleep: peak performance is restored. This finding has led to the universal acknowledgment that sleep serves a restorative function. However, at a more reductionist level, there has been a failure to identify what is restored. Nevertheless, several modern theories of sleep function posit that sleep stabilizes synaptic networks by affecting synaptic efficacy and connectivity (Crick & Mitchinson, 1983; Kavanau, 1994; Krueger & Obál, 1993; Tononi & Cirelli, 2006). Other sleep function theories are closely related, especially those originating from the developmental (Benington & Frank, 2003; Marks, Shaffery, Oksenberg, et al., 1995) and memory (Drosopoulos, Schulze, Fischer, et al., 2007; Maquet, 2004) literatures, in that they posit that sleep is for connectivity. These ideas have their origin in the observation that experience modifies the microcircuitry of the brain. Such use–dependent-induced changes in synaptic efficacy and connectivity have the potential to lead to dysfunction – for example, loss or modification of critical synaptic networks (Kavanau, 1994; Krueger & Obál, 1993). Thus the brain is confronted with the problem of how to maintain synaptic networks that contain both instinctual and learned memories that have proven adaptive (the prima facie evidence is that the organism is alive) in the face of a constantly changing network modified by everyday experience. Herein we presented a sleep regulatory mechanism that simultaneously had the potential to stabilize cell sensitivity by changing receptor populations for excitatory (glutamate) and inhibitory (adenosine) molecules as the need arises due to neuronal activity. This is accomplished via the SRSs that are altered by activity and which, in turn, alter expression of the receptors involved at the sites where the SRSs were induced by use. Although we are far from any comprehensive molecular or genetic understanding of sleep, this view suggests that sleep mechanisms are closely tied to a fundamental sleep function of connectivity. Further, such a mechanism provides insight to the issue of unconsciousness. Thus, if we consider input during waking to induce an environmentally adaptive output, then – after prolonged neuronal use and the activation of those mechanisms shown in Figures 4.2 and 4.3 – the consequent SRS release would induce a new output (state shift). The new output would likely not be relevant to the environmentally driven input and thus be maladaptive if allowed to manifest in motor or cognitive real-time outputs because behavior would not be coordinated in real-time to environmental inputs. There would thus be an adaptive need to prevent the animal from behaving at such times. The local sleep mechanisms are thus not only inseparable from the connectivity/metabolic functions of sleep but also provide the necessity for unconsciousness.

References

Albensi, B. C., & Mattson, M. P. (2000). Evidence for the involvement of TNF and NF-kappaB in hippocampal synaptic plasticity. *Synapse, 35,*151–159.

Basheer, R., Rainnie, D. G., Porkka-Heiskanen, T., Ramesh, V., & McCarley, R. W. (2001). Adenosine, prolonged wakefulness, and A1-activated NF-κB DNA binding in the basal forebrain of the rat. *Neuroscience*, *104*, 731–739.

Basheer, R., Strecker, R. E., Thakkar, M. M., & McCarley, R. W. (2004). Adenosine and sleep-wake regulation. *Progress in Neurobiology*, *73*, 379–396.

Bazhenov, M., Timofeev, I., Steriade, M., & Sejnowski, T. J. (2002). Model of thalamocortical slow-wave sleep oscillations and transitions to activated states. *Journal of Neuroscience*, *22*, 8691–8704.

Beattie, E. C., Stellwagen, D., Morishita, W., Bresnahan, J. C., Ha, B. K., Von Zastrow, et al. (2002). Control of synaptic strength by glial TNF alpha. *Science*, *295*, 2282–2285.

Benington, J. H., & Frank, M. G. (2003). Cellular and molecular connection between sleep and synaptic plasticity. *Progress in Neurobiology*, *69*, 71–101.

Bianco, F., Pravettoni, E., Colombo, A., Schenk, U., Moller, T., Matteoli, M., et al. (2005). Astrocyte-derived ATP induces vesicle shedding and IL-1 beta release from microglia. *Journal of Immunology*, *174*, 7268–7277.

Borbély, A. A., & Tobler, I. (1989). Endogenous sleep-promoting substances and sleep regulation. *Physiological Reviews*, *69*, 605–670.

Brandt, J. A., Churchill, L., Rehman, A., Ellis, G., Mémet, S., Israël, A., et al. (2004). Sleep-deprivation increases activation of nuclear factor kappa B in lateral hypothalamic cells. *Brain Research*, *1004*, 91–97.

Bredow, S., Taishi, P., Guha-Thakurta, N., Obál, F., Jr, & Krueger, J. M. (1997). Diurnal variations of tumor necrosis factor-alpha mRNA and alpha-tubulin mRNA in rat brain. *Journal of Neuroimmunomodulation*, *4*, 84–90.

Cavadini, G., Petrzilka, S., Kohler, P., Jud, C., Tobler, I., Birchler, T., et al. (2007). TNF-alpha suppresses the expression of clock genes by interfering with E-box-mediated transcription. *Proceedings of the National Academy of Sciences of the United States of America*, *104*, 12843–12848.

Chen, Z., Gardi, J., Kushikata, T., Fang, J., & Krueger, J. M. (1999). Nuclear factor kappa B-like activity increases in murine cerebral cortex after sleep deprivation. *American Journal of Physiology*, *45*, R1812–R1818.

Churchill, L., Rector, D., Yasuda, K., Rojas, M. J., Schactler, S., Fix, C., et al. (2006). Tumor necrosis factor α increases surface evoked potentials in the barrel field by whisker deflection during sleep in rats. *Sleep*, *29*, A12–A13.

Churchill, L., Yasuda, K., Yasuda, T., Blindheim, K., Falter, M., Garcia-Garcia, F., et al. (2005). Unilateral cortical application of tumor necrosis factor alpha induces asymmetry in Fos- and interleukin-1 beta-immunoreactive cells within the corticothalamic projection. *Brain Research*, *1055*, 15–24.

Cottone, L. A., Adamo, D., & Squires, N. K. (2004). The effect of unilateral somatosensory stimulation on hemispheric asymmetries during slow-wave sleep. *Sleep*, *27*, 63–68.

Crick, F., & Mitchinson, G. (1983). The function of dream sleep. *Nature*, *304*, 111–114.

Darko, D. F., Miller, J. C., Gallen, C., White, J., Koziol, J., Brown, S. J., et al. (1995). Sleep electroencephalogram delta-frequency amplitude, night plasma levels of tumor necrosis factor alpha, and human immunodeficiency virus infection. *Proceedings of the National Academy of Sciences of the United States of America*, *92*, 12080–12084.

De, A., Krueger, J. M., & Simasko, S. M. (2003). Tumor necrosis factor alpha increases cytosolic calcium response AMPA and KCl in primary cultures of rat hippocampal neurons. *Brain Research*, *981*, 133–142.

De, A., Krueger, J. M., & Simasko, S. M. (2005). Glutamate induces expression and release of tumor necrosis factor alpha in cultured hypothalamic cells. *Brain Research*, *1053*, 54–61.

Dickstein, J. B., Moldofsky, H., Lue, F. A., & Hay, J. B. (1999). Intracerebroventricular injection of TNF-alpha promotes sleep and is recovered in cervical lymph. *American Journal of Physiology*, *276*, R1018–R1022.

Domercq, M., Brambilla, L., Pilati, E., Marchaland, J., Volterra, A., & Bezzi, P. (2006). P2Y1 receptor-evoked glutamate exocytosis from astrocytes control by tumor necrosis factor and prostaglandins. *Journal of Biological Chemistry*, *281*, 30684–30696.

Drosopoulos, S., Schulze, C., Fischer, S., & Born, J. (2007). Sleep's function in the spontaneous recovery and consolidation of memories. *Journal of Experimental Psychology – General*, *136*, 169–183.

Drummond, S. P. A., Brown, G. G., Stricker, J. L., Buxton, R. B., Wong, E. C., & Gillin, J. C. (1999). Sleep deprivation-induced reduction in cortical functional response to serial subtraction. *NeuroReport*, *10*, 3745–3748.

Edelman, G. H. (1987). *Neural Darwinism*. New York: Basic Books.

Eissner, G., Kolch, W., & Scheurich, P. (2004). Ligands working as receptors; Reverse signaling by members of the TNF superfamily enhance the plasticity of the immune system. *Cytokine and Growth Factor Reviews*, *15*, 353–366.

Fang, J., Wang, Y., & Krueger, J. M. (1997). Mice lacking the TNF 55 kD receptor fail to sleep more after TNF alpha treatment. *Journal of Neuroscience*, *17*, 5949–5955.

Farber, K., & Kettenmann, H. (2006). Purinergic signaling and microglia. *Pflügers Archives – European Journal of Physiology*, *452*, 615–621.

Ferrara, M., De Gennaro, L., Curcio, G., Cristiani, R., & Bertini, M. (2002). Interhemispheric asymmetry of human sleep EEG in response to selective slow-wave sleep deprivation. *Behavioral Neuroscience*, *116*, 976–981.

Fix, C., Churchill, L., Hall, S., Kirkpatrick, R., Yasuda, T., & Krueger, J. M. (2006). The number of tumor necrosis factor α-immunoreactive cells increases in layer IV of the barrel field in response to whisker deflection in rats. *Sleep*, *29*, A11.

Floyd, R. A., & Krueger, J. M. (1997). Diurnal variations of TNF alpha in the rat brain. *NeuroReport*, *8*, 915–918.

Foltenyi, K., Greenspan, R. J., & Newport, J. W. (2007). Activation of EGFR and ERK by rhomboid signaling regulates the consolidation and maintenance of sleep in *Drosophila*. *Nature Neuroscience*, *10*, 1160–1167.

Fontaine, V., Mohand-Said, S., Hanoteau, S., Fuchs, C., Pfizenmaier, K., & Eisel, U. (2002). Neurodegenerative and neuroprotective effects of tumor necrosis factor (TNF) in retinal ischemia: Opposite roles of TNF receptor 1 and TNF receptor 2. *Journal of Neuroscience*, *22*, RC216.

Frank, M. G., Issa, N. P., & Stryker, M. P. (2001). Sleep enhances plasticity in the developing visual cortex. *Neuron*, *30*, 275–287.

Franken, P., Thomason, R., Heller, H. C., & O'Hara, B. F. (2007). A non-circadian role for clock genes in sleep homeostasis: A strain comparison. *BMC Neuroscience*, *8*, 87.

Franklin, C. M. (1999). Clinical experience with soluble TNF p75 receptor in rheumatoid arthritis. *Seminars in Arthritis and Rheumatism, 29,* 171–181.

Gabel, C. A. (2007). P2 purinergic receptor modulation of cytokine production. *Purinergic Signalling, 3,* 27–38.

Hendricks, J. C., Finn, S. M., Panckeri, K. A., Chavkin, J., Williams, J. A., Sehgal, A., et al. (2000). Rest in *Drosophila* is a sleep-like state. *Neuron, 26,* 295–298.

Hide, I., Tanaka, M., Inoué, A., Nakajima, K., Kohsaka, S., Inoué, K., et al. (2000). Extracellular ATP triggers tumor necrosis factor-alpha release from rat microglia. *Journal of Neurochemistry, 75,* 965–972.

Huber, R., Ghilardi, M. F., Massimini, M., & Tononi, G. (2004). Local sleep and learning. *Nature, 430,* 78–81.

Inoué, S. (1989). *Biology of sleep substances.* Boca Raton, FL: CRC Press, Inc.

Iwasaki, N., Karashima, A., Tamakawa, Y., Katayama, N., & Nakao, M. (2004). Sleep EEG dynamics in rat barrel cortex associated with sensory deprivation. *NeuroReport, 15,* 2681–2684.

Ji, D., & Wilson, M. A. (2007). Coordinated memory replay in the visual cortex and hippocampus during sleep. *Nature Neuroscience, 10,* 100–107.

Jones, B. (2003). Arousal systems. *Frontiers in Bioscience, 8,* S438–S451.

Jouvet, M. (1984). Neuromediateurs et facteurs hypnogenes [Neurotransmitters and hypnogenetic factors]. *Revue Neurologique (Paris), 140,* 389–400.

Kataoka, T., Enomoto, F., Kim, R., Yokoi, H., Fujimori, M., Sakai, Y., et al. (2004). The effect of surgical treatment of obstructive sleep apnea syndrome on the plasma TNF-alpha levels. *Tohoku Journal of Experimental Medicine, 204,* 267–272.

Kattler, H., Dijk, D. J., & Borbely, A. A. (1994). Effect of unilateral somatosensory stimulation prior to sleep on the sleep EEG in humans. *Journal of Sleep Research, 3,* 1599–1604.

Kavanau, J. L. (1994). Sleep and dynamic stabilization of neural circuitry: A review and synthesis. *Behavioral Brain Research, 63,* 111–126.

Kavanau, J. L. (2005). Evolutionary approaches to understanding sleep. *Sleep Medicine Reviews, 9,* 141–152.

Kavanau, J. L. (2006). Is sleep's "supreme mystery" unraveling? An evolutionary analysis of sleep encounters no mystery; nor does life's earliest sleep, recently discovered in jellyfish. *Medical Hypotheses, 66,* 3–9.

Koch, C. (2004). *The quest for consciousness.* Englewood, CO: Roberts and Company.

Kristiansen, K., & Courtois, G. (1949). Rhythmic activity from isolated cerebral cortex EEG. *Clinical Neurophysiology, 1,* 265–272.

Krueger, J. M., & Obál, F., Jr. (1994). *Sleep Factors.* In N. A. Saunders & C. E. Sullivan (Eds.), *Sleep and breathing* (pp. 79–112). New York: Marcel Dekker, Inc.

Krueger, J. M., & Obál, F., Jr. (1993). A neuronal group theory of sleep function. *Journal of Sleep Research, 2,* 63–69.

Krueger, J. M., & Obál, F., Jr. (2003). Sleep function. *Frontiers in Bioscience, 8,* 511–519.

Krueger, J. M., Rector, D. M., & Churchill, L. (2007). Sleep and cytokines. *Sleep Medicine Clinics, 2,* 161–170.

Kubota, T., Fang, J., Guan, Z., Brown, R. A., & Krueger, J. M. (2001). Vagotomy attenuates tumor necrosis factor-alpha-induced sleep and EEG delta-activity in rats. *American Journal of Physiology, 280,* R1213–R1220.

Kubota, T., Kushikata, T., Fang, J., & Krueger, J. M. (2000). Nuclear factor kappa B (NFκB) inhibitor peptide inhibits spontaneous and interleukin-1β-induced sleep. *American Journal of Physiology*, *279*, R404–R413.

Kubota, T., Li, N., Guan, Z., Brown, R. A., & Krueger, J. M. (2002). Intrapreoptic microinjection of TNF-alpha enhances non-REMS in rats. *Brain Research*, *932*, 37–44.

Kushikata, T., Fang, J., Chen, Z., Wang, Y., & Krueger, J. M. (1998). Epidermal growth factor (EGF) enhances spontaneous sleep in rabbits. *American Journal of Physiology*, *275*, R509–R514.

Ledgerwood, E. C., Pober, J. S., & Bradley, J. R. (1999). Recent advances in the molecular basis of TNF signal transduction. *Laboratory Investigation*, *79*, 1041–1050.

Mahowald, M. W., & Schenck, C. H. (2005). Insights from studying human sleep disorders. *Nature*, *437*, 1279–1285.

Majde, J. A., & Krueger, J. M. (2005). Links between the innate immune system and sleep. *Journal of Allergy & Clinical Immunology*, *116*, 1188–1198.

Malinow, R., & Malenka, R. C. (2002). AMPA receptor trafficking and synaptic plasticity. *Annual Review of Neuroscience*, *25*, 103–126.

Maquet, P. (2004). A role for sleep in the processing of memory traces. Contribution of functional neuroimaging in humans. *Bulletin et mémories de l'Académie royale de médicine de Belgique*, *159*, 167–170.

Maquet, P. (2001). The role of sleep in learning and memory. *Science*, *294*, 1048–1052.

Marks, G. A., & Shaffery, J. P., Oksenberg, A., Speciale, S. G., Roffwarg, H. P. (1995). A functional role for REM sleep in brain maturation. *Behav Brain Res*, *69*, 1–11.

Mascetti, G. G., Rugger, M., Vallortigara, G., Boddo, D. (2007). Monocular-unihemispheric sleep and visual discrimination learning in the domestic chick. *Exp Brain Res 176*, 70–84.

McGinty, D., Szymusiak, R. (2003). Hypothalamic regulation of sleep and arousal. *Front Biosci 8*, d1074–d1083.

Miller, T. B., Goodrich, C. A., Pappenheimer, J. R. (1967). Sleep-promoting effects of cerebrospinal fluid from sleep-deprived goats. *Proc Natl Acad Sci USA 58*, 513–517.

Miyamoto, H., Katagiri, H., & Hensch, T. (2003). Experience-dependent slow-wave sleep development. *Nat Neurosci 6*, 553–554.

Montagna, P. (2005). Fatal familial insomnia: A model disease in sleep physiopathology. *Sleep Med Rev 9*, 339–353.

Mukhametov, L. M., (1984). Sleep in marine mammals. *Expt Brain Res 8*, 227–238.

Mullington, J., Korth, C., Hermann, D. M., Orth, A., Galanos, C., Holsboer, F., et al. (2003). Biochemical regulation of non-rapid eye movement sleep. *Frontiers in Bioscience*, *8*, 520–550.

Mullington, J., Korth, C., Hermann, D. M., Orth, A., Galanos, C., Holsboer, F., et al. (2000). Dose-dependent effects of endotoxin on human sleep. *American Journal of Physiology, 278*, R947–R955.

Obál, F., Jr., & Krueger, J. M. (2003). Biochemical regulation of non-rapid eye movement sleep. *Frontiers in Bioscience, 8*, 520–550.

Opp, M. R. (2005). Cytokines and sleep. *Sleep Medicine Reviews, 9*, 355–364.

Puccioni-Sohler, M., Rieckmann, P., Kitze, B., Lange, P., Albrecht, M., & Flegenhauer, K. (1995). A soluble form of tumor necrosis factor receptor in cerebrospinal fluid and serum of HTVLV-1-associated myelopathy and other neurological diseases. *Neurology, 242*, 239–242.

Rattenborg, N. C., Amlaner, C. J., & Lima, S. L. (2001). Unilateral eye closure and interhemispheric EEG asymmetry during sleep in the pigeon (*Columba livia*). *Brain, Behavior & Evolution*, *58*, 323–332.

Rector, D. M., Topchiy, I., & Rojas, M. (2005a). Local cortical column activity states and localized delta wave differences. *Sleep*, *28*, A26.

Rector, D. M., Topchiy, I. A., Carter, K. M., & Rojas, M. J. (2005b). Local functional state differences between rat cortical columns. *Brain Research*, *1047*, 45–55.

Rehman, A., Taishi, P., Fang, J., Majde, J. A., & Krueger, J. M. (2001). The cloning of a rat peptidoglycan recognition protein (PGRP) and its induction in brain by sleep deprivation. *Cytokine*, *13*, 8–17.

Riha, R. L., Brander, P., Vennelle, M., McArdle, N., Kerr, S. M., Anderson, N. H., et al. (2005). Tumour necrosis factor-alpha (-308) gene polymorphism in obstructive sleep apnoea-hypopnoea syndrome. *European Respiratory Journal*, *26*, 673–678.

Saper, C. B., Scammell, T. E., & Lu, J. (2005). Hypothalamic regulation of sleep and circadian rhythms. *Nature, 437*, 1257–1263.

Sauer, S., Herrmann, E., & Kaiser, W. (2004). Sleep deprivation in honeybees. *Journal of Sleep Research*, *13*, 145–152.

Sauer, S., Kinkelin, M., Herrmann, E., & Kaiser, W. (2003). The dynamics of sleep-like behaviour in honeybees. *Journal of Comparative Physiology, Series A*, *189*, 599–607.

Shoham, S., Davenne, D., Cady, A. B., Dinarello, C. A., & Krueger, J. M. (1987). Recombinant tumor necrosis factor and interleukin 1 enhance slow-wave sleep. *American Journal of Physiology*, *253*, R142–R149.

Siegel, J. M. (2005). Clues to the functions of mammalian sleep. *Nature*, *437*, 1264–1271.

Sookoian, S. C., Gonzalez, C., & Pirola, C. J. (2005). Meta-analysis on the G-308A tumor necrosis factor alpha gene variant and phenotypes associated with the metabolic syndrome. *Obesity Research*, *13*, 2122–2131.

Spiegel, K., Knutson, K., Leproult, R., Tasali, E., & Van Cauter, E. (2005). Sleep loss: A novel risk factor for insulin resistance and Type 2 diabetes. *Journal of Applied Physiology*, *99*, 2008–2019.

Stellwagen, D., & Malenka, R. C. (2006). Synaptic scaling mediated by glial TNF-alpha. *Nature*, *440*, 1054–1059.

Steriade, M. (2003). The corticothalamic system in sleep. *Frontiers in Bioscience*, *8*, d878–d899.

Suzuki, T., Hide, I., Ido, K., Kohsaka, S., Inoué, K., & Nakata, Y. (2004). Production and release of neuroprotective tumor necrosis factor by P2X7 receptor-activated microglia. *Journal of Neuroscience*, *24*, 1–7.

Taishi, P., Churchill, L., Wang, M., Kay, D., Davis, C. J., Guan, X., et al. (2007). TNFα siRNA reduces brain TNF and EEG delta wave activity in rats. *Brain Research*, *1156*, 125–132.

Taishi, P., Gardi, J., Chen, Z., Fang, J., & Krueger, J. M. (1999). Sleep deprivation increases the expression of TNF alpha mRNA and TNF 55kD receptor mRNA in rat brain. *The Physiologist*, *42*, A4.

Takahashi, S., & Krueger, J. M. (1997). Inhibition of tumor necrosis factor prevents warming-induced sleep responses in rabbits. *American Journal of Physiology*, *272*, R1325–R1329.

Terao, A., Matsumura, H., Yoneda, H., & Saito, M. (1998). Enhancement of slow-wave sleep by tumor necrosis factor-alpha is mediated by cyclo-oxygenase-2 in rats. *NeuroReport*, *9*, 3791–3796.

Tobler, I. (2005). Phylogeny of sleep regulation. In M. H. Kryger, T. Roth, & W. C. Dement (Eds.), *Principles and practice of sleep medicine* (pp. 70–90). Philadelphia: Elsevier.

Tononi, G., & Cirelli, C. (2006). Sleep function and synaptic homeostasis. *Sleep Medicine Reviews, 10,* 49–62.

Vgontzas, A. N., Zoumakis, E., Lin, H. M., Bixler, E. O., Trakada, G., & Chrousos, G. P. (2004). Marked decrease in sleepiness in patients with sleep apnea by etanercept, a tumor necrosis factor-α antagonist. *Journal of Clinical Endocrinology & Metabolism, 89,* 4409–4413.

Vyazovskiy, V., Borbély, A. A., & Tobler, I. (2000). Unilateral vibrissae stimulation during waking induces interhemispheric EEG asymmetry during subsequent sleep in the rat. *Journal of Sleep Research, 9,* 367–371.

Walker, A. J., Topchiy, I., Kouptsov, K., & Rector, D. M. (2005). ERP differences during conditioned lick response in the rat. *Sleep, 28,* A15.

Williams, J. A., Sathyanarayanan, S., Hendricks, J. C., & Sehgal, A. (2007). Interaction between sleep and the immune response in *Drosophila*: A role for the NFkB relish. *Sleep, 30,* 389–401.

Yang, L., Lindholm, K., Konishi, Y., Li, R., & Shen, Y. (2002). Target depletion of distinct tumor necrosis factor receptor subtypes reveals hippocampal neuron death and survival through different signal transduction pathways. *Journal of Neuroscience, 22,* 3025–3032.

Yasuda, T., Yasuda, K., Brown, R. A., & Krueger, J. M. (2005). State-dependent effects of light-dark cycle on somatosensory and visual cortex EEG in rats. *American Journal of Physiology: Regulatory, Integrative & Comparative Physiology, 289,* R1083–R1089.

Yoshida, H., Peterfi, Z., Garcia-Garcia, F., Garcia-Garcia, F., Kirkpatrick, R., Yasuda, T., et al. (2004). State-specific asymmetries in EEG slow-wave activity induced by local application of TNF alpha. *Brain Research, 1009,* 129–136.

Yu, Z., Cheng, G., Wen, X., Wu, G. D., Lee, W. T., & Pleasure, D. (2002). Tumor necrosis factor alpha increases neuronal vulnerability to excitotoxic necrosis by inducing expression of the AMPA-glutamate receptor subunit GluR1 via an acid sphingomyelinase- and NF-kappaB-dependent mechanism. *Neurobiology of Disorders, 11,* 199–213.

5

Evolutionary medicine of sleep disorders: Toward a science of sleep duration

PATRICK MCNAMARA AND SANFORD AUERBACH

Introduction

Evolutionary medicine is a relatively new field of inquiry that attempts to apply the findings and principles of evolutionary anthropology and biology to medical disorders (Armelagos, 1991; Cohen, 1989; Nesse & Williams, 1998; Stearns, 1999; Stearns & Koella, 2007; Trevathan, Smith, & McKenna, 1999, 2008; Williams & Nesse, 1991). Although a fair number of medical disorders have been explored from the evolutionary medicine perspective (see the collection of papers in Stearns, 1999, and Trevathan et al., 1999, 2008), sleep disorders have not been among them. This is unfortunate, as application of evolutionary theory to problems of sleep disorders will likely yield significant new insights into both the causes and solutions of all of the major sleep disorders.

In this chapter, we discuss several of these major sleep disorders as well as some of the less common ones. Our choice of which disorders to cover was rather arbitrary: we chose those where, we believe, evolutionary analysis is currently in a position to shed new light on the symptomatology of the disorder as well as on its potential ultimate causes. We were particularly interested in disorders that might also shed light on a potential science of sleep durations.

Why sleep durations? Time spent asleep is one of the most important aspects of sleep, as it is directly linked to the restorative qualities of sleep. If you do not get enough sleep, you do not feel well. Too much or too little sleep relative to the population mean has also been linked to disease and to premature mortality (Bliwise & Young, 2007; Hublin, Partinen, Koskenvuo, et al., 2007; Kripke, 2003; Shankar, Koh, Yuan, et al., 2008; Stranges, Dorn, Shipley, et al., 2008). In addition to its clinical links in humans, sleep duration also appears to contribute to fundamental physiologic and ecologic adaptations of many nonhuman mammalian species. As the chapters in this volume attest, variations in sleep durations across mammalian species are correlated with key physiological and ecological characteristics of species even after correcting for phylogenetic relatedness between species. Clearly sleep duration must be linked in some fundamental way with sleep functions. Unfortunately, we do not yet have a science of sleep duration. Rule-governed changes in sleep durations as a function of sleep disorder therefore might provide unique insight to the normal mechanisms of sleep duration.

Sleep disorders

Sleep disorders can be divided into two very broad classes: dyssomnias and parasomnias. Dyssomnias involve changes in sleep duration such that the patient gets too much or too little sleep. Parasomnias involve partial arousals from within a rapid-eye-movement (REM) or non–rapid-eye-movement (NREM) sleep state. A third class of *sleep-related* disorders involves changes in the circadian pacemaker system such that the daily sleep period is displaced (delayed or advanced) from its normal slot within the 24-hour day. We confine our discussion to the dyssomnias.

Dyssomnias

As mentioned, dyssomnias involve a change in sleep amount from normal reference values. Hypersomnolence is too much sleep and insomnia is too little. Insomnia and excessive daytime sleepiness are, in fact, the most common disorders of sleep. Changes in sleep duration, furthermore, are associated with significant risks to both physical and mental health. Persons with longer REM sleep durations (relative to the population norm), for example, experience greater risks for various medical conditions (e.g., cardiovascular disease, obesity, etc.) and mortality (Brabbins, Dewey, Copeland, et al., 1993; Dew, Hoch, Buysse, et al., 2003; Kripke, 2003). They are also at greater risk for depression. The increased risk for these "comorbidities" remains even after adjusting for age, gender, and previous mental illness and health status (Brabbins et al., 1993; Dew et al., 2003; Kripke, 2003). Moreover, it has become increasingly clear in recent years that the restorative or homeostatic properties of sleep are dependent on an interaction between sleep amounts and sleep intensity parameters, as formalized in Borbély's original two-process model of sleep regulation (Borbély, 1982) and its more recent emendations (Achermann & Borbély, 2003). When they are deprived of sleep, mammals typically exhibit a sleep rebound proportional to the amount of sleep lost (Tobler, 2000), indicating that the amount of sleep, or of some specific intensity component of sleep reflected in "amount of sleep," is physiologically obligatory.

Decades of research into the effects of sleep deprivation suggests that sleep durations can be adjusted upward or downward depending on the animal's ability to sleep intensively or efficiently. The ability to sleep intensively, in turn, appears to be related to both intrinsic physiologic factors such as brain size and extrinsic ecologic factors such as predation pressures and food availability. For mammals, sleep intensity involves getting enough of a certain kind of sleep – particularly slow-wave delta sleep.

Sleep intensity and the homeostatic regulation of sleep

The importance of the intensity dimension of sleep was first uncovered in experiments on effects of sleep deprivation. The most dramatic effect of sleep deprivation in every mammalian species studied thus far has been the phenomenon of "compensatory rebound" or the increase over baseline of sleep times and intensity, where intensity is measured by higher arousal thresholds, enhanced slow-wave activity, enhanced REM frequencies per unit time, and "deeper" as well as longer sleep cycles (Borbély, 1980; Tobler, 2005). After sleep deprivation, mammals attempt to make up for lost sleep by enhancing the intensity of sleep rather than merely the time spent in subsequent sleep. Birds, too, demonstrate a compensatory rebound after sleep deprivation, but the rebound appears to involve increased overall sleep times rather than enhancements in slow-wave activity. On the other hand, only a few avian species have been studied to date, so it is too early to draw any firm conclusions about sleep intensity in birds. Nevertheless, the phenomenon of compensatory rebound phenomena in mammals indicates a very ancient and conserved homeostatic need for a certain set of reference values (amount and intensity) of sleep.

Borbély (1982) first formalized the insight that mammalian sleep durations involved a homeostatic rate-setting mechanism that kept sleep amounts within a "normal" range of values. His proposal for such a mechanism involved interactions between the daily circadian pacemaker, waking durations, and sleep intensity. In his "two-process" model of sleep regulation a sleep need process (process S) increases during waking (or sleep deprivation) and decreases during sleep. This part of the model indexes restorative aspects of sleep and explicitly predicts that sleep is required for some restorative process of the brain or the body or both. Process S is proposed to interact with input from the light-regulated circadian system (process C) that is independent of sleep and wakefulness rhythms. Slow-wave activity (SWA) is taken as an indicator of the time course of process S because SWA is known to correlate with arousal thresholds and to markedly increase during the previous waking period and during the rebound period after sleep deprivation in all mammals studied. After a threshold value of process S is reached (i.e., once the appropriate amount and intensity of slow-wave sleep [SWS] is reached), process C will be activated. Simulations using the model's assumptions show that the homeostatic component of sleep falls in a sigmoidal manner during waking and rises in a saturating exponential manner during sleep.

Both REM and NREM SWS are under homeostatic control and enhancements of sleep intensity over baseline addresses the homeostatic need for sleep. In short, sleep intensity indexes daily functional need.

What then determines functional need for sleep? Tissue repair? Energy requirements? Recent activity levels? Pathogen stress? Brain plasticity or cognitive processing loads? Reproductive competition? Ecological factors such as the availability of food resources or the presence of predators? As the chapters in this volume suggest, comparative data demonstrate that all of these factors play a role in shaping sleep durations and intensity parameters across species. Does any of this information matter for understanding the dyssomnias?

Yes, it does. Let us begin our discussions of the dyssomnias with an evolutionary analysis of that paradigmatic disorder of sleep – insomnia.

Insomnia

Insomnia, of course, involves a restriction in normal amounts of sleep. Sleep duration is decreased relative to the rest of the population or to the individual's normal amount of sleep. Between 1 in 4 and 1 in 3 persons suffer from some amount of insomnia each year. Insomnia tends to increase with age and is associated with incident stressors in the person's life; as stressors decline in their salience, so does the sleep problem. Insomnia in older adults is associated with reduced REM and SWS percentages (Chesson, Hartse, Anderson, et al., 2000; Espiritu, 2008). Insomnia is the most common disorder of sleep. It has been estimated to affect about one-third of the American population.

In our experience, if one simply surveys patients passing through a primary care clinic, almost half will acknowledge some problem with insomnia, even though some may have relatively mild conditions. Although prevalence rates may vary across studies according to the stringency of the definition applied to the population, it is generally agreed that 10% to 15% of the general population will suffer from moderate to severe degrees of insomnia. Furthermore, chronicity is not uncommon. About 90% of those suffering from insomnia will manifest a chronic picture, with symptoms persisting for more than 6 months. Many of these patients will present with symptoms spanning many years or even decades.

The impact of insomnia on people's lives and on the economy in general is enormous by any measure (National Institutes of Health, 2005). In the mid-1990s, the direct costs, in terms of clinical services and medication costs, were estimated to approach $14 billion annually. If one considers indirect costs as generated by accidents, workplace absenteeism, decreased productivity, and increased comorbidity, the cost may increase by another $77 billion to $92 billion. There is an abundant literature supporting the concern that chronic insomnia may be associated with impairments in cognitive function, a decrease in overall health, an increase in accidents, a lowered pain threshold, and an increased susceptibility to accidents and injuries (Roth, Franklin, & Bramley, 2007). In a general way, then, insomnia has been associated with an overall negative impact on quality of life.

The interface between insomnia with other medical and psychiatric disorders is of interest. For instance, several studies suggest that in addition to being a feature of many medical and psychiatric disorders, insomnia may also be a risk factor for predicting the development of future psychiatric disorders. Insomnia may be a risk factor for the development of a relapse after recovery from a mood disorder.

Insomnia may also enhance pain and suffering. The interaction between pain and sleeplessness may be reciprocal. For instance, insomnia may lead to a lowered pain threshold; pain may, in turn, lead to insomnia; treatment of the insomnia may accelerate the treatment of the pain. In a similar fashion, insomnia and depression may be intertwined, and treatment of an underlying depression may lead to improvement in the insomnia; but it seems that targeted treatment of insomnia may also facilitate treatment of the depression.

The importance of insomnia as an independent disorder and its relation to other medical and psychiatric disorders has contributed to a recent paradigm shift in the way clinicians view insomnia. It is not and should not any longer be viewed as a mere symptom of some other problem but rather as an independent "syndrome" requiring independent mechanisms for its production.

Let us now turn to how an evolutionary medicine perspective can shed light on insomnia. The first step is to clarify the definition of insomnia that will be referenced in this discussion. The revised edition of the *International Classification of Sleep Disorders* (American Sleep Disorders Association, 2005) identifies two key features. The first is the presence of at least one feature of sleep disruption despite an adequate opportunity to sleep. There may be difficulties with sleep onset, frequent awakenings, prolonged period(s) of awakening, or an early awake time. The mere acknowledgment of sleep disruption, however, is not adequate for the diagnosis of a sleep disorder. It is also important to identify an impact on daytime function. In some respects, the association of insomnia with daytime symptoms (fatigue, cognitive impairments, or affective symptoms) and other symptoms of distress is an equally important component of the definition of insomnia. Thus daytime affective disturbances associated with sleeplessness are hallmark symptoms of idiopathic insomnia.

It is always tempting to compare insomnia with sleep deprivation. In both cases, there is a disruption of sleep in the context of evidence suggesting stress on the organism. A reader may be quick to point out that the very definition of insomnia requires the presence of distress, even though causality is not specified. Nevertheless, the distinction between insomnia and sleep deprivation becomes evident when one looks at the element of associated sleepiness. Patients suffering from sleep deprivation are generally found to be quite sleepy during the day. There is an abundant literature emphasizing that subjects who are deprived of their usual sleep requirement experience characteristics of sleepiness (Balkin, Rupp, Picchioni, et al., 2008). This is true when subjects are subjected to enforced sleep deprivation or external demands, as may be imposed by the requirements of a disruptive sleep environment, the demands of a busy medical training program, or the influence of a crying infant. All result in an urge to sleep when given the opportunity, a shortened sleep latency (time to fall asleep), or a susceptibility to "microsleeps" – seconds-long dips into a sleep state. Insomniacs, however, are usually not considered to be sleepy even though they may say that they are. Standard "objective" measures of daytime sleepiness, such as the Epworth Sleepiness Scale or the Multiple Sleep Latency Test, generally reveal that insomniacs are no more sleepy than age-matched controls who do not complain of insomnia. In fact, insomniacs may appear to be less prone to sleepiness than the normal controls.

At first, the apparent resistance of the insomniac to the homeostatic drive toward sleepiness may seem paradoxical. If one speaks to the insomniacs arriving at a sleep clinic, it soon becomes clear that the sleepless nights make the sufferers feel terrible the following day. Although they do not feel well, insomniacs are not necessarily sleepy the next day. Despite the claims of problems with sleep initiation and/or maintenance and apparent shortening of total sleep time, these individuals do not display the same degree of sleepiness as the sleep-deprived. In short, insomniacs may resemble to some extent the class of people known as short sleepers, except that these individuals do not complain of feeling terrible the next day, as do the insomniacs (Fichten, Libman, Creti, et al., 2004). Both the insomniacs and the short sleepers exhibit an objective reduction in sleep times relative to the rest of the population. But do both groups of people exhibit resistance to homeostatic drive or the need to make up for lost sleep? It is not clear that the short sleepers do. Although they do not sleep much, short sleepers exhibit either a normal or even an *enhanced* homeostatic drive when deprived of the short amounts of sleep they do get (Aeschbach, Postolache, Sher, et al., 2001). So we are left with insomniacs as the group of individuals who seem to exhibit clear resistance to homeostatic drive.

Is it possible that resistance to the homeostatic drive could be viewed from the perspective of evolutionary medicine? As one reviews the literature on insomnia, it becomes clear that the emphasis has been on the negative impact of insomnia on the individual. As noted, this negative impact has been examined in terms of behavioral, cognitive, social, and even economic terms. Therefore it is unlikely that a clinician would think about a possible advantage of insomnia, even when this idea is phrased in evolutionary terms. A somewhat different approach would be to think that a disorder as common as insomnia may, in certain circumstances, carry an advantage. Perhaps, the answer is in the middle of this same literature. The common theme is that insomnia is typically associated with stress. Although the research often implies that the stress may be a product of the insomnia, causality is usually not addressed. In fact, in the clinical situation, the patient may often blame the symptoms of stress on the insomnia, rather than considering that the insomnia is the product of the stress. Of course, the clinical situation carries a selection bias. Sleep medicine clinics are usually visited by patients who trace the source of their ill feelings to the insomnia and have little insight as to the fact that anxiety or other stressors are the triggers rather than the product of their insomnia. On the other hand, the patient who views his or her insomnia as a product of anxiety or other medical condition will seek assistance outside the sleep clinic and direct attention to the associated disorder.

Perhaps there are two steps we need to take to reformulate insomnia in terms that will permit an analysis appropriate for evolutionary medicine. The first is to note that insomnia is always associated with a state of stress. The second step is to reformulate sleeplessness as a *resistance to homeostatic drive*. Again, the sleep literature has often focused on sleep need and the homeostatic drive to meet that sleep need. This is the drive to sleep that the organism begins to accumulate on awakening. As noted above each organism carries a sleep drive and a daily sleep need. The drive to sleep, in short, is an established fact in the sleep sciences. Much less attention has been directed toward the necessity to resist this homeostatic drive. It is quite easy to consider circumstances where the individual should resist this drive to sleep. Ironically, insomniacs may carry this resistance.

One final point must be made in considering the paradoxical "evolutionary advantage" of insomnia in association with stress. At first glance, one might suspect that the insomniac is at a clear disadvantage because of the impairments in daytime functioning that are frequently reported. Indeed, insomnia may be associated with a decrease in performance. (Whether the drop in performance is a function of the lost sleep or the associated stress is debatable.) Nevertheless, these patients still perform reasonably well. If one starts with predictions of performance based on sleep deprivation studies, one will be surprised by the extent of preserved function. This leads us to the final step in the consideration of insomnia from an evolutionary perspective. Insomnia is a state of homeostatic resistance with a relative preservation of the functions otherwise encountered in the well-rested individual. The advantage of this state in times of stress follows.

How does this formulation apply to the clinician? The first step is that insomnia should be considered as a disorder associated with stress. Both need treatment. The concern about whether to consider one as a symptom or a comorbid disorder is not useful. Even if insomnia is a response to stress, concern over the insomnia adds to stress. Perhaps it is worthwhile to remind the patient that insomnia is likely to be an adaptation serving to handle stress. In fact, if the individual is not experiencing stress, he or she would not seek assistance and the condition would not fulfill the requirements for the definition of insomnia. Simply shifting the assessment of insomnia from the idea that it is only a deficit or burden that the individual is helpless to battle to a new portrayal of the insomniac as particularly good at resisting homeostatic drive might help both the patient and physician to cope with the problem. This conceptual shift might also help sleep specialists shift from treating insomnia with sleeping pills to a strategy of identifying the triggers that activate the inherent capacity to resist homeostatic drive. Among the most potent of these triggers is anxiety.

The link between anxiety and insomnia suggests overactivation or hyperarousal of subcortical structures and of amygdalar circuits as one source of sleeplessness; this supposition has been supported by neuroimaging studies (Carr, Drummond, & Nesthus, 2003; Desseilles, Dang-Vu, Schabus, et al., 2008). The hyperarousal may also lead to an inhibition of SWS and thus the decreased SWS observed in elderly insomniacs.

The culprit in this whole story is, of course, anxiety. Anxiety leads to hyperarousal, which leads to insomnia. Why would it benefit an organism to restrict its time asleep and most especially restrict its time in SWS when it is anxious? The simplest answer, of course, is that anxiety signals danger, and thus vigilance levels need to be maintained until the danger is past. After the danger is past, vigilance can be relaxed and lost sleep made up through the compensatory rebound process described previously. Insomniacs would then be considered humanity's natural sentinels.

There are problems, however, with this explanation for insomnia. If vigilance against "danger" is the ultimate cause of chronic insomnia, one would expect some sort of differential inhibition of the sleep state from which it is most difficult to arouse an individual. But it is not at all clear that it is easier for animals to arouse out of REM versus NREM sleep, especially when considering the lighter stages of NREM sleep. One might think that it would be better to inhibit REM than NREM sleep when faced with danger or predation pressure, because REM makes the animal more vulnerable to predation (muscle paralysis is a typical feature of REM sleep in most species studied). REM sleep is also associated with inhibition or hypoarousal of the prefrontal cortex (PFC) – the cortical region responsible for strategic decision making in most species – surely a skill needed in dealing with danger. Finally, REM sleep is associated with the vivid hallucinatory phenomena called dreams. Failure to inhibit REM sleep when facing danger might predispose the animal to perceptual error. In short, if insomnia is about enhancing vigilance to face danger, one would expect differential inhibition of REM rather than NREM sleep, but what we see in insomnia is greater inhibition or loss of SWS rather than REM sleep.

Thus, if enhanced REM values are a liability when facing danger or threat, why is REM sleep not differentially inhibited in anxious insomniacs? Perhaps what is really going on in both the insomniac and in the case where we face a real threat is preparation for a *fight* rather than a flight. Evolutionary theory and comparative data point to REM sleep and its associated mentation as a system designed to activate aggression circuits and respond to threat (Revonsuo, 2000; McNamara, 2004). Social aggression in many mammalian species including humans has been linked to activation of hypothalamic, amygdalar and limbic sites, and reductions in serotoninergic activity (Crowe & Blair, 2008; Linnoila, De Jong, & Virkkunen, 1989; Linnoila, Virkkunen, George, et al., 1994; Virkkunen, Kallio, Rawlings, et al., 1994). These same brain sites are highly activated in REM sleep, while sertoninergic levels are also reduced. REM-related sleep mentation or dreams are drenched in high levels of aggression. Studies of dream content with standardized scoring methodologies reveal that fully half of all dreams include at least one episode of aggression (Domhoff, 2000; Domhoff & Hall, 1996). REM dreams in particular are saturated with aggression. McNamara, McLaren, Smith, and colleagues (2005), for example, found significantly higher aggression/friendliness percents for REM (65%) versus NREM (33%) and versus wake (23%) reports and the associated effect sizes were moderately large (Cohen's h $= -0.64$ for REM−NREM; -0.88 for REM−wake, and -0.24 for NREM−wake) and were statistically significant. The most dramatic difference between REM versus NREM dreams concerned the aggressor percent, which adjusts the number of dreamer-initiated aggressions by number of instances where the dreamer was a victim of an aggression. The aggressor percent for NREM was 0%, indicating that the dreamer was never reported to be an aggressor in NREM. By contrast, it was 52% for REM ($P < 0.0001$). In short, unlike NREM dreams and daytime reports about social interactions, REM dreams are saturated with scenes of aggression, particularly dreamer-initiated aggressions. Presumably sleep mentation is a reflection of the underlying brain circuits activated in tandem with the mentation. In the case of REM dreams, those circuits must include neural systems that code for and promote aggression.

Thus, in insomnia, hyperarousal and its concomitant high subcortical and amygdalar activation levels would likely lower the threshold for aggressivity in insomniacs. People under threat are indeed anxious, but they are also angry.

Are insomniacs aggressive? Very little data speak directly to this question. Ireland and Culpin (2006) studied levels of aggression in 184 incarcerated juvenile offenders. Aggression was related both to the quantity and quality of sleep reported, with reduced quantity and quality of sleep predicting increased overall aggression.

We have examined one end of the spectrum of changes in sleep-duration insomnia. Now let us look at the other end of the spectrum – disorders involving too much sleep. These include narcolepsy, idiopathic hypersomnia, and Kleine–Levin syndrome.

Narcolepsy and hypersomnolence

It is estimated that as many as 3 million people worldwide are affected by narcolepsy. Narcoleptic symptoms typically are first noticed in teenagers or young adults (Benca, 2007). The age-of-onset profile is similar in some respect to age of onset of schizophrenia and suggests a derailed developmental process of some kind. This derailment implies perhaps that in the typical case, some aspects of sleep continue to develop into adolescence. If we could identify those aspects of sleep that develop in adolescence, we might be better able to identify the process that gets derailed in narcolepsy.

Whatever sleep process is affected by the onset of narcoleptic symptoms, the end result is a disinhibition of REM and the experience of too much sleep. Changes in sleep architecture have been extensively documented in narcolepsy (see Guilleminault & Anagnos, 2000, for review) with reduced REM latencies, either no change or an increased REM percent, increased REM densities, and sleep onset REM (SOREM) instead of the normal process with NREM at onset of sleep. SWS may be reduced or show a rapid decrement in the early part of the sleep period. The individual undergoes what appear to be irresistible "sleep attacks" composed primarily of REM sleep.

In addition to dysregulatory REM attacks, the narcoleptic symptom complex includes cataplexy, sleep paralysis, and hypnogogic hallucinations, all REM-related phenomena. Nighttime sleep is often interrupted by awakenings and terrifying dreams (Guilleminault & Anagnos, 2000; Lee, Bliwise, Lebret-Bories, et al., 1993). Evidence suggests that narcolepsy runs in families; 8% to 12% of people with narcolepsy have a close relative with at least one of the symptoms of narcolepsy. Narcolepsy is strongly associated with the HLA DQB1*0602 genotype. There is also an association with HLA DR2 and HLA DQ1, suggesting autoimmune disorder. Recent findings have implicated loss of hypothalamic orexinergic cells in the hypothalamus of narcoleptics. These cells promote arousal and wakefulness, among other things.

Narcolepsy occurs before age 20 in 80% of cases and before the age of 10 in only 5% to 10% of cases. Thus, like schizophrenia, narcolepsy appears to have its onset typically in adolescence. Whatever the proximate cause of the derailment of the sleep development process at adolescence in narcoleptics, the end result is a disinhibition of REM-related physiologic processes. What developmental processes are occurring at or near puberty that are involved in sleep processes? One such set of processes/structures that come online at or near puberty are the executive functions associated with the PFC.

If REM is necessary for normal PFC development and REM is disrupted during development, PFC functions may be disrupted as well. We know that when REM is disturbed or inhibited in adults, daytime PFC functions are correspondingly impaired (Harrison, Horne, & Rothwell, 2000). Perhaps then, derailment of REM processes during development prevents normal development of prefrontal functions in narcoleptics, and one would notice this problem at adolescence when PFC functions normally come fully into play. If so, this would imply that one functional contribution of sleep, particularly REM, is the development of PFC functions. The functional association between REM sleep and PFC, in turn, would imply correlated evolution of the two traits across mammalian species. This latter evolutionary hypothesis could be tested with comparative data – a hypothesis not yet tested as far as we are aware. But first, is there any evidence that PFC functions are impaired in narcoleptics?

Using functional magnetic resonance imaging (fMRI) techniques, Kaufman, Schuld, Pollmacher, and Auer (2002) demonstrated that patients with narcolepsy showed bilateral reductions in cortical gray matter predominantly in inferior temporal and inferior frontal brain regions. Relative global loss of gray matter was independent of disease duration or medication history. No significant alterations in subcortical gray matter were noted.

In summary, given the evident links between REM and PFC functions revealed by narcolepsy, narcolepsy may give us insights to the correlated evolution of sleep and brain structure – in this case the PFC. The changes in sleep duration associated with narcolepsy may also hold clues for development of a science of sleep duration. To piece those clues together into any kind of coherent framework, we need to examine other disorders of increased sleep duration.

Idiopathic hypersomnia

Idiopathic hypersomnia is excessive sleepiness of unknown cause or origin. It is characterized by persistent complaints of excessive sleepiness and long, unrefreshing daytime naps. Some patients report sleep times of greater than 15 hours per day (Basetti & Aldrich, 2000). Between 20% and 60% of patients report a feeling of "sleep drunkenness," difficulty awaking from sleep, and extreme difficulty "getting going" in the morning. During these periods of sleep drunkenness there may be performance deficits on gait tests and on cognitive tests. There is often a positive family history of excessive daytime sleepiness. Although there is often no change in percentages of REM sleep, there is frequently an increase in SWS parameters.

Kleine–Levin syndrome

Arnult, Lin, Gadoth, et al. (2008) recently reviewed the literature on Kleine–Levin syndrome (KLS). In KLS, some event – a viral infection, a physical trauma, or even seasickness – triggers a hypersomnolent episode. The onset of the sleep attack can be rapid, on the order of hours, and patients may sleep up to 20 hours per day. Associated behavioral changes may include overeating, irritability, hypersexuality, and confusional states. Hypersomnolent bouts may last for weeks and then disappear for long periods of time. Like narcolepsy, KLS onset is more common in adolescence and occurs most often in males. There are typically no significant changes in REM parameters but an enhancement of SWS. There are also sleep-onset REM periods, as in narcolepsy.

In summary, all of the dyssomnias involve a change, either an increase or decrease, relative to the norm in sleep durations. Sleep durations can be adjusted up or down depending on the level of homeostatic drive. If there is resistance to that drive, sleep is adjusted downward. If there is enhancement of that drive, sleep is adjusted upward. Given that homeostatic drive is largely indexed by slow-wave activity in mammalian species, it appears that NREM SWS is driving or linked to the mechanism that adjusts sleep durations. When SWS duration is increased, so is sleepiness; and when sleep is restricted, as in insomnia, so is SWS. These directional relationships do not seem to hold for REM sleep – that is, the effect is selective.

Why might SWS differentially influence sleep times? We mentioned above the special role that SWS plays in sleep intensity and in the restorative aspects of sleep. Apparently, when you get too much or too little SWS, you lose the restorative qualities of sleep and "gain" a feeling of sleepiness or sleeplessness.

In the case of too little sleep or insomnia, we have seen why it might be plausible to think of insomnia as part of a facultative mechanism to enhance the aggressiveness of the organism and to deal with threat. Is there any evidence that the hypersomnias can be viewed as facultative adjustments as well? Too much sleep typically occurs in response to infection (Krueger & Fang, 2000); thus it may be that narcolepsy, KLS, and idiopathic hypersomnia may all be considered adaptive responses to infection.

Most animals, humans included, become sleepy with the onset of infection (Hart, 1990). Experimental work confirms that soon after infectious challenge, animals exhibit an increase in NREM sleep durations and a decrease in REM sleep (Krueger & Fang, 2000). Conversely, sleep loss can render one more vulnerable to infection. After sleep deprivation, several immune system parameters change, including natural killer cell activity, interleukin-1, tumor necrosis factor, prostaglandins, nitric oxide, and adenosine. After prolonged (2 to 3 weeks) sleep deprivation, rats become septicemic (Rechtschaffen & Bergmann, 2001) and die. The relation between sleep states and parasite elimination has not been sufficiently studied but seems a very promising avenue for further research.

All three of the hypersomnia syndromes mentioned here have been linked to infectious disease. In the case of narcolepsy, the strong association with positive HLA findings has suggested autoimmune disease, although this is controversial. In the case of idiopathic hypersomnia and KLS, clinical histories have consistently implicated onset of the disease in response to infection. Whether autoimmunity or response to infection is at issue, it is clear that immune response is altered in all three hypersomnias. It is worth considering this response in evolutionary perspective.

Evolutionary medicine has called attention to the fact that the very successes of modern medicine may create unusual medical disorders. Elimination of common parasitic infections may create ideal conditions for autoimmune inflammatory disorders. Hurtado A., Hurtado I., Sapien, and Hill (1999) pointed out, for example, that prevalence of helminth parasites and incidence of asthma are inversely correlated. They do not typically coexist. Immunoglobulin E (IgE) response is triggered by exposure to helminths. When helminths are eliminated by modern medicine and cleanliness, the immune response is free to overreact to allergens, thus stimulating disorders like asthma. Whether or not we can consider these hypersomnias as adaptive responses to infection or as species of an overreactive immune response and perhaps even autoimmune disorders remains to be seen.

What about the relevance of the dyssomnias for a science of sleep durations? We have seen that duration can be facultatively adjusted up or down depending on need. In the case of insomnia, sleep duration is adjusted downward, we argued, to facilitate aggression. In the case of the hypersomnias, sleep duration is adjusted upward, we argued, to respond to infection. These facultative responses of sleep times to challenges suggest that (1) sleep duration is plastic and of course "adjustable"; (2) SWS drives sleep times more so than does REM sleep; and (3) the two sleep states evolved under differing selective pressures, with slow-wave sleep evolving in response to parasite evolution and REM sleep more tightly linked to brain evolution. These conclusions, however, must be considered tentative and speculative until they can be evaluated more rigorously in the laboratory and the clinic.

References

Achermann, P., & Borbély, A. A. (2003). Mathematical models of sleep regulation. *Frontiers in Bioscience*, *1*(8), S683–S693.

Aeschbach, D., Postolache, T. T., Sher, L., Matthews, J. R., Jackson, M. A., & Wehr, T. A. (2001). Evidence from the waking electroencephalogram that short sleepers live under higher homeostatic sleep pressure than long sleepers. *Neuroscience*, *102*(3), 493–502.

American Sleep Disorders Association, Diagnostic Classification Steering Committee. (2005). *International classification of sleep disorders: Diagnostic and coding manual, ICSD-R*. Westchester, IL: American Academy of Sleep Medicine.

Armelagos, G. J. (1991). Human evolution and the evolution of disease. *Ethnicity & Disease*, *1*(1), 21–25.

Arnult, I., Lin, L., Gadoth, N., File, J., Lecendreux, M., Franco, P., et al. (2008). Kleine–Levin syndrome: A systematic study of 108 patients. *Annals of Neurology*, *63*(4), 482–493.

Balkin, T. J., Rupp, T., Picchioni, D., & Wesensten, N. J. (2008). Sleep loss and sleepiness: Current issues. *Chest*, *134*(3), 653–660.

Basetti, C., & Aldrich, M. (2000). Narcolepsy, idiopathic hypersomnias, and periodic hypersomnias. In A. Culebras (Ed.), *Sleep disorders and neurologic disease* (pp. 323–354), New York: Informa Healthcare.

Benca, R. M. (2007). Narcolepsy and excessive daytime sleepiness: Diagnostic considerations, epidemiology, and comorbidities. *Journal of Clinical Psychiatry*, *68*(Suppl. 13), 5–8.

Bliwise, D. L., & Young, T. B. (2007). The parable of parabola: What the U-shaped curve can and cannot tell us about sleep. *Sleep*, *30*(12), 1614–1615.

Borbély, A. A. (1980). Sleep: Circadian rhythm versus recovery process. In M. Koukkou, D. Lehmann, & J. Angst (Eds.), *Functional states of the brain: Their determinants* (pp. 151–161). Amsterdam: Elsevier.

Borbély, A. A. (1982). A two process model of sleep regulation. *Human Neurobiology*, *1*, 195–204.

Brabbins, C. J., Dewey, M. E., Copeland, R. M., Davidson, I. A., McWilliam, C., Saunders, P., et al. (1993). Insomnia in the elderly: Prevalence, gender differences, and relationships with morbidity and mortality. *International Journal of Geriatric Psychiatry*, *8*, 473–480.

Carr, W., Drummond, S. P., & Nesthus, T. (2003). Neuroimaging sleep debt with fMRI in short- and long-sleepers. *Aviation, Space, and Environmental Medicine*, *74*(8), 902–903.

Chesson, A., Jr., Hartse, K., Anderson, W. M., Davila, D., Johnson, S., Littner, M., et al. (2000). Practice parameters for the evaluation of chronic insomnia. An American Academy of Sleep Medicine report. Standards of practice committee of the American Academy of Sleep Medicine. *Sleep*, *23*(2), 237–241.

Cohen, M. N. (1989). *Health and the rise of civilization*. New Haven, CT: Yale University Press.

Crowe, S. L., & Blair, R. J. (2008). The development of antisocial behavior: What can we learn from functional neuroimaging studies? *Developmental Psychopathology*, *20*(4), 1145–1159.

Desseilles, M., Dang-Vu, T., Schabus, M., Sterpenich, V., Maquet, P., & Schwartz, S. (2008). Neuroimaging insights into the pathophysiology of sleep disorders. *Sleep*, *31*(6), 777–794.

Dew, M. A., Hoch, C. C., Buysse, D. J., Monk, T. H., Begley, A. E., Houck, P. R., et al. (2003). Healthy older adults' sleep predicts all-cause mortality at 4 to 19 years of follow-up. *Psychosomatic Medicine*, *65*, 63–73.

Domhoff, G. W. (2000). Methods and measures for the study of dream content. In M. Kryger, T. Roth, & W. Dement (Eds.), *Principles and practices of sleep medicine* (3rd ed., pp. 463–471). Philadelphia: W. B. Saunders.

Domhoff, G. W., & Hall, C. S. (1996). *Finding meaning in dreams*. New York: Plenum Press.

Espiritu, J. R. (2008). Aging-related sleep changes. *Clinical Geriatric Medicine*, *24*(1), 1–14.

Fichten, C. S., Libman, E., Creti, L., Bailes, S. and Sabourin, S. (2004). Long sleepers sleep more and short sleepers sleep less: A comparison of older adults who sleep well. *Behavioral Sleep Medicine*, *2*, 2–23.

Guilleminault, C., & Anagnos, A. (2000). *Narcolepsy*. In M. H. Kryger, T. Roth, W. C. Dement (Eds.), *Principles and practice of sleep medicine* (3rd ed., pp. 676–686). Philadelphia: W. B. Saunders.

Harrison, Y., Horne, J. A., & Rothwell, A. (2000). Prefrontal neuropsychological effects of sleep deprivation in young adults – A model for healthy again? *Sleep*, *23*(8), 1067–1073.

Hart, B. L. (1990). Behavioral adaptations to pathogens and parasites: Five strategies. *Neuroscience and Biobehavioral Reviews*, *14*, 273–294.

Hublin, C., Partinen, M., Koskenvuo, M., & Kaprio, J. (2007). Sleep and mortality: A population-based 22-year follow-up study. *Sleep 30*(10), 1245–1253.

Hurtado, A. M., Hurtado, I. A., Sapien, R., & Hill, K. (1999). The evolutionary ecology of childhood asthma. In W. R. Trevathan, E. O. Smith, & J. J. McKenna (Eds.), *Evolutionary medicine* (pp. 101–134). New York: Oxford University Press.

Ireland, J. L., & Culpin, V. (2006). The relationship between sleeping problems and aggression, anger, and impulsivity in a population of juvenile and young offenders. *Journal of Adolescent Health*, *38*(6), 649–655.

Kaufman, C., Schuld, A., Pollmacher, T., & Auer, D. P. (2002). Reduced cortical gray matter in narcolepsy: Preliminary findings with voxel-based morphometry. *Neurology*, *58*, 1852–1855.

Kripke, D. F. (2003). Sleep and mortality. *Psychosomatic Medicine*, *65*(1), 74.

Krueger, J. M., & Fang, J. (2000). Sleep and host defense. In M. H. Kryger, T. Roth, & W. Dement (Eds.), *Principles and practice of sleep medicine* (3rd ed., pp. 255–265). Philadelphia: W. B. Saunders.

Lee, J. H., Bliwise, D., Lebret-Bories, E., Guilleminault, C., & Dement, W. C. (1993). Dream-disturbed sleep in insomnia and narcolepsy. *Journal of Nervous & Mental Disease*, *181*(5), 320–324.

Linnoila, M., De Jong, J., & Virkkunen, M. (1989). Monoamines, glucose metabolism, and impulse control. *Psychopharmacology Bulletin*, *25*(3), 404–406.

Linnoila, M., Virkkunen, M., George, T., Eckardt, M., Higley, J. D., Nielsen, D., et al. (1994). Serotonin, violent behavior, and alcohol. *EXS*, *71*, 155–163.

McNamara, P. (2004). *An evolutionary psychology of sleep and dreams*. Westport, CT: Praeger.

McNamara, P., McLaren, D., Smith, D., Brown, A., & Stickgold, R. (2005). A "Jekyll and Hyde" within: Aggressive versus friendly social interactions in REM and NREM dreams. *Psychological Science*, *16*(2), 130–136.

National Institutes of Health. (2005). National Institutes of Health State of the science conference statement on manifestations and management of chronic insomnia in adults, June 13–15, 2005. *Sleep*, *28*(9):1049–1057.

Nesse, R. M., & Williams, G. C. (1998). Evolution and the origin of disease. *Scientific American*, *279*, 86–93.

Revonsuo, A. (2000). The reinterpretation of dreams: An evolutionary hypothesis of the function of dreaming. *Behavioral and Brain Sciences*, *23*(6), 877–901.

Rechtschaffen, A., & Bergmann, B. M. (2001). Sleep deprivation and host defense. *American Journal of Physiology: Regulatory, Integrative and Comparative Physiology*, *280*(2), R602–R603.

Roth, T., Franklin, M., & Bramley, T. J. (2007). The state of insomnia and emerging trends. *American Journal of Managed Care*, *13*(5 Suppl.), S117–S120.

Shankar, A., Koh, W. P., Yuan, J. M., Lee, H. P., & Yu, M. C. (2008). Sleep duration and coronary heart disease mortality among Chinese adults in Singapore: A population-based cohort study. *American Journal of Epidemiology*, *168*(12), 1367–1373.

Stearns, S. C. (1999). *Evolution in health and disease*. New York: Oxford University Press.

Stearns, S. C., & Koella, J. K. (Eds.). (2007). *Evolution in health and disease* (2nd ed.). Oxford: Oxford University Press.

Stranges, S., Dorn, J. M., Shipley, M. J., Kandala, N. B., Trevisan, M., Miller, M. A., et al. (2008). Correlates of short and long sleep duration: A cross-cultural comparison between the United Kingdom and the United States: The Whitehall II Study and the Western New York Health Study. *American Journal of Epidemiology*, *168*(12), 1353–1364.

Tobler, I. (2000). Phylogeny of sleep regulation. In M. H. Kryger, T. Roth, & W. C. Dement (Eds.), *Principles and practice of sleep medicine* (3rd ed., pp. 72–78). Philadelphia: W. B. Saunders.

Tobler, I. (2005). Phylogeny of sleep regulation. In M. H. Kryger, T. Roth, & W. C. Dement (Eds.), *Principles and practice of sleep medicine* (4th ed., pp. 77–90). Philadelphia: W. B. Saunders.

Trevathan, W. R., Smith, E. O., & McKenna, J. J. (1999). *Evolutionary medicine*. New York: Oxford University Press.

Trevathan, W. R., Smith, E. O., & McKenna, J. J. (2008). *Evolutionary medicine and health: New perspectives*. New York: Oxford University Press.

Virkkunen, M., Kallio, E., Rawlings, R., Tokola, R., Poland, R. E., Guidotti, A., et al. (1994). Personality profiles and state aggressiveness in Finnish alcoholic, violent offenders, fire setters, and healthy volunteers. *Archives of General Psychiatry*, *51*, 28–33.

Williams, G. C., & Nesse, R. M. (1991). The dawn of Darwinian medicine. *Quarterly Review of Biology*, *66*(1), 1–22.

6

Primate sleep in phylogenetic perspective

CHARLES L. NUNN, PATRICK MCNAMARA, ISABELLA CAPELLINI, BRIAN T. PRESTON, AND ROBERT A. BARTON

Introduction

The primates comprise a diverse group of eutherian mammals, with between some 200 and 400 species, depending on the taxonomic authority consulted (e.g., Corbet & Hill, 1991; Wilson & Reeder, 2005). Most of these species dwell in tropical forests, but primates also thrive in many other habitats, including savannas, mountainous forests of China and Japan, and even some urban areas. Living primates are divided into two groups, the strepsirrhines (lemurs and lorises) and the haplorrhines (monkeys, apes, and tarsiers). Strepsirrhines include mostly arboreal species and retain several ancestral characteristics, including greater reliance on smell and (in most species) a dental comb that is used for grooming. Most are nocturnal, but some have, in parallel with most haplorrhines, evolved a diurnal niche. They are found only in the Old World tropics. Haplorrhines are more widely distributed geographically, being found in both the New and Old Worlds. They include two groups, the platyrrhines and the catarrhines. Platyrrhines are monkeys native to the New World. Catarrhines include both Old World monkeys and apes. With the exception of owl monkeys in the genus *Aotus*, all monkeys and apes are active during the day (i.e., diurnal), and most live in bisexual social groups that vary in size from 2 to well over 100 adults (Smuts, Cheney, Seyfarth, et al., 1987).

Nonhuman primates are among the best-studied of mammals, in large part because of their close phylogenetic relatedness to humans. Much of the research on wild primates has focused on issues of biomedical importance, such as emerging infectious diseases (Chapman, Gillespie, & Goldberg, 2005; Wolfe, Escalante,

This material is based on work supported in part by the National Institute of Mental Health, Grant 5R01MH070415-03.

123

Karesh, et al., 1998). As a result, we know much about the parasites and pathogens that infect wild primates (Nunn & Altizer, 2005, 2006). An additional goal of studying primates is to gain insight into human evolution (Foley & Lee, 1989; Smuts et al., 1987). Thus a wealth of information is available on primate behavior and ecology (Smuts et al., 1987), primate phylogeny (Disotell, 2008; Purvis, 1995), and the geographical distribution and population sizes of different primate species (Hilton-Taylor, 2002).

Sleep is an important factor in nonhuman primate health, behavior, and ecology and can play a central role in shaping daily activity schedules (Anderson, 1998, 2000). For instance, locating a suitable sleep site can be an important component of individual survival in primates, allowing them to avoid mosquito vectors or predators (Anderson, 1998, 2000; Day & Elwood, 1999; Di Bitetti, Vidal, Baldovino, et al., 2000; Heymann, 1995; Nunn & Heymann, 2005). Thus, hamadryas baboons (*Papio hamadryas*) travel to cliffs to sleep at night (Kummer, 1968), whereas chimpanzees (*Pan troglodytes*) build new nests every night for sleeping (Boesch & Boesch-Achermann, 2000).

Together, these factors make primates particularly valuable for the comparative study of sleep, with the potential to provide critical advances in our understanding of human sleep disorders, the ecology of sleep in nonhuman primates and humans, and the evolution of sleep patterns more generally. Properly controlled comparative studies of sleep in primates and other mammals have been rare, usually requiring comparative biologists to examine variation across mammals rather than within different orders of mammals (Capellini, Barton, McNamara, et al., 2008a; Capellini, Nunn, McNamara, et al., 2008b; Elgar, Pagel, & Harvey, 1988; Lesku, Roth, Amlaner, et al., 2006). A handful of studies have investigated the durations of rapid-eye-movement (REM) and non–rapid-eye-movement (NREM) sleep in primates, however, providing some data for at least initial comparative studies of primate sleep (e.g., Bert & Pegram, 1969; Hsieh, Robinson, & Fuller, 2008; Perachio, 1971).

In this chapter, we review existing knowledge of sleep in primates, focusing in particular on variation in sleep quotas across primate species in relation to ecological and life history traits. Our goals are threefold. First, we aim to identify those aspects of sleep that have changed on the primate lineage. Second, based on our review, we advance selected hypotheses for distinctive characteristics of sleep expression in primates. We test these hypotheses when sufficient data exist. Last, we summarize gaps in our knowledge of primate sleep and identify the primate species that are most important for future data collection. We propose more generally that increased knowledge of sleep expression among nonhuman primates will deepen our understanding of the function of sleep and human sleep disorders.

The measurement of primate sleep quotas and sleep architecture

We focus much of this review on four basic parameters of sleep expression – known as "sleep quotas" – in primates. These simply represent the total time spent asleep per day and the time spent in each of the two major forms of mammalian sleep: active or REM sleep, and quiet or NREM sleep. Periods of REM and NREM sleep alternate throughout the sleeping period, and the mean duration of these *sleep cycles* – measured from initiation of NREM to the end of the subsequent bout of REM sleep – is the fourth parameter of interest.

Despite the apparent validity of using sleep quotas to study sleep expression across primates, these data must be interpreted with caution, given several methodological problems associated with the collection of sleep quotas. Among the most troublesome of these issues has been the necessity for animals to be studied in the laboratory rather than in their natural habitats. When studied in the laboratory, animals may be restrained to record electroencephalographic (EEG) sleep changes. Restraint for animals can be very stressful and can therefore affect the sleep data obtained from the experiment. Data from telemetric recordings are available for only a handful of primates, including baboons (Bert, 1975), lemurs (Vuillon-Cacciuttolo, Balzamo, Petter, et al., 1976), monkeys (Hsieh et al., 2008; Reite, Stynes, Vaughn, et al., 1976), and chimpanzees (Bert, Kripke, & Rhodes, 1970); few would consider these studies to be "natural" with respect to the behavior and ecology of the animals involved. In these studies, the animals undergo surgery to implant electrodes that can measure brain wave activity and transmit the information to a receiver. The transmitter was usually housed in a small box that was affixed to the top of the animal's head, allowing the animal the freedom to move without attached wires. Twenty-four-hour EEG recordings could therefore be obtained from the animal while it moved about and while it slept. There was no need to strap the animal into a chair or restraining device or to drug the animal so that EEG recordings could be secured. In that sense the animal was not restrained when telemetry was used. Recent advances in EEG data acquisition technology offer many promising opportunities for studies of sleep in wild, naturally behaving primates (Rattenborg, Voiren, Vissotski, et al., 2008).

Other factors may also influence sleep measures. For example, the ambient temperature in a laboratory setting may differ from conditions typical in the wild; this is relevant because sleep variables are known to be sensitive to small changes in temperature. Usually laboratories are under constant lighting, which is likely to impact components of the sleep response in some animals. Thus comparative biologists should restrict their analyses to sleep studies that meet a set of basic criteria (e.g., Capellini et al., 2008a; McNamara, Capellini, Harris, et al., 2008).

Table 6.1. *Sleep quotas in primates* [a]

Species	Total sleep	REM duration	NREM duration	Sleep cycle length
Aotus trivirgatus[b]	17	1.82	15.15	
Callithrix jacchus[b]	9.5	1.61	7.9	50
Chlorocebus aethiops[b]	10.1	0.65	9.44	
Erythrocebus patas[b]	10.9	0.86	9.99	
Eulemur macaco	9.4			
Eulemur mongoz[b]	11.9	0.72	11.16	
Homo sapiens[b]	8.5	2.1	6.37	90
Macaca arctoides[b]	9	1.38	7.65	50
Macaca mulatta[b]	10.2	2.05	8.19	
Macaca nemestrina[b]	14	0.92	13	80
Macaca radiata[b]	9.1	1.05	8.06	
Macaca sylvanus[b]	11.7	1.07	10.7	31.1
Microcebus murinus[b]	15.4	0.99	14.4	
Pan troglodytes[b]	11.5	2.06	9.46	90
Papio anubis[b]	9.2	1	8.2	40
Papio papio[b]	10.1	1.06	9	
Perodicticus potto	11			
Phaner furcifer	11.5			
Saguinus oedipus	13.2			19
Saimiri sciureus[b]	9.7	1.77	7.8	12
Theropithecus gelada	10.9			

[a]Blank cells indicate that no data are available.
[b]Based on EEG data.

Specifically, the ideal should be to use data that record the animal's brain activity with an EEG for at least 24 hours under normal (for the animal) light–dark schedules and ambient temperatures and only after the animal has adapted to the laboratory and recording procedures. Unfortunately we cannot yet meet this ideal for comparative studies of primates. In this chapter, we therefore include behavioral measures of sleep for some estimates of total sleep times and limit the data to studies with 12 or more hours of observation (which should be sufficient for most species of primates that sleep in only one block of time per day – i.e., they are monophasic).

Empirical data and general evolutionary patterns

Available data on average sleep quotas in primates are provided in Table 6.1 (McNamara, et al., 2008). We found data on 20 species of nonhuman primates, to which we also added data on humans from Carskadon and Dement

Total Sleep

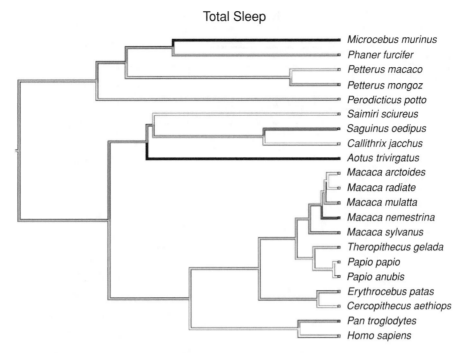

Figure 6.1. The phylogenetic distribution of total sleep time in primates. Darker branches represent longer sleep durations per 24-hour period. Internal nodes were reconstructed using maximum parsimony. Note that the maximum parsimony reconstruction of the root node (11.8) is slightly higher (but within the confidence interval) of the estimate from the Bayesian analysis (11.3, see text and Table 6.2).

(2006). Owl monkeys, cotton-top tamarins (*Saguinus oedipus*), and mouse lemurs (*Microcebus murinus*) appear to be the "marathon sleepers" among the primate species studied thus far. Their average total sleep time per day ranges from 13 to 17 hours. Interestingly, the two longest-sleeping species are nocturnal (owl monkeys and mouse lemurs). Our evolutionarily closest relative, the chimpanzee, sleeps an average of 11.5 hours per day, quite close to the phylogenetic average for all primates. The short sleepers in our dataset sleep for 8 to 10 hours and are phylogenetically diverse; they include humans, a handful of cercopithecine monkeys, a lemur, and some New World primates.

With these data, we can investigate evolutionary patterns – for example, by reconstructing ancestral states of sleep traits on primate phylogeny. Before conducting such tests, however, it is important to assess whether primate sleep quotas exhibit what evolutionary biologists call "phylogenetic signal" (Blomberg & Garland, 2002; Blomberg, Garland, & Ives, 2003). This concept simply captures whether more closely related species exhibit more similar values in their sleep quotas; such an effect would indicate that the trait is shared through common

NREM Sleep

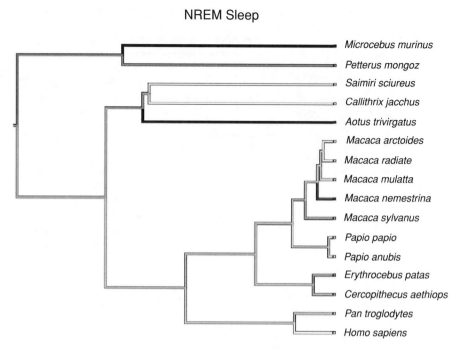

Figure 6.2. The phylogenetic distribution of NREM sleep in primates. Darker branches represent increased duration of NREM sleep. See Table 6.1 for details. Internal nodes were reconstructed using maximum parsimony.

descent, suggesting that it is an evolved trait. In addition, phylogenetic signal would show that data vary sufficiently across species for comparative study, such that measurements available for different species can be linked to ecological, life history, or behavioral traits of those species. Phylogenetic signal in sleep traits has been shown to exist across a wide range of mammalian species (Capellini et al., 2008a), but this appears to be structured mostly at the order level (e.g., with rodents exhibiting more similarity to other rodents than to carnivores), and we lack an understanding of patterns of phylogenetic signal within orders, including primates. Figures 6.1 to 6.3 indicate that more closely related primate species exhibit similar trait values, and a statistical test for phylogenetic signal (Blomberg et al., 2003) reveals some evidence for greater similarity among more closely related primate species, especially for REM and NREM sleep durations examined separately (Table 6.2). Although these test results are not statistically significant, they approach significance; hence, given the small sample sizes and the probable measurement error involved with estimating sleep quotas, it is reasonable to suggest that these results point toward the existence of phylogenetic signal (Blomberg et al., 2003; Ives, Midford, & Garland, 2007). These analyses thus suggest that

Table 6.2. *Phylogenetic signal in primate sleep parameters*[a]

Variable	K statistic	P value
Total sleep	0.241	0.108
NREM sleep	0.297	0.072
REM sleep	0.218	0.061

[a]The *P* value does not reflect a significant level of the K statistic; rather, it indicates whether the mean square error (MSE) for the dataset is significantly lower than the mean MSE on permuted datasets because such a result would indicate significant phylogenetic signal (see Blomberg et al., 2003).

REM Sleep

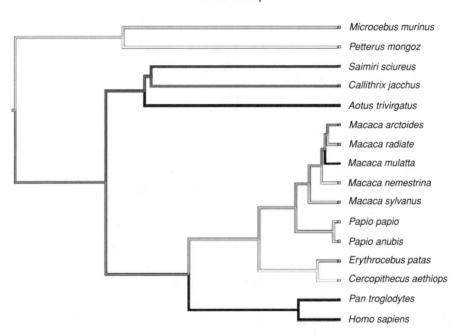

Figure 6.3. The phylogenetic distribution of REM sleep in primates. Darker branches represent increased duration of REM sleep. See Table 6.1 for details. Internal nodes were reconstructed using maximum parsimony.

evolutionary history explains some of the variation in primate sleep traits, as was found more convincingly in a larger set of mammals (Capellini et al., 2008a).

Assuming that evidence for phylogenetic signal is strengthened as more species of primates are studied, we can also use phylogenetic methods to reconstruct sleep characteristics of the ancestral primate and examine trends in sleep characteristics

Table 6.3. *Bayesian estimates of ancestral states and evolutionary parameters*[a]

	Log-likelihood (harmonic mean)	Ancestral value	Lower 95% CI	Upper 95% CI	Mean λ	Mean κ	Mean δ
Total sleep	−47.5	11.3	9.4	13.4	0.32	2.39	1.23
NREM	−38.7	10.0	7.3	13.2	0.38	2.42	1.09
REM	−12.1	1.3	0.7	1.9	0.57	1.24	1.26

[a]Results from 10,001 samples of a Bayesian posterior probability distribution, calculated in Bayes Traits (Pagel & Meade, 2007). Ancestral value reflects the mean estimate, and lower and upper 95% confidence interval (CI) reflects the distribution of values obtained from the Bayesian analysis.

over primate evolution. To reconstruct sleep in the ancestral primate, we used a Bayesian approach, as this provides a means to put confidence intervals on the reconstructed ancestral node (Pagel, Meade, & Barker, 2004). To implement these tests, we used the program BayesTraits (Pagel & Meade, 2007) with Purvis's (1995) "supertree" for the estimate of primate phylogeny. From this analysis, we estimate that the ancestral primate slept an average of just over 11 hours per day, with 10 hours of NREM and 1.3 hours of REM sleep. The confidence intervals are provided in Table 6.3.

This methodology also provides another way to assess phylogenetic signal (Freckleton, Harvey, & Pagel, 2002) and models of evolutionary change (Pagel, 1997, 1999). For example, the program BayesTraits calculates a parameter known as λ, ranging between 0 (no phylogenetic signal) and 1 (phylogenetic signal consistent with a Brownian motion model of evolution); higher values thus indicate greater phylogenetic signal (Freckleton et al., 2002). The values presented in Table 6.3 suggest that REM sleep exhibits more phylogenetic signal than other traits. Conversely, the parameter κ investigates how evolutionary rate varies in relation to branch length (i.e., the time separating speciation events on the phylogeny). We find that κ is lower for REM than for NREM. This indicates that especially for NREM sleep, more change in sleep times occurs on longer branches, thus suggesting that sleep has not evolved according to an adaptive radiation model (with large changes early in a clade and occurring on short branches). Similarly, the values of $\delta > 1$ indicate that more change occurs later in evolution than in the early stages and confirms a pattern of species adaptation rather than of early adaptive radiation. From this analysis, it appears that the two forms of sleep have undergone somewhat different evolutionary trajectories in primates. We should note, however, that most of these parameters had extremely wide confidence intervals that encompassed 1, probably owing to the small sample sizes; thus we consider these analyses to be exploratory.

Primates also exhibit variability in the amount of time devoted to NREM sleep and REM sleep (Table 6.1). For example, monkeys spend between 7 and 15 hours in NREM sleep (Figure 6.2). Time devoted to REM sleep varies from a little over 30 minutes per day in the vervet monkey (*Cercopithecus aethiops*) to 2 hours per day in the chimpanzee and human (Figure 6.3). REM sleep may have increased in the great apes, although this is based on only two ape species in the dataset. Increases in REM sleep can be seen in other lineages.

In general, it appears that total sleep duration in primates is most sensitive to the amount of NREM sleep. Thus, in analyses of independent contrasts that control for the nonindependence of species values (Felsenstein, 1985; Garland, Harvey, & Ives, 1992; Nunn & Barton, 2001), we found that evolutionary increases in NREM sleep correlate strongly with evolutionary increases in total sleep among primates, whereas REM sleep shows no such association (Figure 6.4, panels a and b). In contrast to work across mammals more generally (Capellini et al., 2008a), we failed to find a significant association between NREM and REM sleep in primates (Figure 6.4c). Although this may again reflect low statistical power, it is noteworthy that the slope of this nonsignificant relationship is in fact negative (see legend of Figure 6.4), whereas previous work demonstrated positive associations across mammals more generally (Capellini et al., 2008a). Additional data collection on primate sleep may reveal a difference in this regard in primates compared to other mammals.

Sleep in relation to biological characteristics of primates

In the previous section we showed how the evolutionary history of a species helps explain why closely related species have similar sleep durations; but what explains the remaining variation in primate sleep? Here we consider factors that might account for variability in sleep patterns. In this section, we first review features of primates that might be related to sleep characteristics and then, when sufficient data exist, present tests that investigate some of these predictions.

It is generally agreed that one of the major evolutionary transitions in the primate order involved a shift from a nocturnal to a diurnal activity period, which has occurred more than once (Martin, 1990). The shift from nocturnality to diurnality was associated with dramatic changes in ecology and behavioral capacities. Prominent among ecological changes was increased predation pressure from diurnal predators, such as raptors, leading to a need to monitor the environment visually for these and other predators. The shift to a diurnal lifestyle may also have played a role in the evolution of larger, permanent social groups in most primate lineages, as this would have afforded greater safety from predators (Janson, 1992; van Schaik, 1983). Moreover, some diurnal lineages became more terrestrial and

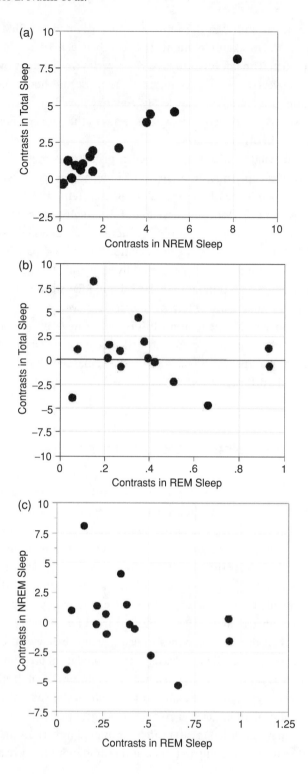

moved into more open habitats (Nunn & Barton, 2001). Living on the ground may have exposed these animals to greater predation pressure and thus led to selection for even larger social groups (Nunn & van Schaik, 2002). Finally, the shift to diurnality is associated with greater sexual selection, especially among terrestrial species (e.g., greater body mass dimorphism; Plavcan & van Schaik, 1997).

In terms of brain evolution, there emerged a tendency among diurnal primates toward reduction of the olfactory region of the brain and expansion of the cerebral cortex, associated with an increasing reliance on visual sensory modalities (Barton, Purvis, & Harvey, 1995). Diurnal primates, for example, have either dichromatic or trichromatic color vision and fields of view that significantly overlap, resulting in true three-dimensional depth perception (Cartmill, 1974; Martin & Ross, 2005). The combination of greater predation pressures and enhanced social interaction likely also promoted expansions in brain volumes linked to visual processing and management of social capacities (Barton, 1998; Barton & Dunbar, 1997). In primates, living in a larger social group is associated with increased neocortex size (Dunbar, 1992, 1998), and increasing sexual selection covaries with the size of brain structures involved in sensorimotor skills (Lindenfors, Nunn, & Barton, 2007).

The evolutionary shift to a diurnal lifestyle, therefore, had an enormous impact on primate behavior and life history strategies. In what follows, we consider how these and other features may have influenced primate sleep. When possible, we test these proposals using data from *The Phylogeny of Sleep* research group (McNamara et al., 2008).

Evolution of monophasic sleep

One of the most fundamental descriptors of sleep concerns whether it occurs in one bout per 24-hour period of time (i.e., is monophasic) or encompasses multiple bouts in a daily cycle (i.e., is polyphasic) (Ball, 1992). The occurrence of sleep relative to the daily photoperiod varies across mammals, with most species

---

Figure 6.4. Relationships among sleep states and total sleep. Total sleep is highly correlated with NREM sleep (a) but shows no obvious association with REM sleep in primates (b), suggesting that NREM accounts for most of the variation in total sleep. Results for NREM and REM are, respectively: $t_{14} = 23.9, P < 0.0001$; $t_{14} = 0.07, P = 0.94$. (c) NREM and REM sleep durations are not significantly correlated ($t_{14} = -0.51, P = 0.62$). Analyses were based on independent contrasts calculated with the PDAP module (Midford et al., 2005) in Mesquite (Maddison & Maddison, 2006), using Nee transformed branch lengths (Purvis, 1995) to better meet the assumptions of independent contrasts (Garland et al., 1992). The phylogeny matches that used in other studies of sleep in mammals (Capellini et al., 2008a).

exhibiting a polyphasic form, characterized by short bouts of sleep throughout the day and night. This polyphasic pattern is most likely to be the ancestral state in mammals, based on a maximum likelihood analysis of 56 species (Capellini et al., 2008b). Some lemur species, such as *Eulemur*, show an interesting pattern of cathemerality, meaning that they are active during both day and night (Tattersall, 1987). Sleep is not so rigidly restricted to night or day in these species, potentially resulting in a tendency toward polyphasic sleep patterns, and the same appears to be true of the nocturnal mouse lemur (Capellini et al., 2008b). In anthropoid primates, however, sleep is clearly monophasic, with one sleep period during the 24-hour cycle (Capellini et al., 2008b).

Thus we conclude that monophasic sleep is a derived trait in anthropoid primates, with an origin either at the base of the primate clade (with subsequent reversals in some strepsirrhines), or probably originating on the lineage leading to monkeys and apes.

Reductions in sleep among diurnal primates

Our analyses of the phylogenetic distribution of total sleep, NREM sleep, and REM sleep suggest that the diurnal activity period exerted a major influence on patterns of sleep in primates, with diurnal primates sleeping less than nocturnal primates. Total sleep duration appears to vary according to whether the species is nocturnal or diurnal, with longer sleep durations in nocturnal primate species (see Figure 6.1). This could reflect the fact that many small-bodied nocturnal species, such as the mouse lemur, seek protected sleep sites to reduce predation pressure, which in turn allows them to sleep for longer periods. Conversely, the shorter sleep durations in diurnal species may reflect increased sleep efficiency associated with monophasic sleep (see below), increased foraging needs that limit the time available for sleep, or thermoregulatory costs associated with inactivity during the night, when temperatures are lowest. In addition, the increased needs for social interactions in these species may also constrain time available for sleeping.

We examined these patterns using phylogenetically based statistical methods. Among the species in our dataset, total sleep time decreases over the three independent transitions in activity period from nocturnality to diurnality. In treating activity period as a continuously varying character, the association between diurnality and total sleep time is significant ($t_{17} = -2.94$, $P = 0.009$), although it is based on only five informative contrasts (i.e., the others exhibit no variation in the activity period variable). Some phylogenetic assumptions are violated in this analysis, likely due to treating activity period as if it were a continuous character. Further information on sleep in other lineages of primates – as well as other mammals – would help to address the effects of nocturnality more directly.

Increased sleep intensity

Primate sleep exhibits a differentiation of NREM sleep into at least two distinctive types: a light form characterized by spindling activity and a deep form characterized by slow-wave activity (SWA) (Balzamo, Santucci, Seri, et al., 1977; Hsieh et al., 2008). In the great apes (including humans), NREM sleep can be differentiated into four substages, with substages III and IV marked predominantly by SWA or delta activity, which indexes sleep intensity in humans (Tobler, 2005). Stages I and II are characterized mainly by spindling activity. Interestingly, whereas in most primates spindling activity is more characteristic of juvenile than of adult sleep, marked spindling activity appears to persist into the adult state in the great apes as well as in humans (Bert, Balzamo, Chase, et al., 1975). The differentiation of NREM into substages, with their concomitant spindling activity and greater SWA, may have helped primates to achieve enhanced sleep intensity for each bout of sleep. Indeed, monophasic sleep in mammals may be more efficient because it involves less time in light sleep and monophasic sleepers spend less time asleep per day (Capellini et al., 2008b).

It could also be that nocturnal predation risk on diurnal primates favored lighter sleep as a means of detecting predators (Lima, Rattenborg, Lesku, et al., 2005). Lighter sleep might also be favored in the context of social sleeping – for example, to monitor competitive interactions, mating opportunities, or risks to infants from other individuals in the group (especially male infanticide) (van Schaik & Janson, 2000).

Altered developmental sleep patterns

The shift to a diurnal activity pattern exposed primates to a different suite of predation pressures. With increased predation, one might expect that primate young would be born in a precocial state. Perhaps consistent with this prediction, we see a transition from "parking" infants in a safe place, among some strepsirrhines, to being carried by the mother, in haplorrhines. REM sleep has been implicated in brain development owing to the age-related changes in its expression, with REM sleep dominating the sleep of juveniles (Carroll, Denenberg, & Thoman, 1999; Reite et al., 1976). In addition, previous comparative analyses have suggested that mammalian species that give birth to immature (altricial) infants – those requiring the highest degree of subsequent brain development – exhibit longer durations of REM sleep (Zepelin, Seigel, & Tobler, 2005). However, a recent study called into question the generality of this result in phylogeny-based tests (Capellini et al., 2008a), as did a more focused study of three mammalian species (Thurber, Jha, Coleman, et al., 2008). Thurber et al. showed that when sleep times are compared at a common developmental landmark (age at eyes opening), ferret kits – considered among the most altricial neonates among mammals – do not show longer sleep times than other species (cats and rats), as predicted by the hypothesis that REM sleep helps the developing neonatal brain. In addition, ferrets have less REM sleep (55% of total sleep time) than the other two species (75% in both cats and rats).

Nonetheless, primate juveniles, as compared to other mammalian juveniles, may devote less of their sleep time to REM, in accordance with this general association between REM and altriciality. According to Bert and colleagues (Bert, 1975; Bert, Pegram, & Balzamo, 1972), the amount of time spent in REM is reduced in juvenile primates relative to other mammals. For example, vervet monkey juveniles spend 5.6% of their sleep time in REM; patas monkey 7.9%; wild baboons, 5.9%; *Macaca radiata*, 11.5%; *Macaca nemestrina*, 11.1%; and *Macaca mulatta*, 15.5%. This overall primate trend in reduction of juvenile REM sleep quotas is partially reversed in chimpanzees and humans, where infants spend between 22% and 50% of their time in REM (Balzamo, Bradley, & Rhodes, 1972; Balzamo et al., 1977; Salzarulo & Ficca, 2002). This could perhaps reflect their relatively more altricial state at birth (at least for humans), or it could reflect differences in cognitive demands in these species and the need for investment in brain tissue.

Sociality and primate sleep

Increased predation pressures and sociality in general may also promote the practice of cosleeping between mother and infant and sleeping in "huddles" with kin and nonkin in the social group. The primate infant extracts metabolic resources from the mother throughout the night, and cosleeping could protect infants from competitive interactions within groups, including infanticide attempts by males. Koyama (1973) and Vessey (1973) described sleeping huddles in free-ranging bonnet (*M. radiata*) and rhesus macaques (*M. mulatta*), respectively. In both species, the most frequent huddle size was two, and huddles were composed primarily of mother–infant pairs, same-sex individuals, or male–female sexual consortships. The cosleeping pattern is established in infancy but persists into adulthood. In infancy, cosleeping involves nursing during the night. Juveniles who are weaned regularly return to sleep in contact with their mothers at night – for example, baboons (Altmann J., Altmann S., & Hausfater, 1981), gorillas (Goodall, 1979), and orangutans (Horr, 1977).

In addition to protection against predation, the practice of the sleep huddle or cosleeping among adults may also serve other functions. Some of these functions may include a thermoregulatory function (e.g., Altmann, 1980; Anderson & McGrew, 1984; Gartlan & Brain, 1968; Gaulin & Gaulin, 1982; Suzuki, 1965) or a sexual function (Anderson & McGrew, 1984; Fruth & Hohmann, 1993). It is also known, however, that sleeping in a larger group can increase the risk of acquiring

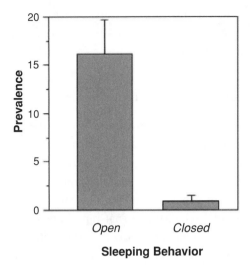

Figure 6.5. Malaria prevalence in neotropical primates in relation to sleeping behavior. Mean prevalence in nonphylogenetic tests is lower among genera that sleep in closed microhabitats (Nunn & Heymann, 2005). Results were also significant in the majority of phylogenetic tests using independent contrasts, although analyses were based on only two or three evolutionary transitions, depending on the phylogeny used.

vector-borne infections, particularly malaria. Thus, in New World primates, the prevalence of malaria increases when the number of animals sleeping in a group increases (Figure 6.5) (Davies, Ayres, Dye, et al., 1991; Nunn & Heymann, 2005). This probably reflects that larger groups of animals emit more of the cues used by mosquitoes to locate hosts, and it suggests that sleeping in larger groups and reuse of sleep sites (Hausfater & Meade, 1982) might have costs to primates (Nunn & Altizer, 2006).

An intriguing topic for the future concerns the possibility of links between sociality and sleep quotas. A recent study of *Drosophila* found that when flies lived in socially enriched environments with many conspecifics, they exhibited increased sleep times (Ganguly-Fitzgerald, Donlea, & Shaw, 2006). This increased sleep time affected daytime but not nighttime sleep. A previous study conducted across mammals found that increased social sleeping correlates with shorter sleep durations (Capellini et al., 2008a). This could indicate the existence of trade-offs between time for social interactions and sleep, or it could reflect increased sleep efficiency in social species, possibly because they gain safety in numbers and can thus spend more time in deep sleep; but these two hypotheses are not mutually exclusive. We tested whether primates that live in larger networks of other individuals sleep for a longer period of time each day, but we found no significant association (independent contrasts: $t_{18} = -0.13$, $P = 0.90$; humans were removed from the analysis due to uncertainty in measuring group size). Other measures of sociality, such as the time spent grooming or the number of grooming partners, would be interesting to examine in future research.

Conclusions and recommendations for future research

Based on a review of the existing literature and new analyses conducted for this chapter, we identified five major hypothesized characteristics of primate sleep, many of which were associated with transitions to increased diurnality. These characteristics include the following: (1) *Consolidation of sleep into a single long bout*, possibly to achieve greater sleep intensities, but this also could be a side effect of typically strict activity periods (nocturnal versus diurnal lifestyles). (2) *Reductions in sleep times among diurnal primate species*, which could reflect a number of different advantages or constraints associated with diurnality. (3) *Increased sleep intensity*, possibly associated with differentiation of NREM sleep stages into lighter and deeper stages of sleep and testable once more data have accumulated on sleep intensity measures. (4) *Developmental shifts in sleeping patterns*, including less REM among juveniles. (5) *Maintenance of social contact during sleep*, which likely has advantages in terms of infant care, predation risk, and thermoregulation, but also costs in terms of parasitism (Nunn & Altizer, 2006).

Even when data are available to test these hypotheses, our conclusions must remain tentative, given the small number of primate species studied by sleep scientists. In many ways, we see this chapter as an illustration of the questions that remain to be answered and of the phylogenetic approaches that can be brought to bear on these questions. In this context, it is instructive to point out the species and clades that are missing from current comparative research and would be most important to include in the future. First, we have information on only two species of apes – the chimpanzee and humans. A high priority for future research should be to collect sleep data in the other great apes, specifically gorillas (*Gorilla gorilla*), orangutans (*Pongo pygmaeus*), and bonobos (*Pan paniscus*), along with one or more species of gibbons (*Hylobates* spp.).

Second, few diurnal strepsirrhines have been sampled, and only two have been sampled for REM and NREM sleep (one nocturnal and one diurnal species). Key species to test in this regard are those with good behavioral and ecological sampling, including the ring-tailed lemur (*Lemur catta*), the sifaka (*Propithecus* spp.), and the brown lemur (*Eulemur fulvus*).

Third, among the monkeys, notable sampling gaps include the diurnal Cebidae in the New World (especially *Alouatta*, *Cebus*, *Callicebus*, and *Ateles*), and the colobines in the Old World (especially one or more species of *Presbytis* and *Colobus* monkeys). In addition, although many studies have investigated sleep in macaques (*Macaca*),

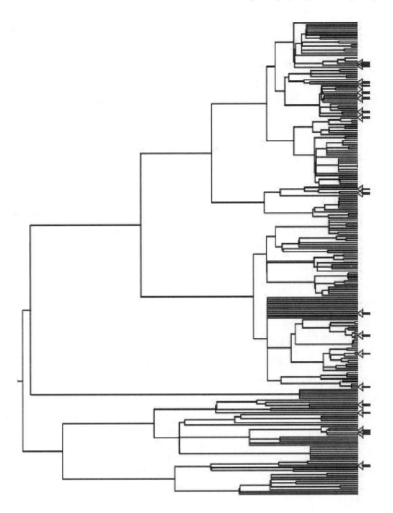

Figure 6.6. Sampling gaps in primates. Species that have been studied for sleep are indicated with arrows next to terminal tips from a recent estimate of mammalian phylogeny (Bininda-Emonds, et al., 2007). Species with arrows, in order from top to bottom, are *Cercopithecus aethiops*, *Erythrocebus patas*, *Papio hamadryas* (representing the two species of *Papio* in our dataset), *Theropithecus gelada*, *Macaca arctoides*, *M. radiata*, *M. mulatta*, *M. nemestrina*, *M. sylvanus*, *Homo sapiens*, *Pan troglodytes*, *Aotus trivirgatus*, *Callithrix jacchus*, *Saguinus oedipus*, *Saimiri sciureus*, *Microcebus murinus*, *Phaner furcifer*, *Eulemur macaco*, *E. mongoz*, and *Perodicticus potto*. Not all of these species have been studied using EEG.

no species of arboreal guenons (*Cercopithecus*) have been studied for their sleep. Moreover, *Cercocebus* and *Mandrillus* are also missing from our understanding of primate sleep, and folivores (leaf eaters) are largely absent. Obtaining data on folivores will be critically important for understanding the links between diet, metabolic rate, and sleep patterns. For example, folivorous primates may spend more time resting (thus including both sleep and quiet resting time) relative to frugivorous species (Oates, 1987), but whether such a pattern affects sleep time or only quiet resting time, and the factors that drive these differences (e.g., more time needed to find fruits or more constraints due to digestion), remains to be clarified. As noted above, it would help to have better sampling of nocturnal primates. However, given the small number of transitions in activity period in primates, it may be necessary to investigate the effect of activity period across mammals more generally.

Figure 6.6 reveals that sampling has occurred at a generally "shallower" depth in Old World monkeys, meaning that taxa with sleep data tend to share a common ancestor more recently, as compared to sampling for sleep in New World monkeys and strepsirrhines. This might be due to higher rates of diversification in these lineages (Purvis, Nee, & Harvey, 1995), but it also indicates somewhat broader phylogenetic sampling, in terms of longer phylogenetic branches covered, for all groups except Old World monkeys. All of the previous points essentially relate to aspects of the sampling "depth" shown in Figure 6.6.

In summary, the evolution of primate sleep patterns is as yet little understood, with many sampling gaps related to the phylogenetic distribution of primates and the ecological characteristics that are important for testing hypotheses for the function of sleep. Nonetheless, it is possible to formulate clear hypotheses for how major transitions associated with primate evolution have impacted primate sleep patterns and in some cases to test these hypotheses using existing data. Given the scientific and health benefits of studying primate sleep, it is critically important to fill some of the gaps in our knowledge of primate sleep and to do so in a way that provides the strongest tests of comparative hypotheses.

References

Altmann, J. (1980). *Baboon mothers and infants*. Cambridge, MA: Harvard University Press.

Altmann, J., Altmann, S., & Hausfater, G. (1981). Physical maturation and age estimates of yellow baboons, *Papio cynocephalus*, in Amboseli National Park. *American Journal of Primatology, 1*, 389–399.

Anderson, J. R. (1998). Sleep, sleeping sites, and sleep-related activities: Awakening to their significance. *American Journal of Primatology, 46*, 63–75.

Anderson, J. R. (2000). Sleep-related behavioural adaptations in free-ranging anthropoid primates. *Sleep Medicine Reviews, 4*, 355–373.

Anderson, J. R., & McGrew, W. C. (1984). Guinea baboons (*Papio papio*) at a sleeping site. *American Journal of Primatology, 6*, 1–14.

Ball, N. J. (1992). The phasing of sleep in mammals. In C. Stampi (Ed.), *Why we nap: Evolution, chronobiology, and functions of polyphasic and ultrashort sleep* (pp. 31–49). Boston: Birkhauser.

Balzamo, E., Bradley, R. J., & Rhodes, J. M. (1972). Sleep ontogeny in the chimpanzee: From two months to forty-one months. *Electroencephalography and Clinical Neurophysiology, 33*, 47–60.

Balzamo, E., Santucci, V., Seri, B., Vuillon-Cacciuttolo, G., & Bert, J. (1977). Nonhuman-primates: Laboratory animals of choice for neurophysiologic studies of sleep. *Laboratory Animal Science*, *27*, 879–886.

Barton, R. A. (1998). Visual specialization and brain evolution in primates. *The Royal Society of London Series B–Biological Sciences*, *265*, 1933–1937.

Barton, R. A., & Dunbar, R. I. M. (1997). Evolution of the social brain. In A. Whiten & R. W. Byrne (Eds.), *Machiavellian intelligence II: Extensions and evalutations* (pp. 240–263). Cambridge: Cambridge University Press.

Barton, R. A., Purvis, A., & Harvey, P. H. (1995). Evolutionary radiation of visual and olfactory brain systems in primates, bats, and insectivores. *Philosophical Transactions of the Royal Society, London, Series B*, *348*, 381–392.

Bert, J. (1975). Generic characteristics and specific characteristics of the ponto-geniculo-occipital spike activity (PGO) in 2 baboons, *Papio hamadryas* and *Papio papio*. *Brain Research*, *88*(2), 362–366.

Bert, J., Balzamo, E., Chase, M., & Pegram, V. (1975). The sleep of the baboon, *Papio papio,* under natural conditions and in the laboratory. *Electroencephalography and Clinical Neurophysiology*, *39*, 657–662.

Bert, J., Kripke, D. F., & Rhodes, J. (1970). Electroencephalogram of the mature chimpanzee: Twenty-four hour recordings. *Electroencephalography and Clinical Neurophysiology*, *28*(4), 368–373.

Bert, J., & Pegram, V. (1969). The sleep electroencephalogram in Cercopithecinae: *Erythrocerbus patas* and *Cercopithecus aethiops sabaeus*. *Folia Primatologica*, *11*(1), 151–159.

Bert, J., Pegram, V., & Balzamo, E. (1972). Comparison of sleep between 2 *Macaca* species (*Macaca radiata* and *Macaca mulatta*). *Folia Primatologica*, *17*(3), 202–208.

Bininda-Emonds, O. R. P., Cardillo, M., Jones, K. E., MacPhee, R. D. E., Beck, R. M. D., Grenyer, R., et al. (2007). The delayed rise of present-day Mammals. *Nature*, *446*, 507–512.

Blomberg, S. P., & Garland, T. (2002). Tempo and mode in evolution: Phylogenetic inertia, adaptation and comparative methods. *Journal of Evolutionary Biology*, *15*, 899–910.

Blomberg, S. P., Garland, T., & Ives, A. R. (2003). Testing for phylogenetic signal in comparative data: Behavioral traits are more labile. *Evolution*, *57*, 717–745.

Boesch, C., & Boesch-Achermann, H. (2000). *The chimpanzees of the Tai Forest*. Oxford: Oxford University Press.

Capellini, I., Barton, R. A., McNamara, P., Preston, B., & Nunn, C. L. (2008a). Ecology and evolution of mammalian sleep. *Evolution*, *62*, 1764–1776.

Capellini, I., Nunn, C. L., McNamara, P., Preston, B., & Barton, R. A. (2008b). Sleep cycles, predators, and energetics in mammals. *Functional Ecology*, *22*, 847–853.

Carroll, D. A., Denenberg, V. H., & Thoman, E. B. (1999). A comparative study of quiet sleep, active sleep, and waking on the first 2 days of life. *Developmental Psychobiology*, *35*(1), 43–48.

Carskadon, M. A., & Dement, W. C. (2006). Normal human sleep: An overview. In M. H. Kryger, T. Roth, & W. C. Dement (Eds.), *Principles and practice of sleep medicine* (4th ed., pp. 13–23). Philadelphia: W. B. Saunders.

Cartmill, M. (1974). Rethinking primate origins. *Science*, *184*(135), 436–443.

Chapman, C. A., Gillespie, T. R., & Goldberg, T. L. (2005). Primates and the ecology of their infectious diseases: How will anthropogenic change affect host–parasite interactions? *Evolutionary Anthropology*, *14*,134–144.

Corbet, G. B., & Hill, J. E. (1991). *A world list of mammalian species.* Oxford: Oxford University Press.

Davies, C. R., Ayres, J. M., Dye, C., & Deane, L. M. (1991). Malaria infection rate of Amazonian primates increases with body weight and group size. *Functional Ecology 5*, 655–662.

Day, R. T., & Elwood, R. W. (1999). Sleeping site selection by the golden-handed tamarin *Saguinus midas midas*: The role of predation risk, proximity to feeding sites, and territorial defence. *Ethology, 105*,1035–1051.

Di Bitetti, M. S., Vidal, E. M. L., Baldovino, M. C., & Benesovsky, V. (2000). Sleeping site preferences in tufted capuchin monkeys (*Cebus apella nigritus*). *American Journal of Primatology, 50*, 257–274.

Disotell, T. R. (2008). *Primate Phylogenetics.* In: *Encyclopedia of Life Sciences.* Chinchester: John Wiley and Sons.

Dunbar, R. I. M. (1992). Neocortex size as a constraint on group size in primates. *Journal of Human Evolution, 20*, 469–493.

Dunbar, R. I. M. (1998). The social brain hypothesis. *Evolutionary Anthropology, 6*, 178–190.

Elgar, M. A., Pagel, M. D., & Harvey, P. H. (1988). Sleep in mammals. *Animal Behavior, 36*, 1407–1419.

Felsenstein, J. (1985). Phylogenies and the comparative method. *The American Naturalist, 125*, 1–15.

Foley, R. A., & Lee, P. C. (1989). Finite social space, evolutionary pathways, and reconstructing hominid behavior. *Science, 243*, 901–906.

Freckleton, R. P., Harvey, P. H., & Pagel, M. (2002). Phylogenetic analysis and comparative data: A test and review of evidence. *The American Naturalist, 160*, 712–726.

Fruth, B., & Hohmann, G. (1993). Comparative analyses of nest building behavior in bonobos and chimpanzees. In R. W. Wrangham, W. C. McGrew, F. B. M. de Waal, & P. G. Heltne (Eds.), *Chimpanzee cultures* (pp. 109–128). Cambridge, MA: Harvard University Press.

Ganguly-Fitzgerald, I., Donlea, J., & Shaw, P. J. (2006). Waking experience affects sleep need in *Drosophila. Science, 313*, 1775–1781.

Garland, T., Harvey, P. H., & Ives, A. R. (1992). Procedures for the analysis of comparative data using phylogenetically independent contrasts. *Systematic Biology, 4*, 18–32.

Gartlan, J. S., & Brain, C. K. (1968). Ecology and social variability in *Cercopithecus aethiops* and *C. mitis.* In P. C. Jay (Ed.), *Primates: Studies in adaptation and variability* (pp. 253–292). New York: Holt, Rinehart and Winston.

Gaulin, S. J. C., & Gaulin, C. K. (1982). Behavioral ecology of *Alouatta seniculus* in Andean cloud forest. *International Journal of Primatology, 3*(1), 1–32.

Goodall, A. (1979). *The wandering gorillas.* London: William Collins & Sons.

Hausfater, G., & Meade, B. J. (1982). Alternation of sleeping groves by yellow baboons (*Papio cynocephalus*) as a strategy for parasite avoidance. *Primates, 23*, 287–297.

Heymann, E. W. (1995). Sleeping habits of tamarins, *Saguinus mystax*, and *Saguinus fuscicollis* (Mammalia: Primates; Callitrichidae), in northeastern Peru. *Journal of Zoology, London, 237*, 211–226.

Hilton-Taylor, C. (2002). *IUCN Red List of threatened species.* Morges: IUCN.

Horr, D. A. (1977). Orangutan maturation, growing up in a female world. In S. Chevalier-Skolnikoff & F. E. Poirier (Eds.), *Primate bio-social development* (pp. 289–321). New York: Garland Publishing.

Hsieh, K. C., Robinson, E. L., & Fuller, C. A. (2008). Sleep architecture in unrestrained rhesus monkeys (*Macaca mulatta*) synchronized to 24-hour light-dark cycles. *Sleep, 31*(9), 1239–1250.

Ives, A. R., Midford, P. E., & Garland, T. (2007). Within-species variation and measurement error in phylogenetic comparative methods. *Systematic Biology*, *56*, 252–270.

Janson, C. H. (1992). Evolutionary ecology of primate social structure. In E. A. Smith, & B. Winterhalder (Eds.), *Evolutionary ecology and human behavior* (pp. 95–130). New York: Aldine de Gruyter.

Koyama, N. (1973). Dominance, grooming, and clasped-sleeping relationships among bonnet monkeys in India. *Primates*, *14*, 225–244.

Kummer, H. (1968). Social organization of Hamadryas baboons: A field study. *Bibliotheca Primatologica*, *6*.

Lesku, J. A., Roth II, T. C., Amlaner, C. J., Lima, S. L. (2006). A phylogenetic analysis of sleep architecture in mammals: The integration of anatomy, physiology, and ecology. *The American Naturalist*, *168*, 1–13.

Lima, S. L., Rattenborg, N. C., Lesku, J. A., & Amlaner, C. J. (2005). Sleeping under the risk of predation. *Animal Behavior*, *70*, 723–726.

Lindenfors, P., Nunn, C. L., & Barton, R. A. (2007). Primate brain architecture and selection in relation to sex. *BioMed Central Biology*, *5* (20). *doi:10.1186/1741-7007-5-20*.

Maddison, W. P., & Maddison, D. R. (2006). *Mesquite: A modular system for evolutionary analysis*, version 1.1. See http://mesquiteproject.org

Martin, R. D. (1990). *Primate origins and evolution*. London: Chapman & Hall.

Martin, R. D., & Ross, C. F. (2005). The evolutionary and ecological context of primate vision. In J. Kremers (Ed.), *The primate visual system: A comparative approach* (pp. 1–36). Chichester: John Wiley and Sons.

McNamara, P., Capellini, I., Harris, E., Nunn, C. L., Barton, R. A., & Preston, B. (2008). The phylogeny of sleep database: A new resource for sleep scientists. *The Open Sleep Journal*, *1*, 11–14.

Midford, P. E., Garland, Jr., T., & Maddison, W. P. (2005). *PDAP package of Mesquite*, version 1.07.

Nunn, C. L., & Altizer, S. (2005). The global mammal parasite database: An online resource for infectious disease records in wild primates. *Evolutionary Anthropology*, *14*, 1–2.

Nunn, C. L., & Altizer, S. M. (2006). *Infectious diseases in primates: Behavior, ecology, and evolution*. Oxford: Oxford University Press.

Nunn, C. L., & Barton, R. A. (2001). Comparative methods for studying primate adaptation and allometry. *Evolutionary Anthropology*, *10*, 81–98.

Nunn, C. L., & Heymann, E. W. (2005). Malaria infection and host behaviour: A comparative study of neotropical primates. *Behavioral Ecology and Sociobiology*, *59*, 30–37.

Nunn, C. L., & van Schaik, C. P. (2002). Reconstructing the behavioral ecology of extinct primates. In J. M. Plavcan, R. F. Kay, W. L. Jungers, & C. P. van Schaik (Eds.), *Reconstructing behavior in the fossil record* (pp. 159–216). New York: Kluwer Academic/Plenum.

Oates, J. F. (1987). Food distribution and foraging behavior. In B. B. Smuts, D. L. Cheney, R. M. Seyfarth, R. W. Wrangham, and T. T. Struhsaker (Eds.), *Primate Societies* (pp. 197–209). Chicago: University of Chicago Press.

Pagel, M. (1997). Inferring evolutionary processes from phylogenies. *Zoologica Scripta*, *26*, 331–348.

Pagel, M. (1999). Inferring the historical patterns of biological evolution. *Nature*, *401*, 877–884.

Pagel, M., & Meade, A. (2007). *BayesTraits Version 1.0*. http://www.evolution.rdg.ac.uk. Reading, UK.

Pagel, M., Meade, A., & Barker, D. (2004). Bayesian estimation of ancestral character states on phylogenies. *Systematic Biology*, *53*, 673–684.

Perachio, A. A. (1971). Sleep in the nocturnal primate, *Aotus trivirgatus*. *Proceedings of the 3rd International Congress on Primates* (vol. *2*, pp. 54–60). Basel: Karger.

Plavcan, J. M., & van Schaik, C. P. (1997). Intrasexual competition and body weight dimorphism in anthropoid primates. *American Journal of Physical Anthropology*, *103*, 37–68.

Purvis, A. (1995). A composite estimate of primate phylogeny. *Philosophical Transactions of the Royal Society, London, Series B*, *348*, 405–421.

Purvis, A., Nee, S., & Harvey, P. H. (1995). Macroevolutionary inferences from primate phylogeny. *Proceedings of the Royal Society London Series B*, *260*, 329–333.

Rattenborg, N. C., Voiren, B., Vyssotski, A. L., Kays, R. W., Spoelstra, K., Kuemmeth, F., et al. (2008). Sleeping outside the box: Electroencephalographic measues of sleep in sloths inhabiting a rainforest. *The Royal Society Biology Letters*. doi:10.1098/rsbl.2008.0203.

Reite, M., Stynes, A. J., Vaughn, L., Pauley, J. D., & Short, R. A. (1976). Sleep in infant monkeys: Normal values and behavioral correlates. *Physiology and Behavior*, *16*(3), 245–251.

Salzarulo, P., & Ficca, G. (2002). *Awakening and sleep-wake cycle across development*. Philadelphia: J. Benjamins.

Smuts, B. B., Cheney, D. L., Seyfarth, R. M., Wrangham, R. W., & Struhsaker, T. T. (1987). *Primate societies*. Chicago: University of Chicago Press.

Suzuki, K. (1965). The pattern of mammalian brain gangliosides: Evaluation of the extraction procedures, postmortem changes and the effect of formalin preservation. *Journal of Neurochemistry*, *12*(7), 629–638.

Tattersall, I. (1987). Cathemeral activity in primates: A definition. *Folia Primatologica*, *49*, 200–202.

Thurber, A., Jha, S. K., Coleman, T., & Frank, M. G. (2008). A preliminary study of sleep ontogenesis in the ferret (*Mustela putorius furo*). *Behavioural Brain Research*, *189*, 41–51.

Tobler, I. (2005). Phylogeny of sleep regulation. In M. Kryger, T. Roth, & W. Dement (Eds.), *Principles and practice of sleep medicine* (4th ed., pp. 77–90). Philadelphia: W. B. Saunders.

van Schaik, C. P. (1983). Why are diurnal primates living in groups? *Behaviour*, *87*, 120–143.

van Schaik, C. P., & Janson, C. (2000). *Infanticide by males and its implications*. Cambridge: Cambridge University Press.

Vessey, S. H. (1973). Night observations of free-ranging rhesus monkeys. *American Journal of Physical Anthropology*, *38*(2), 613–619.

Vuillon-Cacciuttolo, G., Balzamo, E., Petter, J. J., & Bert, J. (1976). Wakefulness-sleep cycle studied by telementry in a lemurian (*Lemur macaco fulvus*). *Revue d'électroencéphalographie et de neurophysiologie clinique*, *6*(1), 34–36.

Wilson, D. E., & Reeder, D. M. (2005). *Mammal species of the world*. Baltimore: Johns Hopkins University Press.

Wolfe, N. D., Escalante, A. A., Karesh, W. B., Kilbourn, A., Spielman, A., Lal, A. A. (1998). Wild primate populations in emerging infectious disease research: The missing link? *Emerging Infectious Diseases*, *4*, 149–158.

Zepelin, H., Siegel, J. M., & Tobler, I. (2005). Mammalian sleep. In M. H. Kryger, T. Roth, & W. C. Dement (Eds.), *Principles and practice of sleep medicine* (4th ed, pp. 91–100). Philadelphia: W. B. Saunders.

7

A bird's-eye view of the function of sleep

NIELS C. RATTENBORG AND CHARLES J. AMLANER

Introduction

Sleep has been detected in every animal that has been adequately studied (Cirelli & Tononi, 2008). The ubiquitous nature of sleep suggests that it evolved early in the course of evolution and therefore may serve a conserved function essential to all animals. This hypothesis forms the rationale behind the development of "simple" animal models of sleep (Allada & Siegel, 2008; Mignot, 2008). By studying sleep in animals such as the fruit fly (*Drosophila melanogaster*), where the power of genetic techniques can be readily employed, we may gain insight into the initial (perhaps cellular) function of sleep, a function that may still be relevant to understanding sleep in humans. Indeed, recent studies have already demonstrated remarkable similarities between sleep in *Drosophila* and sleep in mammals (Hendricks, Finn, Panckeri, et al., 2000; Shaw, Cirelli, Greenspan, et al., 2000; reviewed in Cirelli & Bushey, 2008). Although the utility of studying sleep in "simple" animal models is undeniable, it is unlikely that this approach alone will tell the whole story, especially given that *Drosophila* do not exhibit brain states comparable to mammalian slow-wave sleep (SWS) and rapid eye-movement (REM) sleep (Cirelli, 2006; Cirelli & Bushey, 2008; Hendricks & Sehgal, 2004; Nitz, van Swinderen, Tononi, et al., 2002). Indeed, the heterogeneous nature of mammalian sleep suggests that the specific changes in brain activity that accompany SWS and REM sleep might serve secondarily evolved functions not found in simple animals. Interestingly, birds, as the only nonmammalian taxonomic group to exhibit unequivocal SWS and REM sleep (Klein, Michel, & Jouvet, 1964; Ookawa & Gotoh, 1964; reviewed in Amlaner & Ball, 1994), provide a largely unrecognized opportunity to glean insight into the functions of these states by revealing overriding principles common to both lineages (Rattenborg, Martinez-Gonzalez, Lesku, et al., 2008a).

145

This chapter summarizes our current understanding of avian sleep. We first describe the basic changes in brain activity and physiology that accompany avian SWS and REM sleep. Although we emphasize the similarities between avian and mammalian sleep, potentially meaningful differences are also discussed. Finally, we summarize our recent proposal that the convergent evolution of similar sleep states in mammals and birds is linked to the convergent evolution of relatively large and highly interconnected brains capable of complex cognition in each group (Rattenborg, 2006a; Rattenborg, Martinez-Gonzalez, & Lesku, 2009).

Avian sleep

Slow-wave sleep

As in mammals, when compared to wakefulness, the electroencephalogram (EEG) during SWS is characterized by increased slow-wave activity (SWA, 0.5 to 4 hertz power density) (Figure 7.1). However, the degree to which SWA increases between wakefulness and SWS is smaller in birds when compared to mammals (Tobler & Borbély, 1988). Moreover, as in some mammals, states intermediate between wakefulness and SWS are often observed. In some studies, such "drowsiness" is scored as a state distinct from SWS (Figure 7.1), whereas in others a distinction between drowsiness and SWS is not made. Thalamocortical spindles, a hallmark of the lighter stages of SWS in mammals (Steriade, 2006), have not been observed during avian SWS, although recordings from the thalamus have not been reported. Early reports of sleep spindles in birds were later determined to be artifacts originating from intermittent, brief oscillations of the eyes that occur during SWS in birds (Zepelin, Hartzer, & Pendergast, 1998) (Figure 7.1). Hippocampal sharp waves similar to those occurring during mammalian SWS (Hahn, Sakmann, & Mehta, 2006; Isomura, Sirota, Ozen, et al., 2006; Mölle, Yeshenko, Marshall, et al., 2006; Sirota, Csicsvari, Buhl, et al., 2003) have not been reported during avian SWS, although, as with spindles, this phenomenon has not been investigated sufficiently. Metabolism (Ball, Amlaner, Shaffery, et al., 1988; Stahel, Megirian, & Nicol, 1984), brain temperature (Szymczak, 1989), and heart and respiratory rate (van Twyver & Allison, 1972) are all usually lower during SWS when compared to wakefulness. Thermoregulatory responses – such as shivering, ptiloerection (feather raising), and panting – can occur during SWS (Heller, Graf, & Rautenberg, 1983; Hohtola, Rintamaki, & Hissa, 1980). Time spent in SWS may decrease, increase, or remain unchanged across the major sleep period (see Szymczak, Helb, & Kaiser, 1993). In pigeons (*Columba livia*), both the time spent in SWS and the duration of SWS episodes decrease across the night, due largely to an increase in the incidence and duration of REM sleep episodes (Martinez-Gonzalez, Lesku, & Rattenborg, 2008). SWS episodes last 50 seconds early in the night and 25 seconds

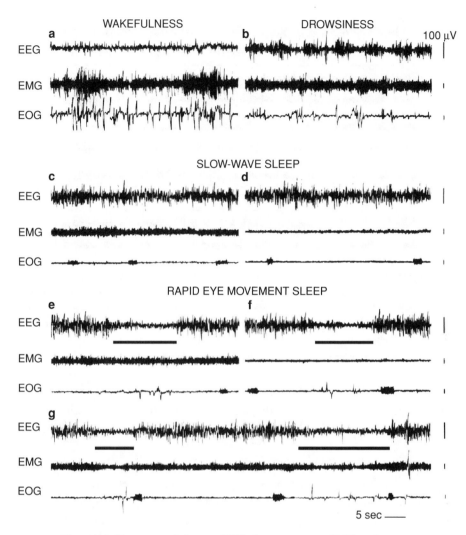

Figure 7.1. Electroencephalogram (EEG), electromyogram (EMG), and electro-occulogram (EOG) recordings of wakefulness (a), drowsiness (b), slow-wave sleep (SWS) (c, d), and rapid-eye-movement (REM) sleep (e to g) in an emperor penguin (*Aptenodytes forsteri*). During drowsiness, the EEG alternates rapidly between a pattern typical of wakefulness and that typical of SWS. SWS can occur with high (c) or low (d) tonic EMG activity. The intermittent increases in EOG amplitude occurring during SWS reflect brief, rapid oscillations of the eyes. Transitions between SWS and REM sleep are shown in (e to g); the horizontal black bars mark REM sleep. As with SWS, REM sleep can occur with high (e) or low (f) tonic EMG activity. In rare cases, however, EMG activity may decrease during episodes of REM sleep (g). The states depicted in the penguin are typical of those reported in other bird species. (Modified from Buchet, Dewasmes, & Le Maho, 1986.)

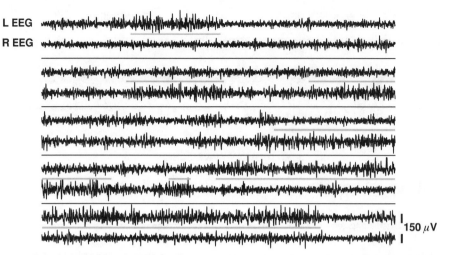

L EEG

R EEG

150 μV

Figure 7.2. Unihemispheric slow-wave sleep (USWS) in the European quail (*Coturnix coturnix*). Five consecutive minutes of electroencephalograms (EEG) recorded from the left (L) and right (R) hemispheres. The thin black lines separate each minute. Episodes of interhemispheric asymmetry are marked by the gray lines. Note that the quail alternates between sleeping primarily with left or right hemisphere. The last episode of asymmetry lasts 77 seconds. (Unpublished data, Rattenborg & Derénaucourt.)

toward the end of the night. Episodes of sleep (SWS and REM sleep combined) last approximately 4 minutes early in the night and 1 minute by the end of the night. Such relatively brief sleep episodes are typical of birds (Amlaner & Ball, 1994).

Unihemispheric slow-wave sleep

Birds often sleep with one eye open. Unilateral eye closure is associated with an interhemispheric asymmetry in the level of SWA. The hemisphere contralateral to the closed eye exhibits levels of SWA typical of SWS with both eyes closed, and the hemisphere contralateral to the open eye exhibits SWA intermediate between SWS and wakefulness (Ball et al., 1988; Ookawa & Gotoh, 1965; Peters, Vonderahe, & Schmid, 1965; reviewed in Ookawa, 2004; Rattenborg, Amlaner, & Lima, 2000). In contrast to mammals, where unihemispheric SWS (USWS) is known to occur only in marine mammals (cetaceans, seals in the Order Otariidae, walruses, and manatees) (reviewed in Lyamin, Manger, Ridgway, et al., 2008), sleep with one eye open has been observed in several avian orders and therefore may be an ancestral trait (Rattenborg et al., 2000). Although the degree of asymmetry occasionally approaches that occurring during USWS in marine mammals (Ball et al., 1988), the asymmetry is typically less pronounced in birds (Rattenborg, Amlaner, & Lima, 2001; Rattenborg, Lima, & Amlaner, 1999a). Figure 7.2 shows USWS alternating between the left and right hemispheres in a European quail

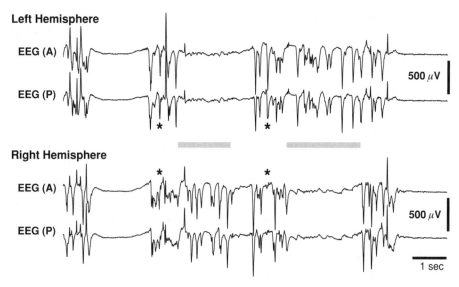

Left Hemisphere

EEG (A)

EEG (P)

500 μV

*   *

Right Hemisphere

*   *

EEG (A)

EEG (P)

500 μV

1 sec

Figure 7.3. Asymmetric burst suppression in a pigeon (*Columba livia*) recorded under 2.0% isoflurane anesthesia. The electroencephalograms (EEG) were simultaneously recorded from the anterior (A) and posterior (P) hyperpallia of each hemisphere. Although bursting (high-amplitude slow waves) and suppression (low-amplitude EEG) often occurred simultaneously in the two hemispheres, bursting also occurred in one hemisphere while the other showed suppression. Episodes of asymmetrical burst suppression are marked with the gray lines. During such episodes, the slight variation in amplitude in the suppressed hemisphere (not evident when both hemispheres showed suppression) may reflect volume conduction from the bursting hemisphere. Note that even when the hemispheres were bursting at the same time, the individual slow waves were coherent only within a given hemisphere and not between hemispheres (*see asterisks*). This pharmacological condition demonstrates the capacity for the two hemispheres to function largely as independent units in birds. (Unpublished data, Rattenborg.)

(*Coturnix coturnix*), a species in which the degree of asymmetry is particularly evident. Note that the last episode of USWS in the left hemisphere lasted well over 1 minute. The capacity for the two hemispheres to function independently (even in the absence of unilateral visual input) is most evident under anesthesia, where burst suppression, a phenomenon thought to share mechanisms in common with SWS (Steriade, Amzica, & Contreras, 1994), can occur independently in the two hemispheres (Figure 7.3). The limited interhemispheric connections in the avian brain may contribute to such hemispheric asymmetries during anesthesia and natural sleep (Rattenborg et al., 2000).

Sleeping with one eye open may be useful under circumstances that simultaneously require sleep and wakefulness. For instance, mallard ducks (*Anas platyrhynchos*) sleeping exposed at the edge of a group spend proportionately more time sleeping with one eye open when compared to ducks sleeping safely flanked by other birds (Rattenborg, Lima, & Amlaner, 1999b). Moreover, ducks at the edge of a group direct the open eye away from the other birds, as if watching for approaching predators. These findings indicate that unilateral eye closure serves a predator detection function. More generally, the preference for sleeping with both eyes closed and both hemispheres simultaneously under safe conditions, indicates that USWS is a less efficient form of SWS, presumably because sleep processes associated with SWA are reduced in the hemisphere contralateral to the open eye.

In addition to antipredator vigilance, it has also been suggested that birds utilize USWS during prolonged nonstop flights (reviewed in Rattenborg, 2006b). Although birds that engage in nonstop flights lasting several days, weeks, or longer, such as European swifts (*Apus apus*) and frigatebirds (*Fregata* sp.) (Weimerskirch, Chastel, Barbraud, et al., 2003; Weimerskirch, Le Corre, Jaquemet, et al., 2004), would seemingly need to sleep on the wing, the electrophysiological recordings required to measure sleep have not been performed. Consequently, it remains unclear whether such birds sleep in flight or have evolved novel mechanisms that allow them to suspend sleep temporarily during long flights (see Rattenborg, Mandt, Obermeyer, et al., 2004). The recent development of miniaturized EEG devices (Vyssotski, Serkov, Itskov, et al., 2006) that can be used to record sleep in the wild (Rattenborg, Voirin, Vyssotski, et al., 2008b) may provide a means to answer this question.

REM sleep

As in mammals, avian REM sleep is characterized by a low-amplitude, high-frequency (or activated) EEG pattern similar to that occurring during wakefulness (Figure 7.1). Other electrophysiological correlates of mammalian REM sleep (i.e., irregular brainstem neuronal activity or ponto-geniculo-occipital waves) have not been investigated sufficiently to determine whether they are present or absent in birds. Van Twyver and Allison (1972) made specific attempts to record a mammal-like hippocampal theta rhythm during REM sleep in pigeons but did not find evidence of this phenomenon. In contrast to SWS, REM sleep is not known to occur unihemispherically. Although nuchal electromyographic recordings only occasionally show reductions in muscle tone during REM sleep (Figure 7.1), behavioral signs of reduced tone (e.g., head drooping) are often observed (Dewasmes, Cohen-Adad, Koubi, et al., 1985). Rapid eye movements (conjugate and disconjugate) and occasional twitches, such as bill movements, also accompany avian REM sleep. As in mammals, thermoregulatory responses, such as shivering, are reduced during avian REM sleep (Heller et al., 1983). Slight increases in brain temperature have been observed during REM sleep (Szymczak, 1989). Heart rate may increase, decrease, or remain unchanged during individual episodes of REM sleep (see Dewasmes et al., 1985). Episodes of REM sleep in birds, when compared to those in mammals, are short, typically lasting less than 10 seconds. In many species, REM sleep increases across the night, owing largely to an increase in the incidence of REM sleep episodes (Ayala-Guerrero, Mexicano, & Ramos, 2003; Rattenborg et al., 2004; Szymczak et al., 1993; Tobler & Borbély, 1988), although a decline across the night has also been reported (Szymczak, 1987). In some studies, the incidence *and* duration of REM sleep episodes increases across the night in a manner comparable to that in humans (Fuchs, 2006; Low, Shank, Sejnowski, et al., 2008; Martinez-Gonzalez et al., 2008). Although earlier studies suggested that birds only spend on average 43 minutes per day in REM sleep (i.e., approximately 8% of the total time spent sleeping [Roth, Lesku, Amlaner, et al., 2006], recent studies suggest that Passerines (songbirds) and perhaps all birds have more REM sleep than previously recognized (Fuchs, 2006; Low et al., 2008; Rattenborg et al., 2004; Szymczak et al., 1993).

The reasons for the short duration of avian REM sleep episodes remain unclear. Given that muscle tone declines during REM sleep, it has been suggested that the short duration may protect birds from falling out of trees. However, several lines of evidence argue against this idea. Perching birds have tendons in their legs that passively pull their feet closed around a branch when they perch. Because this mechanism does not require muscular effort, it is not susceptible to REM sleep–related reductions in muscle tone. Moreover, birds that sleep supported on the ground, such as geese, also have short episodes of REM sleep, even when their heads are supported on their backs (Dewasmes et al., 1985).

The short duration of REM sleep episodes may represent a fundamental difference between birds and mammals. Alternatively, it is conceivable that mammals and birds differ only in the amount of cortical activation occurring during REM sleep. This idea stems from studies in monotremes, where REM sleep–related neuronal activity occurs in the brainstem while the cortex exhibits EEG activity characteristic of SWS (Siegel, Manger, Nienhuis, et al., 1996, 1999). Apparently, REM sleep with cortical activation evolved after the monotreme lineage diverged from the lineage leading to marsupial and placental (or therian) mammals (but see Nicol, Andersen, Phillips, et al., 2000). Moreover, this suggests that REM sleep–related cortical activation evolved independently in birds and therian mammals. Given the difference between REM sleep in monotremes and therian mammals, it is conceivable that birds exhibit a REM sleep stage intermediate between that exhibited by monotreme and therian mammals. Specifically, birds may exhibit more REM sleep at the level of the brainstem than is reflected by cortical activation; the short periods of cortical activation may simply reflect the most intense (phasic) periods of REM sleep. This hypothesis is consistent with the observation that episodes of REM sleep often occur in clusters separated by only several seconds of SWS, particularly toward the end of the night in diurnal birds. Clearly, brainstem recordings are needed to test this hypothesis. Even if more REM sleep is found at the level of the brainstem, however, the difference in the amount of cortical activation is still potentially very interesting, especially when viewed in the context of hypotheses that implicate REM sleep–related cortical activation in memory consolidation (reviewed in Stickgold & Walker, 2007). If avian and mammalian REM sleep–related cortical activation serve similar functions, such functions must be achievable within bouts of REM sleep lasting less than 10 seconds or through the cumulative action of many short bouts.

Sleep homeostasis

Despite the gross similarities in the EEGs of mammals and birds, early studies suggested that mammalian and avian SWS were regulated differently. In mammals, SWS-related SWA increases and decreases as a function of time spent awake and asleep, respectively, and therefore appears to reflect a homeostatically regulated process (process S) that accumulates during wakefulness and declines during SWS (Tobler, 2005). Until recently, the only EEG-based sleep deprivation studies to examine sleep homeostasis in birds were conducted on pigeons. In contrast to mammals, pigeons deprived of sleep for 24 hours did not show an increase in SWA during recovery SWS (Tobler & Borbély, 1988). A subsequent study also failed to detect a compensatory response to light-induced sleep deprivation reportedly lasting several weeks in pigeons (Berger & Phillips, 1994). Collectively, these studies suggested that avian SWS is not homeostatically regulated, possibly because birds lack the neural substrate necessary for SWS homeostasis, such as a neocortex (Zepelin, Siegel, & Tobler, 2005). However, two factors prompted us to re-evaluate avian SWS homeostasis. First, because SWA increases in only some mammals following relatively short periods of sleep deprivation (Tobler & Jaggi, 1987), it was possible that birds would show a homeostatic response to shorter, presumably more ecologically realistic periods of sleep loss. Second, although pigeons sleep more at night, our experience (Martinez-Gonzalez et al., 2008) and that of others (Tobler & Borbély, 1988) indicates that pigeons spend 40% to 45% of the day in SWS, a finding that calls into question the report of long-term light-induced sleep suppression. In fact, although scored SWS was greatly reduced under constant light in Berger and Phillips' study, the overall level of SWA (regardless of scored state) actually increased above that occurring during the light phase of the 12:12 light–dark photoperiod. Consequently it appears that their pigeons were actually compensating for the lost SWS that would have occurred at night by increasing SWA during constant light. Furthermore, the difference between the amounts of scored SWS during light in general, when compared to the other studies of pigeons (Martinez-Gonzalez et al., 2008; Tobler & Borbély, 1988), may reflect the use of a stricter threshold for scoring SWS in Berger and Phillips' (1994) study.

We designed a sleep deprivation protocol that directly tested both explanations for the absence of a homeostatic response to sleep deprivation in the previous studies in pigeons (Martinez-Gonzalez et al., 2008). To determine whether the duration of sleep deprivation may account for the absence of a homeostatic response following 24 hours of sleep deprivation, we deprived pigeons of sleep for only 8 hours, using the gentle handling technique. To determine whether SWS, or SWA according to Berger and Phillips (1994), occurring during the light is homeostatically regulated in pigeons, we conducted the deprivation during the last 8 hours of the light phase of the 12:12 light–dark photoperiod. Under this protocol, pigeons lost approximately 3 hours of SWS. We then compared SWS-related SWA during recovery at night to that occurring during the corresponding hours of the preceding baseline night. Although the time spent in SWS did not change during recovery, SWA increased significantly during recovery SWS and progressively declined thereafter in a manner comparable to that observed in mammals (Figure 7.4). This study thus provided the first experimental evidence for SWS homeostasis in birds (reviewed in Rattenborg, et al., 2009), and demonstrated that the neocortex per se is not necessary for this process. Furthermore, it suggested that the absence of a homeostatic response following 24 hours of sleep deprivation was due to the longer duration of the deprivation (Tobler & Borbély, 1988), although the influence of other factors cannot be excluded. Our results also demonstrate that SWA (or SWS) occurring during the light does in fact reflect homeostatically regulated sleep-related processes. Consequently the absence of an increase in SWA following constant light appears to indicate simply that light is not an effective form of sleep deprivation in pigeons. Subsequent to our study, similar results were obtained in another avian Order (Passeriformes) (Jones, Vyazovskiy, Cirelli, et al., 2008a), suggesting that, as in mammals, SWS homeostasis may be a trait shared by the avian Class (reviewed in Rattenborg et al., 2009).

In addition to exhibiting a compensatory increase in SWA following sleep deprivation, additional aspects of sleep homeostasis are similar in birds and mammals. As in rodents (Borbély, Tobler, & Hanagasioglu, 1984; Huber, Deboer, & Tobler, 2000; Tobler & Jaggi, 1987), the power density of higher frequencies (approximately 10 to 20 hertz) also increased during recovery SWS in pigeons (Figure 7.4) (Martinez-Gonzalez et al., 2008) and sparrows (*Zonotrichia leucophrys gambelii*) (Jones et al., 2008a). Although the functional significance of this phenomenon remains unclear, it nonetheless represents another parallel between avian and mammalian sleep. Also as in mammals, REM sleep increases following 8 and 24 hours of sleep

154   Niels C. Rattenborg and Charles J. Amlaner

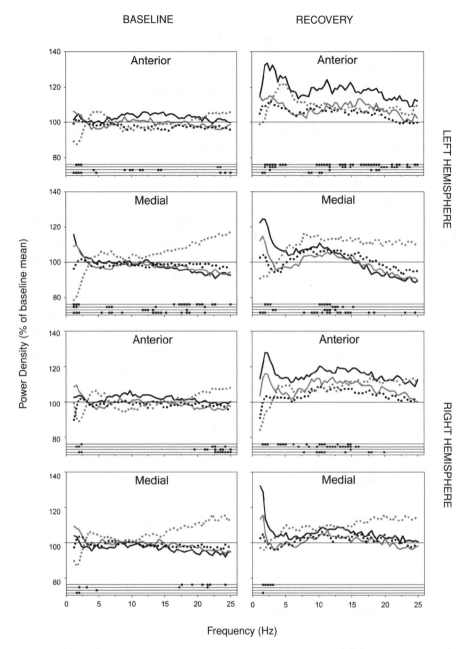

BASELINE   RECOVERY

Figure 7.4. Electroencephalogram power density (0.78 to 25 hertz) during slow-wave sleep (SWS) on a baseline night (*left column*) and a recovery night (*right column*) immediately following 8 hours of sleep deprivation. The power density for each quarter (first, solid black line; second, solid gray line; third, dotted black line; fourth, dotted gray line) of each night is expressed as a percent of the entire baseline night SWS mean (i.e., the 100% line) for each frequency bin and brain region (left and right,

deprivation in pigeons (Martinez-Gonzalez et al., 2008; Tobler & Borbély, 1988; see also Newman, Paletz, Rattenborg, et al., 2008). Interestingly, although REM sleep increased following 8 hours of sleep deprivation, the duration of individual episodes of REM sleep (and SWS) actually decreased. The shorter duration of REM sleep and SWS episodes appears to reflect more frequent switching between sleep states, because the duration of episodes of wakefulness was unchanged during recovery. This pattern is unlike that observed in mammals, where the duration of sleep bouts usually increases following sleep deprivation (Vyazovskiy, Achermann, & Tobler, 2007). The reason for this interesting difference remains unclear. Nevertheless, despite this difference, the fundamental features of mammalian SWS and REM sleep homeostasis are present in birds.

Convergent evolution of SWS and REM sleep in mammals and birds

SWS and REM sleep were either inherited from the common ancestor to mammals and birds or evolved independently in their respective ancestors. Most studies of sleep in poikilothermic vertebrates (reptiles, amphibians, and fish) indicate that the latter is the case. Although reptiles clearly sleep, they do not exhibit the high-amplitude slow waves characteristic of SWS in mammals and birds (reviewed in Hartse, 1994; Rattenborg, 2006a). Slow waves have been reported in awake, stimulated reptiles and amphibians (Rial, Nicolau, Gamundi, et al., 2007), but these slow waves appear to reflect neural processes unlike those that instantiate slow waves during sleep in mammals and birds, and therefore reflect superficially similar but unrelated phenomena (Rattenborg, 2007). Most studies report an association between sleep and the occurrence of high-amplitude sharp waves in the dorsal cortex arising from background EEG activity with an amplitude lower than that occurring during alert wakefulness. Several lines of evidence suggest that these sharp waves may be homologous to similar sharp waves occurring in the mammalian hippocampus during SWS (Hahn et al., 2006; Isomura et al., 2006; Mölle et al., 2006; Sirota et al., 2003). Notably, both sharp waves

←──

Figure 7.4 (*continued*) anterior and medial pallia) in each pigeon (N = 5). The mean percent is plotted at the end of each frequency bin. For the baseline night, values for each quarter and frequency bin were compared to the baseline night average. Significant differences ($P < 0.05$, two-tailed paired t-test after significant repeated measures ANOVA) are indicated by filled squares on the lines at the bottom of each plot; statistical data for the first through fourth quarters is presented on the first (*top*) through fourth (*bottom*) lines, respectively. For the recovery night, values for each quarter and frequency bin were compared to the corresponding quarter of the baseline night, with significant differences similarly indicated at the bottom of each plot. (Modified from Martinez-Gonzalez et al., 2008.)

increase following sleep deprivation and respond similarly to various pharmacological agents (Hartse & Rechtschaffen, 1982; reviewed in Hartse, 1994; Rattenborg, 2007). Indeed, more recent neurophysiological studies indicate that these sharp waves originate in the reptilian hippocampus and propagate to the dorsal cortex (Gaztelu, García-Austt, & Bullock, 1991; Lorenzo, Macadar, & Velluti, 1999; Lorenzo & Velluti, 2004; Velluti, Russo, Simini, et al., 1991). Thus, although reptiles appear to exhibit components of the SWS state, they lack the large-scale, slow, synchronous network oscillations that define SWS in mammals and birds.

In addition, the evidence for REM sleep in poikilothermic vertebrates is equivocal. Although REM sleep has been reported in reptiles and fish, based largely on eye and limb movements occurring during periods of sleep (reviewed in Hartse, 1994; Rattenborg, 2006a), these events may simply reflect brief arousals from sleep rather than REM sleep–related phasic twitching. Moreover, a study in turtles failed to find signs of REM sleep in recordings of brainstem neuronal activity similar to that observed in monotreme or placental mammals during REM sleep (Eiland, Lyamin, & Siegel, 2001). Taken together, the available evidence argues against the presence of SWS or REM sleep in poikilothermic vertebrates and thereby suggests that both states evolved independently in mammals and birds.

Why do they sleep similarly?

Why are mammals and birds the only animals known to exhibit SWS and REM sleep? One obvious explanation is that it has something to do with the independent evolution of homeothermy in each group. For instance, early functional hypotheses proposed that SWS evolved in mammals and birds to offset the energy costs associated with a higher metabolic rate by reducing the metabolic rate during SWS and preventing animals from expending energy during times of the day when they are unable to forage effectively (Berger & Phillips, 1995; van Twyver & Allison, 1972; Walker & Berger, 1980; Zepelin et al., 2005). However, the energy savings associated with engaging in SWS, rather than wakefulness, are relatively small (Stahel et al., 1984). Moreover, for animals such as sloths (*Bradypus variegates*), with sensory systems and foraging strategies that enable them to forage effectively during the day and night (Rattenborg, et al., 2008b), engaging in SWS at the expense of foraging would be costly rather than frugal in terms of energy. Indeed, in contrast to earlier comparative studies, recent studies using modern statistical methods have shown that animals with higher relative metabolic rates actually engage in *less* SWS, not more, as predicted by the energy conservation hypothesis for SWS (Capellini, Barton, McNamara, et al., 2008; Lesku, Roth, Amlaner, et al., 2006; Lesku, Roth, Rattenborg, et al., in press); presumably, mammals with higher relative metabolic rates have to spend more time foraging to acquire sufficient energy to fuel their metabolism. The energy conservation hypothesis also does not explain the presence of USWS in marine mammals and birds. Instead, the fact that SWS can occur locally in the brain suggests that it serves a function for the brain itself (Huber, Ghilardi, Massimini, et al., 2004). Finally, the fact that animals engage in SWS (and REM sleep), despite the inherent risk of predation (Lima & Rattenborg, 2007; Lima, Rattenborg, Lesku, et al., 2005), indicates that sleep serves an essential function for the brain that is incompatible with significant sensory processing.

Although energy conservation may not be the primary function of sleep, the convergent evolution of SWS and REM sleep in mammals and birds may nonetheless be interrelated with the evolution of homeothermy. For instance, it has been suggested that the convergent evolution of homeothermy contributed to the convergent evolution of large brains in mammals and birds (Allman, 1999; Jerison, 2001; Shimizu, 2008). Moreover, as expected based on brain size, it has recently become apparent that mammals and birds independently evolved complex cognitive abilities not observed in reptiles (reviewed in Butler, 2008; Butler & Cotterill, 2006; Jarvis, Güntürkün, Bruce, et al., 2005). In an extreme case, corvids (e.g., crows, ravens, magpies, and jays) exhibit cognition comparable to that found in nonhuman primates, including the manufacture and use of tools, object permanence (Piagetian stage 6), theory of mind, and retrospective as well as prospective cognition (reviewed in Emery & Clayton, 2004, 2005; Kirsch, Güntürkün, & Rose, 2008), and possibly self-recognition (Prior, Schwartz, & Güntürkün, 2008). Gray parrots (*Psittacus erithacus*) also exhibit rudiments of referential language and abstract categorical reasoning (Pepperberg, 2002). Even birds not commonly thought to be particularly smart, such as pigeons, can memorize up to 725 visual patterns (von Fersen & Delius, 1989), learn to categorize images based on abstract features (Lubow, 1974; Watanabe, Sakamoto, & Wakita, 1995; Yamazaki, Aust, Huber, et al., 2007), and engage in tasks requiring transitive inference (von Fersen, Wynne, Delius, et al., 1992) and working memory (Güntürkün, 2005a, 2005b). Given the advanced cognitive abilities of birds and recent work implicating mammalian sleep in various aspects of learning and cognition (reviewed in Stickgold & Walker, 2007), it is conceivable that birds also evolved mammalian-like sleep states to maintain their heavily interconnected brains and advanced cognition.

We recently proposed that the independent evolution of SWS, in particular, was linked to the independent evolution of large, heavily interconnected brains capable of performing complex cognition in mammals and birds (Rattenborg, 2006a; Rattenborg et al., 2009). There are both mechanistic and functional components to this hypothesis. We first discuss the mechanistic basis of this hypothesis and then the functional implications. In mammals, SWS-related SWA reflects the large-scale, slow (<1 hertz) alternation of neocortical neuronal membrane potentials

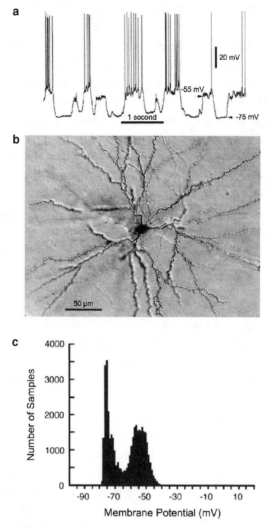

Figure 7.5. Spontaneous slow oscillations of the membrane potential in pallial neurons of an anesthetized pigeon. (a) Slow oscillations of the membrane potential between "up states" (−55 millivolts) with frequent action potentials and "down states" (−75 millivolts) with no action potentials recorded from a neuron (b) in the external pallium. (c) Frequency histogram of membrane potentials showing the tendency of a neuron in the nidopallium caudolateral to be either in the down state (−75 millivolts) or up state (−55 millivolts). (Modified from Reiner et al., 2001.)

between a hyperpolarized "down state" with no action potentials and a depolarized "up state" with action potentials occurring at a rate comparable to wakefulness (Steriade, 2006). Similar slow oscillations also occur in pigeons (Figure 7.5) (Reiner, Stern, & Wilson, 2001) and presumably form the basis for SWS-related SWA in birds.

The mammalian slow oscillation originates in the neocortex and is synchronized across large neuronal populations via corticocortical connections. In humans, the slow oscillation is most prominent in the heavily interconnected frontal cortex and propagates posteriorly as a traveling wave, presumably via corticocortical connections (Massimini, Huber, Ferrarelli, et al., 2004; Murphy, Riedner, Huber, et al., 2009). Multiple slow oscillations can originate simultaneously from different foci, thereby resulting in complex wave interactions that give rise to EEG spectral power across the SWA bandwidth (i.e., 0.5 to 4 hertz). Several lines of evidence indicate that the slow oscillation is synchronized via corticocortical connections. The synchronizing role of corticocortical connections has been shown through lesion studies, both in vitro and in vivo (Amzica & Steriade, 1995; Sanchez-Vives & McCormick, 2000; Timofeev, Grenier, Bazhenov, et al., 2000). Moreover, conditions that increase or decrease corticocortical connectivity in specific regions also increase or decrease SWA, respectively, in those regions (Miyamoto, Katagiri, & Hensch, 2003; Vyazovskiy & Tobler, 2005). Finally, in computer simulations of SWS, the removal of corticocortical connections (Hill & Tononi, 2005) or weakening of their strength (Esser, Hill, & Tononi, 2008) causes simulated neurons to oscillate more asynchronously, with the result that SWA is reduced.

Given the role of corticocortical connectivity in generating EEG SWA in mammals, the level of connectivity in the dorsal cortex of reptiles may explain why they lack high-amplitude slow waves during sleep. Indeed, their three-layered dorsal cortex is thought to be homologous only with layers I, V, and VI of the six-layered mammalian neocortex; reptiles apparently lack layers II and III, the layers with the most extensive corticocortical projections in mammals (reviewed in Medina & Reiner, 2000). Interestingly, during development, layers II and III arise from a subventicular zone (SVZ) of neuronal proliferation that is not present in reptiles (Martinez-Cerdeño, Noctor, & Kreigstein, 2006; Molnár, Métin, Stoykova, et al., 2006). Consequently this evolutionary innovation may explain why the mammalian neocortex shows extensive corticocortical connectivity and SWS-related SWA, whereas the reptilian dorsal cortex does not. If this hypothesis is correct, then birds, which show SWS-related SWA, should also show a high degree of connectivity. Indeed, although the avian brain does not exhibit a truly laminar cortical structure similar to the neocortex or dorsal cortex, the hyperpallium, which develops from the same embryonic pallial neural tissue as the neocortex and dorsal cortex (Jarvis et al., 2005), shows extensive "palliopallial" connectivity (Medina & Reiner, 2000). Other pallial regions, such as the mesopallium and nidopallium of the dorsal ventricular ridge, also show extensive interconnectivity when compared to homologous structures in reptiles (Butler & Cotterill, 2006; Tömböl, 1995). Like the neurons of layers II and III in the mammalian neocortex, neurons in this heavily interconnected region of the avian brain appear to originate from an SVZ of proliferation not found in reptiles (Martinez-Cerdeño et al., 2006; Molnár et al., 2006). Moreover, although additional comparative work is needed, the avian and mammalian SVZs seem to have evolved independently in each group. Consequently these differences in development and resulting inter-connectivity may explain why birds and mammals exhibit SWS-related SWA and reptiles do not. Furthermore, this interconnectivity is most likely responsible for mediating the complex cognitive abilities of birds and mammals.

The preceding discussion has focused on mechanistic reasons for the presence of SWS-related SWA in mammals and birds. This raises several compelling questions. Are slow waves simply functionless epiphenomena of heavily interconnected brains occurring in a sleep state otherwise no different from that in reptiles and other animals? Or, are slow waves an emergent property of heavily interconnected brains that serve secondarily evolved functions unique to mammals and birds? Certainly it is also possible that slow waves reflect a secondarily evolved *mechanism* to achieve an evolutionarily conserved function mediated by different mechanisms in animals lacking slow waves (Gilestro, Tononi, & Cirelli, 2009; Rattenborg et al., 2009). One intriguing possibility is that along with the evolution of complex brains and cognition, SWS evolved in mammals and birds to maintain their complex brains. Indeed, several hypotheses suggest that rather than being mere epiphenomena, slow waves and associated cellular processes play an active role in maintaining adaptive brain function during wakefulness (Benington, 2000; Benington & Frank, 2003; Jha, Jones, Coleman, et al., 2005; Krueger & Obál, 1993; Krueger, Rector, Roy, et al., 2008; Marshall, Helgadóttir, Mölle, et al., 2006; Sejnowski & Destexhe, 2000; Steriade & Timofeev, 2003; Stickgold & Walker, 2007; Tononi & Cirelli, 2003, 2006). Particularly appealing are hypotheses that specifically account for the presence of SWS homeostasis, as reflected in the relationship between SWA and time spent awake and asleep (i.e., process S) (Benington, 2000). For instance, Tononi and Cirelli's "synaptic homeostasis hypothesis" proposes that slow waves are directly involved in maintaining adaptive levels of synaptic strength in the neocortex, a process reflected in the homeostatic regulation of SWA (Tononi & Cirelli, 2003, 2006). According to this hypothesis, interaction with the environment during wakefulness causes an overall increase in synaptic strength and density that, if left unchecked, would saturate, leading to reduced space, increased energy demands, and a reduced ability to acquire new information. The slow network oscillations occurring during SWS are thought to remedy this problem in the following manner. Synaptic strength accumulated during wakefulness causes neurons to oscillate more synchronously and SWA to be greatest at the beginning of sleep. During SWS, however, the slow oscillations cause long-term depression and an overall downscaling of synaptic strength. Although overall synaptic strength declines during sleep, the relative strength of most synapses is retained, thereby preserving previously learned information. However, weak, newly formed synapses may be removed during downscaling. This process may reduce the signal-to-noise ratio in neural networks, and thereby account for some of the improvements in cognitive performance observed following sleep (Hill, Tononi, & Ghilardi, 2008; Huber et al., 2004). Finally, as synaptic strength declines during sleep, neurons oscillate less synchronously, with the result that SWA and downscaling diminish in a self-limiting manner.

The synaptic homeostasis hypothesis for SWS is supported by behavioral (Huber, et al., 2004), neurophysiological (Vyazovskiy, Cirelli, Pfister-Genskow, et al., 2008), and molecular evidence in mammals (Cirelli, 2006; Cirelli, Gutierrez, & Tononi, 2004; Cirelli, LaVoute, & Tononi, 2005; Huber, Tononi, & Cirelli, 2007; Vyazovskiy, et al., 2008). Compared to mammals, the evidence linking avian SWS to synaptic downscaling is limited. Nonetheless, the presence of (1) SWS homeostasis in birds (reviewed in Rattenborg et al., 2009), (2) evidence linking SWS to imprinting in chicks (*Gallus domesticus*) (Jackson, McCabe, Nicol, et al., 2008) and sleep in general to song learning in young male zebra finches (*Taeniopygia guttata*) (Derégnaucourt, Mitra, Fehér, et al., 2005; Margoliash, 2005; Shank & Margoliash, 2009), and (3) recent molecular data showing similar changes between wakefulness and sleep in the expression of plasticity-related genes in the brains of mammals and birds (Jones, Pfister-Genskow, Benca, et al., 2008b) is at least consistent with the notion that SWS may serve a similar function in birds. Moreover, given the absence of large-scale, slow, synchronous neuronal activity during sleep in reptiles and other animals, this function (or mechanisms for achieving this function) may have evolved independently in mammals and birds to maintain their large, heavily interconnected brains and associated complex cognitive abilities (Figure 7.6) (Rattenborg, 2006a; Rattenborg et al., 2009).

Finally, as with SWS, the convergent evolution of REM sleep may also be related to the convergent evolution of complex brains and cognition in mammals and birds. Although the role of REM sleep in synaptic downscaling, if one exists, is unclear, other mechanisms occurring during REM sleep appear to play a role in sleep-dependent plasticity in mammals (reviewed in Stickgold & Walker, 2007). Perhaps owing to a lack of research on the subject, evidence linking REM sleep and learning is minimal in birds. Nonetheless, an increase in REM sleep has been reported following imprinting in chicks (Solodkin, Cardona, & Corsi-Cabrera, 1985). As mentioned earlier, sleep in general also appears to be involved in song learning in zebra finches (Derégnaucourt et al., 2005; Shank & Margoliash, 2009). Additional research is clearly needed to determine the extent to which avian SWS and REM sleep are involved in plasticity.

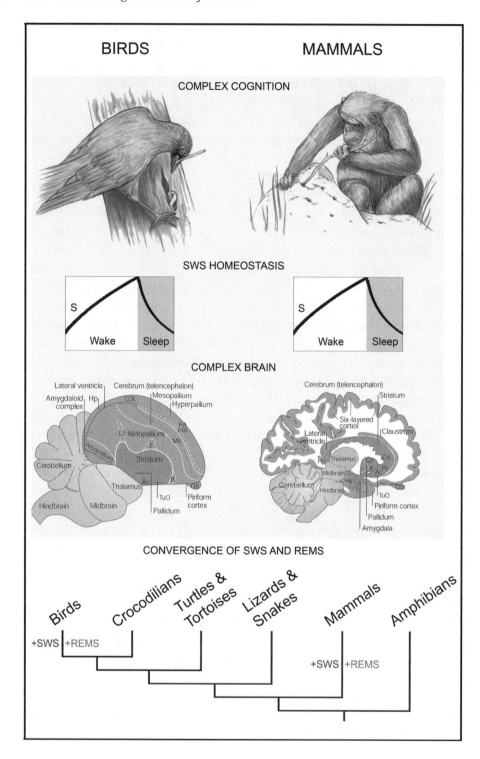

BIRDS   MAMMALS

COMPLEX COGNITION

SWS HOMEOSTASIS

COMPLEX BRAIN

CONVERGENCE OF SWS AND REMS

Concluding remarks

In addition to the similarities between avian and mammalian sleep, the points where they differ may also inform our understanding of the evolution and function of SWS and REM sleep. For instance, SWS-related spindles and hippocampal sharp waves and the hippocampal theta rhythm that occurs during mammalian REM sleep have not been observed in birds. Although differences in pallial architecture may make it more difficult to detect these phenomena in the avian EEG, it is also possible that they simply do not occur in birds. If confirmed with more targeted approaches, such differences are potentially very interesting, especially given that each of these electrophysiological events has been implicated in sleep-dependent plasticity in mammals (reviewed in Stickgold & Walker, 2007). Furthermore, if similar forms of sleep-dependent plasticity occur in birds despite the absence of electrophysiological phenomena comparable to those implicated in these processes in mammals, conclusions drawn from mammals may require reevaluation; or mammalian and avian sleep may perform the same functions but via different mechanisms. Comparisons of such divergent mechanisms may reveal

---

Figure 7.6. The proposed link between the evolution of complex brains, slow-wave sleep (SWS) homeostasis (process S), and complex cognition in birds and mammals. (See color Plate 1.) The phylogenetic tree (*bottom figure*) shows the convergent evolution of slow-wave sleep (SWS) and rapid-eye-movement (REM) sleep in birds and mammals. The proposed link between the convergent evolution of complex brains, SWS homeostasis (process S), and complex cognition in birds and mammals is depicted in the shaded panel (see text for details). The brain diagrams show the modern consensus view that the majority of the avian telencephalon (green) is derived from the same pallial embryonic neural tissue that gives rise to the mammalian neocortex (green) and functions in a similar manner despite the absence of the laminar cytoarchitecture found in the neocortex. Abbreviations: Ac, accumbens; B, basorostralis; Cd, caudate nucleus; CDL, dorsal lateral corticoid area; E, entopallium; GP, globus pallidus (i, internal segment; e, external segment); HA, hyperpallium apicale; Hp, hippocampus; IHA, interstitial hyperpallium apicale; L2, field L2; LPO, lobus parolfactorius; MV, mesopallium ventrale; OB, olfactory bulb; Pt, putamen; TuO, olfactory tubercle. Solid white lines are lamina (cell-sparse zones separating brain subdivisions). Large white areas in the human cerebrum are axon pathways called white matter. Dashed gray lines divide regions that differ by cell density or cell size; dashed white lines separate primary sensory neuron populations from adjacent regions. (The brain diagrams were adapted by permission from Macmillan Publishers Ltd: *Nature Reviews Neuroscience,* Jarvis et al., 2005. The drawings of tool use in a New Caledonian crow [*Corvus moneduloides*] and a chimpanzee [*Pan troglodytes*] by C. Cain are from Emery & Clayton [2004]. The mentality of crows: Convergent evolution of intelligence in corvids and apes, *Science, 306,* 1903–1907. Reprinted with permission from AAAS. Entire figure modified from Rattenborg et al., 2009.)

overriding principles that might otherwise remain obscure under an exclusively mammal-based research approach. Clearly additional comparative work on sleep in birds and other nonmammalian vertebrates may provide further insight into the functions of sleep in mammals, including humans.

References

Allada, R., & Siegel, J. M. (2008). Unearthing the phylogenetic roots of sleep. *Current Biology*, *18*, R670–R679.

Allman, J. M. (1999). *Evolving brains.* New York: W. H. Freeman.

Amlaner, C. J., & Ball, N. J. (1994). Avian sleep. In M. H. Kryger, T. Roth, & W. C. Dement (Eds.), *Principles and practice of sleep medicine* (2nd ed., pp. 81–94). Philadelphia: W. B. Saunders.

Amzica, F., & Steriade, M. (1995). Disconnection of intracortical synaptic linkages disrupts synchronization of a slow oscillation. *Journal of Neuroscience*, *15*, 4658–4677.

Ayala-Guerrero, F., Mexicano, G., & Ramos, J. I. (2003). Sleep characteristics in the turkey *Meleagris gallopavo*. *Physiology & Behavior*, *78*, 35–40.

Ball, N. J., Amlaner, C. J., Shaffery, J. P., & Opp, M. R. (1988). Asynchronous eye closure and unihemispheric quiet sleep of birds. In W. P. Koella, F. Obál, H. Schulz, & P. Visser (Eds.), *Sleep '86* (pp. 151–153). New York: Gustav Fischer.

Benington, J. H. (2000). Sleep homeostasis and the function of sleep. *Sleep*, *23*, 959–966.

Benington, J. H., & Frank, M. G. (2003). Cellular and molecular connections between sleep and synaptic plasticity. *Progress in Neurobiology*, *69*, 71–101.

Berger, R. J., & Phillips, N. H. (1994). Constant light suppresses sleep and circadian rhythms in pigeons without consequent sleep rebound in darkness. *American Journal of Physiology*, *267*, R945–R952.

Berger, R. J., & Phillips, N. H. (1995). Energy conservation and sleep. *Behavioral Brain Research*, *69*, 65–73.

Buchet, C., Dewasmes, G., & Le Maho, Y. (1986). An electrophysiological and behavioral study of sleep in emperor penguins under natural ambient conditions. *Physiology and Behavior*, *38*(3), 331–335.

Butler, A. B. (2008). Evolution of brains, cognition, and consciousness. *Brain Research Bulletin*, *75*, 442–449.

Butler, A. B., & Cotterill, R. M. (2006). Mammalian and avian neuroanatomy and the question of consciousness in birds. *Biological Bulletin*, *211*, 106–127.

Borbély, A. A., Tobler, I., & Hanagasioglu, M. (1984). Effect of sleep deprivation on sleep and EEG power spectra in the rat. *Behavioural Brain Research*, *14*, 171–182.

Capellini, I., Barton, R. A., McNamara, P., Preston, B. T., & Nunn, C. L. (2008). Phylogenetic analysis of the ecology and evolution of mammalian sleep. *Evolution*, *62*, 1764–1776.

Cirelli, C. (2006). Cellular consequences of sleep deprivation in the brain. *Sleep Medicine Reviews*, *10*, 307–321.

Cirelli, C., & Bushey, D. (2008). Sleep and wakefulness in *Drosophila melanogaster*. *Annals of the New York Academy of Sciences*, *1129*, 323–329.

Cirelli, C., & Tononi, G. (2008). Is sleep essential? *Public Library of Science Biology*, *6*, E216.

Cirelli, C., Gutierrez, C. M., & Tononi, G. (2004). Extensive and divergent effects of sleep and wakefulness on brain gene expression. *Neuron*, *41*, 35–43.

Cirelli, C., LaVaute, T. M., & Tononi, G. (2005). Sleep and wakefulness modulate gene expression in *Drosophila*. *Journal of Neurochemistry*, *94*, 1411–1419.

Derégnaucourt, S., Mitra, P. P., Fehér, O., Pytte, C., & Tchernichovski, O. (2005). How sleep affects the developmental learning of bird song. *Nature*, *433*, 710–716.

Dewasmes, G., Cohen-Adad, F., Koubi, H., & Le Maho, Y. (1985). Polygraphic and behavioral study of sleep in geese: Existence of nuchal atonia during paradoxical sleep. *Physiology and Behavior*, *35*, 67–73.

Eiland, M. M., Lyamin, O. I., & Siegel, J. M. (2001). State-related discharge of neurons in the brainstem of freely moving box turtles (*Terrapene carolina major*). *Archives of Italian Biology*, *139*, 23–36.

Emery, N. J., & Clayton, N. S. (2004). The mentality of crows: Convergent evolution of intelligence in corvids and apes. *Science*, *306*, 1903–1907.

Emery, N. J., & Clayton, N. S. (2005). Evolution of the avian brain and intelligence. *Current Biology*, *15*, R946–R950.

Esser, S. K., Hill, S. L., & Tononi, G. (2008). Sleep homeostasis and cortical synchronization: I. Modeling the effects of synaptic strength on sleep slow waves. *Sleep*, *30*, 1617–1630.

Fuchs, T. (2006). *Brain-behavior adaptations to sleep loss in the nocturnally migrating Swainson's thrush* (*Catharus ustulatus*). Unpublished doctoral dissertation, Bowling Green State University, Bowling Green, OH.

Gaztelu, J. M., García-Austt, E., & Bullock, T. H. (1991). Electrocorticograms of hippocampal and dorsal cortex of two reptiles: Comparison with possible mammalian homologs. *Brain, Behavior, and Evolution*, *37*, 144–160.

Gilestro, G. F., Tononi, G., & Cirelli, C. (2009). Widespread changes in synaptic markers as a function of sleep and wakefulness in *Drosophila*. *Science*, *324*, 109–112.

Güntürkün, O. (2005a). The avian "prefrontal cortex" and cognition. *Current Opinion in Neurobiology*, *15*, 686–693.

Güntürkün, O. (2005b). Avian and mammalian "prefrontal cortices": Limited degrees of freedom in the evolution of the neural mechanisms of goal-state maintenance. *Brain Research Bulletin*, *66*, 311–316.

Hahn, T. T., Sakmann, B., & Mehta, M. R. (2006). Phase-locking of hippocampal interneurons' membrane potential to neocortical up-down states. *Nature Neuroscience*, *9*, 1359–1361.

Hartse, K. M. (1994). Sleep in insects and nonmammalian vertebrates. In M. H. Kryger, T. Roth, & W. C. Dement (Eds.), *Principles and practice of sleep medicine* (2nd ed., pp. 95–104). Philadelphia: W. B. Saunders.

Hartse, K. M., & Rechtschaffen, A. (1982). The effect of amphetamine, nembutal, alpha-methyl-tyrosine, and parachlorophenylalanine on the sleep-related spike activity of the tortoise (*Geochelone carbonaria*) and on the cat ventral hippocampus spike. *Brain, Behavior, and Evolution*, *2*, 199–222.

Heller, H. C., Graf, R., & Rautenberg, W. (1983). Circadian and arousal state influences on thermoregulation in the pigeon. *American Journal of Physiology*, *245*, R321–R328.

Hendricks, J. C., & Sehgal, A. (2004). Why a fly? Using *Drosophila* to understand the genetics of circadian rhythms and sleep. *Sleep*, *27*, 334–342.

Hendricks, J. C., Finn, S. M., Panckeri, K. A., Chavkin, J., Williams, J. A., Sehgal, A., et al. (2000). Rest in *Drosophila* is a sleep-like state. *Neuron, 25,* 129–138.

Hill, S., & Tononi, G. (2005). Modeling sleep and wakefulness in the thalamocortical system. *Journal of Neurophysiology, 93,* 1671–1698.

Hill, S., Tononi, G., & Ghilardi, M. F. (2008). Sleep improves the variability of motor performance. *Brain Research Bulletin, 76,* 605–611.

Hohtola, E., Rintamaki, H., & Hissa, R. (1980). Shivering and ptiloerection as complementary cold defense responses in the pigeon during sleep and wakefulness. *Journal of Comparative Physiology, 136,* 77–81.

Huber, R., Deboer, T., & Tobler, I. (2000). Effects of sleep deprivation on sleep and sleep EEG in three mouse strains: Empirical data and simulations. *Brain Research, 857,* 8–19.

Huber, R., Tononi, G., & Cirelli, C. (2007). Exploratory behavior, cortical BDNF expression, and sleep homeostasis. *Sleep, 30,* 129–139.

Huber, R., Ghilardi, M. F., Massimini, M., & Tononi, G. (2004). Local sleep and learning. *Nature, 430,* 78–81.

Isomura, Y., Sirota, A., Ozen, S., Montgomery, S., Mizuseki, K., Henze, D. A., et al. (2006). Integration and segregation of activity in entorhinal-hippocampal subregions by neocortical slow oscillations. *Neuron, 52,* 871–882.

Jackson, C., McCabe, B. J., Nicol, A. U., Grout, A. S., Brown, M. W., & Horn, G. (2008). Dynamics of a memory trace: Effects of sleep on consolidation. *Current Biology, 18,* 393–400.

Jarvis, E. D., Güntürkün, O., Bruce, L., Csillag, A., Karten, H., Kuenzel, W., et al. (2005). Avian brains and a new understanding of vertebrate brain evolution. *Nature Reviews Neuroscience, 6,* 151–159.

Jerison, H. J. (2001). The evolution of neural and behavioral complexity. In G. Roth & M. F. Wullimann (Eds.), *Brain evolution and cognition* (pp. 523–553). New York: Wiley.

Jha, S. K., Jones, B. E., Coleman, T., Steinmetz, N., Law, C. T., Griffin, G., et al. (2005). Sleep-dependent plasticity requires cortical activity. *Journal of Neuroscience, 25,* 9266–9274.

Jones, S. G., Vyazovskiy, V. V., Cirelli, C., Tononi, G., & Benca, R. M. (2008a). Homeostatic regulation of sleep in the white-crowned sparrow (*Zonotrichia leucophrys gambelii*). *BioMed Central Neuroscience, 9,* 47.

Jones, S., Pfister-Genskow, M., Benca, R. M., & Cirelli, C. (2008b). Molecular correlates of sleep and wakefulness in the brain of the white-crowned sparrow. *Journal of Neurochemistry, 105,* 46–62.

Kirsch, J. A., Güntürkün, O., & Rose, J. (2008). Insight without cortex: Lessons from the avian brain. *Consciousness and Cognition, 17,* 475–483.

Klein, M., Michel, F., & Jouvet, M. (1964). Etude polygraphique du sommeil chez les oiseaux. [Polygraphic study of sleep with birds]. *Comptes Rendus des Seances de la Societe de Biologie et de ses Filiales, 158,* 90–103.

Krueger, J. M., & Obál, F. Jr. (1993). A neuronal group theory of sleep function. *Journal of Sleep Research, 2,* 63–69.

Krueger, J. M., Rector, D. M., Roy, S., Van Dongen, H. P. A., Belenky, G., & Panksepp, J. (2008). Sleep as a fundamental property of neuronal assemblies. *Nature Reviews Neuroscience, 9,* 910–919.

Lesku, J. A., Roth, II, T. C., Amlaner, C. J., & Lima, S. L. (2006). A phylogenetic analysis of sleep architecture in mammals: The integration of anatomy, physiology, and ecology. *American Naturalist*, *168*, 441–453.

Lesku, J. A., Roth, T. C., Rattenborg, N. C., Amlaner, C. J., & Lima, S. L. (in press). History and future of comparative analyses in sleep research. *Neuroscience and Biobehavioral Reviews*, doi:10.1016/j.neubiorev.2009.04.002.

Lima, S. L., & Rattenborg, N. C. (2007). A behavioural shutdown can make sleeping safer: A strategic perspective on the function of sleep. *Animal Behaviour*, *74*, 189–197.

Lima, S. L., Rattenborg, N. C., Lesku, J. A., & Amlaner, C. J. (2005). Sleeping under the risk of predation. *Animal Behaviour*, *70*, 723–736.

Lorenzo, D., & Velluti, J. C. (2004). Noradrenaline decreases spike voltage threshold and induces electrographic sharp waves in turtle medial cortex in vitro. *Brain, Behavior, and Evolution*, *64*, 104–114.

Lorenzo, D., Macadar, O., & Velluti, J. C. (1999). Origin and propagation of spontaneous electrographic sharp waves in the in vitro turtle brain: A model of neuronal synchronization. *Clinical Neurophysiology*, *110*, 1535–1544.

Low, P. S., Shank, S. S., Sejnowski, T. J., & Margoliash, D. (2008). Mammalian-like features of sleep structure in zebra finches. *Proceedings of the National Academy of Sciences of the United States of America*, *105*, 9081–9086.

Lubow, R. E. (1974). High-order concept formation in the pigeon. *Journal of Experimental Analysis of Behavior*, *21*, 475–483.

Lyamin, O. I., Manger, P. R., Ridgway, S. H., Mukhametov, L. M., & Siegel, J. M. (2008). Cetacean sleep: An unusual form of mammalian sleep. *Neuroscience and Biobehavioral Reviews*, *32*, 1451–1484.

Margoliash, D. (2005). Song learning and sleep. *Nature Neuroscience*, *8*, 546–548.

Marshall, L., Helgadóttir, H., Mölle, M., & Born, J. (2006). Boosting slow oscillations during sleep potentiates memory. *Nature*, *444*, 610–613.

Martínez-Cerdeño, V., Noctor, S. C., & Kriegstein, A. R. (2006). The role of intermediate progenitor cells in the evolutionary expansion of the cerebral cortex. *Cerebral Cortex*, *16*(1), i152–i161.

Martinez-Gonzalez, D., Lesku, J. A., & Rattenborg, N. C. (2008). Increased EEG spectral power density during sleep following short-term sleep deprivation in pigeons (*Columba livia*): Evidence for avian sleep homeostasis. *Journal of Sleep Research*, *17*, 140–153.

Massimini, M., Huber, R., Ferrarelli, F., Hill, S., & Tononi, G. (2004). The sleep slow oscillation as a traveling wave. *Journal of Neuroscience*, *24*, 6862–6870.

Medina, L., & Reiner, A. (2000). Do birds possess homologues of mammalian primary visual, somatosensory, and motor cortices? *Trends in Neuroscience*, *23*, 1–12.

Mignot, E. (2008). Why we sleep: The temporal organization of recovery. *Public Library of Science Biology*, *6*, E106.

Miyamoto, H., Katagiri, H., & Hensch, T. (2003). Experience-dependent slow-wave sleep development. *Nature Neuroscience*, *6*, 553–554.

Mölle, M., Yeshenko, O., Marshall, L., Sara, S. J., & Born, J. (2006). Hippocampal sharp wave-ripples linked to slow oscillations in rat slow-wave sleep. *Journal of Neurophysiology*, *96*, 62–70.

Molnár, Z., Métin, C., Stoykova, A., Tarabykin, V., Price, D. J., Francis, F., et al. (2006). Comparative aspects of cerebral cortical development. *European Journal of Neuroscience*, *23*, 921–934.

Murphy, M., Riedner, B. A., Huber, R., Massimini, M., Ferrarelli, F., & Tononi, G. (2009). Source modeling sleep slow waves. *Proceedings of the National Academy of Sciences of the USA*, *106*, 1608–1613.

Newman, S. M., Paletz, E. M., Rattenborg, N. C., Obermeyer, W. H., & Benca, R. M. (2008). Sleep deprivation in the pigeon (*Columba livia*) using the disk-over-water method. *Physiology and Behavior*, *93*, 50–58.

Nicol, S. C., Andersen, N. A., Phillips, N. H., & Berger, R. J. (2000). The echidna manifests typical characteristics of rapid eye movement sleep. *Neuroscience Letters*, *283*, 49–52.

Nitz, D. A., van Swinderen, B., Tononi, G., & Greenspan, R. J. (2002). Electrophysiological correlates of rest and activity in *Drosophila melanogaster*. *Current Biology*, *12*, 1934–1940.

Ookawa, T. (2004). The electroencephalogram and sleep in the domestic chicken. *Avian and Poultry Biology Reviews*, *15*, 1–8.

Ookawa, T., & Gotoh, J. (1964). Electroencephalographic study of chickens; Periodic recurrence of low voltage and fast waves during behavioral sleep. *Poultry Science*, *43*, 1603–1604.

Ookawa, T., & Gotoh, J. (1965). Electroencephalogram of the chicken recorded from the skull under various conditions. *Journal of Comparative Neurology*, *124*, 1–14.

Pepperberg, I. M. (2002). In search of King Solomon's ring: Cognitive and communicative studies of grey parrots (*Psittacus erithacus*). *Brain Behavior and Evolution*, *59*, 54–67.

Peters, J., Vonderahe, A., & Schmid, D. (1965). Onset of cerebral electrical activity associated with behavioral sleep and attention in the developing chick. *Journal of Experimental Zoology*, *160*, 255–262.

Prior, H., Schwarz, A., & Güntürkün, O. (2008). Mirror-induced behavior in the magpie (*Pica pica*): Evidence of self-recognition. *Public Library of Science Biology*, *6*, E202.

Rattenborg, N. C. (2006a). Evolution of slow-wave sleep and palliopallial connectivity in mammals and birds: A hypothesis. *Brain Research Bulletin*, *69*, 20–29.

Rattenborg, N. C. (2006b). Do birds sleep in flight? *Naturwissenschaften*, *93*, 413–425.

Rattenborg, N. C. (2007). Response to commentary on evolution of slow-wave sleep and palliopallial connectivity in mammals and birds: A hypothesis. *Brain Research Bulletin*, *72*, 187–193.

Rattenborg, N. C., Lima, S. L., & Amlaner, C. J. (1999a). Facultative control of avian unihemispheric sleep under the risk of predation. *Behavioural Brain Research*, *105*, 163–172.

Rattenborg, N. C., Lima, S. L., & Amlaner, C. J. (1999b). Half-awake to the risk of predation. *Nature*, *397*, 397–398.

Rattenborg, N. C., Amlaner, C. J., & Lima, S. L. (2000). Behavioral, neurophysiological, and evolutionary perspectives on unihemispheric sleep. *Neuroscience and Biobehavioral Reviews*, *24*, 817–842.

Rattenborg, N. C., Amlaner, C. J., & Lima, S. L. (2001). Unilateral eye closure and interhemispheric EEG asymmetry during sleep in the pigeon (*Columba livia*). *Brain, Behavior, and Evolution*, *58*, 323–332.

Rattenborg, N. C., Mandt, B. H., Obermeyer, W. H., Winsauer, P. J., Huber, R., Wikelski, M., et al. (2004). Migratory sleeplessness in the white-crowned sparrow (*Zonotrichia leucophrys gambelii*). *Public Library of Science Biology*, *2*, E212.

Rattenborg, N. C., Martinez-Gonzalez, D., Lesku, J. A., & Scriba M. (2008a). A bird's-eye view on the function of sleep. *Science*, *322*, 527.

Rattenborg, N. C., Voirin, B., Vyssotski, A. L., Kays, R. W., Spoelstra, K., Kuemmeth, F., et al. (2008b). Sleeping outside the box: Electroencephalographic measures of sleep in sloths inhabiting a rainforest. *Biology Letters*, 4, 402–405.

Rattenborg, N. C., Martinez-Gonzalez, D., & Lesku, J. A. (2009). Avian sleep homeostasis: Convergent evolution of complex brains, cognition, and sleep functions in mammals and birds. *Neuroscience and Biobehavioral Reviews*, 33, 253–270.

Reiner, A., Stern, E. A., & Wilson, C. J. (2001). Physiology and morphology of intratelencephalically projecting corticostriatal-type neurons in pigeons as revealed by intracellular recording and cell filling. *Brain, Behavior, and Evolution*, 58, 101–114.

Rial, R. V., Nicolau, M. C., Gamundi, A., Akaârir, M., Garau, C., Aparicio, S., et al. (2007). Comments on evolution of sleep and the palliopallial connectivity in mammals and birds. *Brain Research Bulletin*, 72, 183–186.

Roth, II, T. C., Lesku, J. A., Amlaner, C. J., & Lima, S. L. (2006). A phylogenetic analysis of the correlates of sleep in birds. *Journal of Sleep Research*, 15, 395–402.

Sanchez-Vives, M. V., & McCormick, D. A. (2000). Cellular and network mechanisms of rhythmic recurrent activity in neocortex. *Nature Neuroscience*, 3, 1027–1034.

Sejnowski, T. J., & Destexhe, A. (2000). Why do we sleep? *Brain Research*, 886, 208–223.

Shank, S. S., & Margoliash, D. (2009). Sleep and sensorimotor integration during early vocal learning in a songbird. *Nature*, 458, 73–77.

Shaw, P. J., Cirelli, C., Greenspan, R. J., & Tononi, G. (2000). Correlates of sleep and waking in *Drosophila melanogaster*. *Science*, 287, 1834–1837.

Shimizu, T. (2008). The avian brain revisited: Anatomy and evolution of the telencephalon. In S. Watanabe & M. A. Hofman (Eds.), *Integration of comparative neuroanatomy and cognition* (pp. 55–73). Tokyo: Keio University Press.

Siegel, J. M., Manger, P. R., Nienhuis, R., Fahringer, H. M., & Pettigrew, J. D. (1996). The echidna *Tachyglossus aculeatus* combines REM and non-REM aspects in a single sleep state: Implications for the evolution of sleep. *Journal of Neuroscience*, 16, 3500–3506.

Siegel, J. M., Manger, P. R., Nienhuis, R., Fahringer, H. M., Shalita, T., & Pettigrew, J. D. (1999). Sleep in the platypus. *Neuroscience*, 91, 391–400.

Sirota, A., Csicsvari, J., Buhl, D., & Buzsáki, G. (2003). Communication between neocortex and hippocampus during sleep in rodents. *Proceedings of the National Academy of Sciences of the United States of America*, 100, 2065–2069.

Solodkin, M., Cardona, A., & Corsi-Cabrera, M. (1985). Paradoxical sleep augmentation after imprinting in the domestic chick. *Physiology and Behavior*, 35, 343–348.

Stahel, C. D., Megirian, D., & Nicol, S. C. (1984). Sleep and metabolic rate in the little penguin (*Eudyptula minor*). *Journal of Comparative Physiology, B: Biochemical, Systemic, and Environmental Physiology*, 154, 487–494.

Steriade, M. (2006). Grouping of brain rhythms in corticothalamic systems. *Neuroscience*, 137, 1087–1106.

Steriade, M., & Timofeev, I. (2003). Neuronal plasticity in thalamocortical networks during sleep and waking oscillations. *Neuron*, 37, 563–576.

Steriade, M., Amzica, F., & Contreras, D. (1994). Cortical and thalamic cellular correlates of electroencephalographic burst-suppression. *Electroencephalography and Clinical Neurophysiololgy*, 90, 1–16.

Stickgold, R., & Walker, M. P. (2007). Sleep-dependent memory consolidation and reconsolidation. *Sleep Medicine, 8,* 331–343.

Szymczak, J. T. (1987). Daily distribution of sleep states in the rook (*Corvus frugilegus*). *Journal of Comparative Physiology, A: Sensory, Neural and Behavioral Physiology, 161,* 321–327.

Szymczak, J. T. (1989). Influence of environmental temperature and photoperiod on temporal structure of sleep in corvids. *Acta Neurobiologiae Experimentalis, 49,* 359–366.

Szymczak, J. T., Helb, H. W., & Kaiser, W. (1993). Electrophysiological and behavioral correlates of sleep in the blackbird (*Turdus merula*). *Physiology and Behavior, 53,* 1201–1210.

Timofeev, I., Grenier, F., Bazhenov, M., Sejnowski, T. J., & Steriade, M. (2000). Origin of slow cortical oscillations in deafferented cortical slabs. *Cerebral Cortex, 10,* 1185–1199.

Tobler, I. (2005). Phylogeny of sleep regulation. In M. H. Kryger, T. Roth, & W. C. Dement (Eds.), *Principles and practice of sleep medicine* (4th ed., pp. 77–90). Philadelphia: W. B. Saunders.

Tobler, I., & Borbély, A. A. (1988). Sleep and EEG spectra in the pigeon (*Columba livia*) under baseline conditions and after sleep-deprivation. *Journal of Comparative Physiology, A: Sensory, Neural and Behavioral Physiology, 163,* 729–738.

Tobler, I., & Jaggi, K. (1987). Sleep and EEG spectra in the Syrian hamster (*Mesocricetus auratus*) under baseline conditions and following sleep deprivation. *Journal of Comparative Physiology, A: Sensory, Neural and Behavioral Physiology, 161,* 449–459.

Tömböl, T. (1995). *Golgi structure of telencephalon of chicken.* Budapest, Hungary: Semmelweis University Medical School.

Tononi, G., & Cirelli, C. (2003). Sleep and synaptic homeostasis: A hypothesis. *Brain Research Bulletin, 62,* 143–150.

Tononi, G., & Cirelli, C. (2006). Sleep function and synaptic homeostasis. *Sleep Medicine Reviews, 10,* 49–62.

van Twyver, H., & Allison, T. (1972). A polygraphic and behavioral study of sleep in the pigeon (*Columba livia*). *Experimental Neurology, 35,* 138–153.

Velluti, J. C., Russo, R. E., Simini, F., & Garcia-Austt, E. (1991). Electroencephalogram in vitro and cortical transmembrane potentials in the turtle (*Chrysemys d'orbigny*). *Brain, Behavior, and Evolution, 38,* 7–19.

von Fersen, L., Wynne, C. D. L., Delius, J. D., & Staddon, J. E. R. (1992). Transitive inference formation in pigeons. *Journal of Experimental Psychology: Animal Behavior Processes, 17,* 334–341.

von Fersen, L., & Delius, J. D. (1989). Long-term retention of many visual patterns by pigeons. *Ethology, 82,* 141–155.

Vyazovskiy, V. V., & Tobler, I. (2005). Regional differences in NREM sleep slow-wave activity in mice with congenital callosal dysgenesis. *Journal of Sleep Research, 14,* 299–304.

Vyazovskiy, V. V., Achermann, P., & Tobler, I. (2007). Sleep homeostasis in the rat in the light and dark period. *Brain Research Bulletin, 74,* 37–44.

Vyazovskiy, V. V., Cirelli, C., Pfister-Genskow, M., Faraguna, U., & Tononi, G. (2008). Molecular and electrophysiological evidence for net synaptic potentiation in wake and depression in sleep. *Nature Neuroscience, 11,* 200–208.

Vyssotski, A. L., Serkov, A. N., Itskov, P. M., Dell'Omo, G., Latanov, A. V., Wolfer, D. P., et al. (2006). Miniature neurologgers for flying pigeons: Multichannel EEG and action and field potentials in combination with GPS recording. *Journal of Neurophysiology, 95,* 1263–1273.

Walker, J. M., & Berger, R. J. (1980). Sleep as an adaptation for energy conservation functionally related to hibernation and shallow torpor. *Progress in Brain Research, 53*, 255–278.

Watanabe, S., Sakamoto, J., & Wakita, M. (1995). Pigeons' discrimination of paintings by Monet and Picasso. *Journal of the Experimental Analysis of Behavior, 63*, 165–174.

Weimerskirch, H., Chastel, O., Barbraud, C., & Tostain, O. (2003). Frigatebirds ride high on thermals. *Nature, 421*, 333–334.

Weimerskirch, H., Le Corre, M., Jaquemet, S., Potier, M., & Marsac, F. (2004). Foraging strategy of a top predator in tropical waters: Great frigatebirds in the Mozambique Channel. *Marine Ecology Progress Series, 275*, 297–308.

Yamazaki, Y., Aust, U., Huber, L., Hausmann, M., & Güntürkün, O. (2007). Lateralized cognition: Asymmetrical and complementary strategies of pigeons during discrimination of the "human concept." *Cognition, 104*, 315–344.

Zepelin, H., Hartzer, M. K., & Pendergast, S. (1998). Saccadic oscillations in the sleep of birds. *Sleep, 21*, 6.

Zepelin, H., Siegel, J. M., & Tobler, I. (2005). Mammalian sleep. In M. H. Kryger, T. Roth, & W. C. Dement (Eds.), *Principles and practice of sleep medicine* (4th ed., pp. 91–100). Philadelphia: W. B. Saunders.

8

The evolution of wakefulness: From reptiles to mammals

RUBEN V. RIAL, MOURAD AKAÂRIR, ANTONI GAMUNDÍ,
M. CRISTINA NICOLAU, AND SUSANA ESTEBAN

Introduction

The evolution of sleep has been the subject of several studies and reviews (Allison & Cicchetti, 1976; Allison & Van Twyver, 1970; Hartse, 1994; Karmanova, 1982; Meddis, 1983; Monnier, 1980; Tauber, 1974). However, corresponding studies on the evolution of wakefulness have been fewer (Esteban, Nicolau, Gamundí, et al., 2005; Nicolau, Akaârir, Gamundí, et al., 2000), despite a number of reasons supporting the greater importance of waking in animal adaptation. First of all, waking and sleep are inseparable, an obvious assertion that, notwithstanding, has been ignored in most reviews (see, for instance, Zepelin, 1994; Zepelin & Rechtschaffen, 1974; Zepelin, Siegel, & Tobler, 2005). These reviews compute correlations between the main traits of sleep and ecological variables while forgetting that high correlations of a given trait with total sleep time also imply high correlations with waking time. The present review proposes a change of paradigm from sleep centeredness to waking centeredness.

Let us give an example of the paradigm change: there might be two possible viewpoints related to the high danger of a particular species' exposure within an environment, namely:

1. Sleep is a dangerous state. Therefore natural selection must have reduced sleep in dangerous environments.
2. Alertness is necessary to cope with danger. Therefore natural selection must have increased waking time in dangerous environments.

This research was supported by a grant of the Spanish Ministerio de Ciencia y Tecnologia, BFI 2002-04583-C02-029 and SAF2007-66878-C02-02. Author's contribution: This report was written under the direction of R. Rial. All remaining authors have equally contributed to its final form.

The difference between the two alternatives might seem subtle: but the former focuses on sleep as a key adaptive factor, while the latter is waking-centered. From one perspective, the waking-centered viewpoint should be preferred, as animals have the most interesting behavior during waking, while the most conspicuous phenotypic sleep trait consists in resting immobile during (almost) 99% of the time.

Nevertheless, sleep scholars have a lot of good reasons for their narrowly focused interest, as sleep troubles constitute a deep medical problem. However, the sleep-centered viewpoint probably causes some significant distortions. We discuss some of them in the remainder of this chapter.

It may be that the scarcity of studies devoted to the evolution of wakefulness is due to the obvious advantages of being awake. However, this is not quite evident. As D. C. Dennett (1995) has stated, "...but why does sleep need a 'clear biological function' at all? It is being awake that needs an explanation. Mother Nature economizes where she can. If we could get away with it, we'd sleep our entire life. That is what trees do, after all..." (p. 339). So, there are live beings that do well without waking at all, and the need of waking is not so obvious. But animals – unlike trees – have selected the ability to move and make an active search for food, reproductive partners, and escape from predation. Thus two solutions of the problem of being alive exist: to "sleep" during the whole life and to be awake. Plants selected the former solution; animals, the latter.

The animal solution has its limitations. It is known that the maximal thermo-dynamic stability always produces cycles in complex systems (Prigogine & Nicolis, 1977), and this can be related to the known exuberance of rhythms observed in live beings. Moreover, the biological rhythms allow for a better use of the environmental resources and an optimal distribution of the territory. Indeed, two species with a similar lifestyle may share a territory without competing if they have different time schedules. These properties of rest–activity cycles determine their well-recognized adaptiveness (Aschoff, 1964). In fact, all live beings are under high evolutionary pressure to find a unique ecological niche to minimize predation and competition, a pressure that leads to the distribution of biological time into periods of activity and rest. Animals that failed to be specialists in a particular temporal niche would disappear, as they would die without descendants. Thus the cyclic organization of rest and activity has an undisputed adaptive value.* Both rest and activity appear to be coupled to geophysical cycles; this is particularly true in ectothermic animals, whose rest is obligatory during the cold nocturnal phase of the cycle.

Mammalian sleep occupies most of the resting period and shares with rest a number of features. Therefore, if rest is an essential part of the life in all animals, and if sleep is essentially rest, rest and sleep should share the same primary function, which we have called the trivial function of sleep (Rial, Nicolau, Gamundi, et al., 2007). However, the sleep of mammals differs from that of birds and both are different from mere rest, because sleep has a number of additional features. As the function of rest is well recognized, we advocate that the search for the function of sleep should be restricted to the differential traits present in sleeping animals and absent in those that only rest. What is needed is an explanation for the high brain activity observed during sleep, the two phases, the eye movements and twitches, the changes in the homeostatic regulation, the dreams, and so on. In other words, what should be explained is why the resting state of mammals and birds has been supplemented by the unexpected and complex features of sleep.

Most likely, the complexity of sleep was the cause of the famous statement, currently attributed to A. Rechtschaffen (1971), that "if sleep does not serve an absolute vital function, then it is the biggest mistake the evolutionary process ever made." Thirty-seven years later, this statement remains without a widely accepted answer. However, the lack of an adequate answer might be due to the inappropriateness of the question instead of the difficulty in finding an answer. It is obvious that adaptationism pervades the statement. According to Gould and Levontin (1979), the extreme of adaptationism – panglossianism – considers that every anatomical and functional detail observable in a live being is an adaptation. This simplistic viewpoint has now been abandoned, as many of the traits observed are acknowledged to be neutral (Kimura, 1983). Moreover, it is well accepted that adaptation is an onerous concept (Williams, 1966), as it is too easy to imagine adaptations for a given trait and even for its opposite. Therefore the burden of the proof of a particular adaptation must always be put on its defender and, in the absence of such a proof, adaptation should be considered as nonexistent.

The argument of complexity has a long history in evolutionary thinking (Dawkins, 1986). However, complexity by itself does not allow one to presume a vital function, as argued by Rial et al. (2007); they used noncoding DNA as a counterexample, which is undoubtedly complex and probably lacks any primary function. Let us remark that noncoding DNA shares with sleep an extremely variable phylogenetic distribution. Despite the complexity of sleep, most traits distinguishing sleep from simple motor rest could have no function at all without implying any evolutionary mistake.

Let us go back to the example of the paradigm change we introduced earlier, where there is an implicit misconception: the alleged dangerousness of sleep. Many sleep researchers have stated that a sleeping animal is helpless and have deduced from that claim the proposition that to be asleep is risky. However, it is well known that victims of predation are the very young, the sick, and the old, even when they are fully awake. This is consistent with the well-accepted coevolution of predator and prey (Abrams, 2000). No prey animal can afford to sleep when its predators are active. The alleged vulnerability of an animal during sleep confuses cause and effect: primarily, sleep does not *cause* vulnerability; quite the opposite, vulnerability *causes* reduced sleep – indeed, we should formulate it even better as "vulnerability *leads to* increased waking."

Another alleged negative consequence of sleep is that the total sleeping time must be subtracted from other more useful tasks, such as finding food or reproductive partners. However, this is another instance of confusion between cause and effect: no live being would trade eating or reproductive efficiency for sleep; as soon as an essential need arises, sleep is put aside. Humans who reduce their sleep to achieve their goals constitute proof that sleep is trivial in face of the superior need to be awake, which should become the center of the paradigm.

The sleep rebound effect observed after deprivation provides a seemingly powerful argument that sleep fills an essential function. However, this could also be a mechanism to maintain the rest–activity cycles that might have been jeopardized after the development of homeothermy. In nature, cold-blooded animals never have the possibility of missing a resting period. This possibility arose only from the thermal regulatory freedom acquired by homeothermic animals, and it can be postulated that powerful mechanisms should have appeared to avoid the neglect of rest. Although the homeostatic regulation of sleep could in principle work against the circadian one (Benington, 2000), this never happens in practice, as the impossibility of a permanent adaptation to inverted light–dark activity cycles has been shown to occur in humans. Hence, the relative weakness of the homeostatic regulation, when compared with the circadian one, could mean that the sleep rebound could be a short-term signal to restore the most essential long-term periodic organization of activity and rest.

A result of the overconfidence of the people who rely on the complexity argument, the alleged inconveniences of sleep and the well-recognized need to recover lost sleep, is that many researchers feel that some benefit should be obtained during sleep. Many hypotheses have been proposed to show such benefit. Among them, the supposed consolidation of memory and learning is gaining strength. Certainly some empirical results show impairments in memory retention after sleep loss and even improvements after a good night of sleep. However, the accumulation of evidence may never serve to achieve a definitive proof. The number of empirical results showing that the earth is flat is uncountable, but a single counterexample is enough to falsify the hypothesis. In spite of whatever evidence may have been previously accumulated, we know that the earth is not flat; similarly, nobody will convince us that humans who sleep just 8 hours a day have less learning capacity than a cat that sleeps 16 hours daily, not to mention the platypus. Meanwhile, the empirical evidence thus far obtained is weak and even contradictory (Coenen & Van Luijtelaar, 1985; Siegel, 2001; Vertes 2004; Vertes & Eastman, 2000); in many cases one may suspect that the results are due to the stress accompanying the deprivation (Rial et al., 2007).

Next, we claim that the importance of sleep-distinctive traits is minimal compared to those of waking. Sleep has few, if any, traits on which natural selection could have applied pressure, and it seems difficult to imagine how Mother Nature could have selected the traits observed in a polysomnographic study (Rial et al., 2007). The situation is quite the opposite for wakefulness, where an overwhelming part of the waking behavior has obvious consequences for survival. Sleep traits, which must be recorded with a sophisticated artificial machine, do not resist the comparison with waking traits readily observable by means of the naked eye.

At the end of this introduction we do not expect all readers to be convinced by our argument in favor of a change of paradigm from sleep centeredness to waking centeredness. But we believe that we have given some reasonable evidence that this paradigm change does have significant advantages. We have argued that (1) the absence of function for a given trait must always be considered the null hypothesis; (2) complexity is not a firm proof to invoke an essential function; (3) animals sleep only after having fulfilled all vital functions; (4) deprivation-produced sleep debts may point not to a mysterious benefit obtained during sleep but rather to the maintenance of rest–activity cycles; (5) no adaptation has been unequivocally demonstrated for sleep; and (6) waking has, by far, more adaptive importance than sleep.

The following pages are devoted to a study of the evolution of waking in the transition from reptiles to mammals; we hope that the reader will understand the known facts better from this more appropriate approach. We will begin by comparing the anatomical and functional substrates of wakefulness in the two groups. As a consequence, an important functional discontinuity in the transition of wakefulness from cold- to warm-blooded animals is observed. This discontinuity was probably the result of the transition from diurnal (reptilian) to nocturnal (mammalian) lifestyles, with the accompanying changes in thermoregulation. The final consequence of those changes was the development of the cortical-based waking state, with unprecedented capacities that allowed the development of consciousness. A side effect of these evolutionary changes was the relegation of reptilian subcortical wakefulness to an inactive state, which resulted in sleep. This sequence of events stresses the greater importance of wakefulness over the "triviality" of sleep and explains the probable absence of an adaptive value for sleep: it is a side result of another true adaptation.

Homology and analogy

The following paragraphs will show that the apparent similarity of the waking states of different animal groups is superficial. The concepts of homology and analogy (Campbell & Hodos, 1970), basic in evolutionary studies, will be used to support the lack of continuity in the evolution of wakefulness. Two traits are homologous if their origin can be traced back to a common ancestor irrespective of their actual morphology and/or function. A paradigmatic example is the chain of ossicles of the mammalian middle ear, which is homologous to the bones forming the jaw articulation of fish, with a well-accepted common origin (Tumarkin, 1948). On the contrary, analogy occurs when two traits have a similar structure and/or function but have been developed after convergent evolution. The eyes of cephalopods and vertebrates serve as an example of analogy; they were developed through adaptive convergence with no common ancestor from which the two groups could have inherited them, in spite of serving the same function and having strikingly similar structures. This said, one should be aware that a similar structure and function commonly persists in homologous structures and that the number of examples of homology, with common ancestry and without morphological and functional difference, is large. This prompted Hennig (1966) to propose the auxiliary principle of analogy–homology: "Never assume convergent or parallel evolution; always assume homology in the absence of contrary evidence" (p. 121).

It is clear that the behavioral and functional aspects of waking are the same in every zoological group and every animal species. Hence, following Hennig's principle, waking must be homologous in all animals unless contrary evidence is demonstrated. Restricting the field to the subject of this review, the waking state in reptiles and mammals should be assumed to be homologous; therefore those sleep scholars who assumed homology between the two waking types were apparently right. However, a second look at this question is taken in the following paragraphs.

Waking and sensory processing in reptiles and mammals

An efficient waking state depends on the ability to get environmental information through adequate senses, analyze it by means of appropriate nerve centers, and produce adaptive behavioral responses. The homology between the main sensory organs of vertebrates is beyond doubt, but the homology of the systems analyzing the sensory input and ordering the appropriate output may be questioned. The following paragraphs review the similarities and differences in the sensory performance of different parts of the brains of reptiles and mammals. Attention is given particularly to the visual system, in view of its importance.

Structure of the visual system in reptiles and mammals

The thalamic sensory projections in terrestrial vertebrates have been classified into two main neuronal groups: lemnothalamic and collothalamic (Butler,

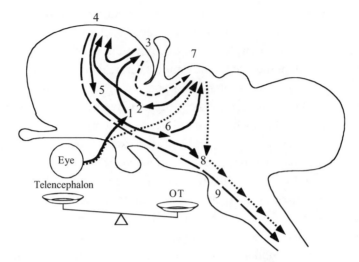

Figure 8.1. Schematic drawing of the amniote brain showing the collothalamic and lemnothalamic organization of the visual pathways in amniotes. (See color Plate 6.) (1) lemnothalamic nuclear group; (2) collothalamic nuclear group; (3) striate (visual) cortex (in mammals), dorsal cortex in reptiles; (4) extrastriate (secondary visual) cortex in mammals, DVR in reptiles; (5) striatal complex; (6) substantia nigra; (7) optic tectum (OT); (8) reticular formation; (9) corticospinal tract, exclusive of mammals (red dashed line). In blue: corticotectal projections. See further explanation in the text.

1974). The lemnothalamic system comprises the lemniscal pathways, with direct connections between the sensory cells and the dorsal thalamus, while the sensory cells of the collothalamic system have a direct projection to the mesencephalic colliculi and then send projections ascending up to the dorsal thalamus and down to the reticular formation (Butler & Hodos, 2005). The two systems are schematized in Figure 8.1.

Taking the collothalamic system first, a group of retinal fibers ends in the optic tectum (OT) in a precise topographical organization. The OT relays to the reticular formation through the tectoreticular and tectobulbar multisynaptic pathways (Butler & Hodos, 2005). Also, a tectothalamic pathway projects from the OT up to a collothalamic nucleus, which is the nucleus rotundus in reptiles and the visual part of the pulvinar in mammals (Aboitiz, Morales, & Montiel, 2003; de la Iglesia & López-García, 1997; Guirado, Dávila, Real, et al., 2000). The collothalamic ascending pathway connects with the dorsal ventricular ridge (DVR) in reptiles (Gonzalez & Ruschen, 1988; Hoogland & Vermeulen Van Der Zee, 1989) and the extrastriate visual cortex in mammals (Chalupa, 1991). In the two groups, both structures send projections to subpallial regions (the striatal complex), with final output to the substantia nigra, pars reticulata (SNr), through the nigrostriatal bundle (Smeets & Medina, 1995; ten Donkelaar & De Boer-Van Huizen, 1981; Voneida & Sligar, 1979).

The SNr has extensive projections to the reticular formation, the final common pathway in reptiles (green line of short arrows in Figure 8.1) (ten Donkelaar, 1998), but also with motor functions in mammals (Habaguchi, Takakusaki, Sitoh, et al., 2002; Siegel & McGinty, 1977; Takakusaki, Habaguchi, Ohtinata-Sugimoto, et al., 2003). However, the SNr also sends important GABAergic inhibitory lines to the OT, whose importance is further described in the next paragraphs.

Regarding the lemnothalamic system, the retinal ganglion cells connect to the lateral geniculate nucleus and the striate cortex in mammals (Ramón y Cajal, 1909/1952), which sends output to the extrastriate cortex (Montero, 1993; Rosa & Kubitzer, 1999) and the basal ganglia. Another interesting pathway connects several sensory cortical areas to the OT to control the ocular movements (Butler & Hodos, 2005) (see Figure 8.1, blue line).

The previous description shows that the general structure of the visual system is similar in mammals and reptiles and probably reflects the primitive pattern in all vertebrates (Butler & Hodos, 2005; ten Donkelaar, 1998). According to Sewards and Sewards (2002), however, some important differences exist and can be shown after telencephalic and tectal lesions.

The ablation of the telencephalon causes important visual deficits in mammals, although their severity varies in relation with the importance of the visual system in different groups. For instance, the rodent's telencephalic lesions cause impairments in the capacity to detect fine details of the visual image, but it is evident that the animals are not blind. On the contrary, similar lesions in primates cause a complete loss of visual awareness, although some simple visual tasks may continue with the so-called blind (tectal) vision. The telencephalic ablations have few consequences in the observable behavior of a reptile, while the tectal lesions cause deep impairments (Peterson, 1980).

These relationships also have a clear anatomical counterpart in the relative size of the telencephalic and tectal regions involved in the visual analysis. The reptilian OT has an important development, while the telencephalic visual areas are quite small. Reciprocally, the mammalian telencephalon has had a huge development in total surface; in addition, the three-layered reptilian pallium has been transformed into the six-layered mammalian isocortex, while the tectum has a reduced extension, a difference that is particularly evident in primates. Therefore the balance between tectal and forebrain analysis is characterized by a displacement toward the telencephalon in the transition from reptiles to mammals (again, see Figure 8.1).

Functional properties of tectal and telencephalic processing

Both collothalamic and lemnothalamic ascending lines (red ascending lines in Figure 8.1) may reach the telencephalon, where they provoke responses that probably result from the recognition of complex details in visual images (Sewards & Sewards, 2002). It should be noted, however, that the telencephalic processing depends on both collo- and lemnothalamic systems (Sewards & Sewards, 2002), whereas the descending tectoreticular and tectobulbar output is dependent on the tectal analysis only.

The collothalamic telencephalic elaboration is performed in the dorsal ventricular ridge (DVR) in reptiles and in the extrastriate cortex in mammals, whereas the lemnothalamic analysis depends on the dorsal cortex in reptiles and the striate cortex in mammals. The two telencephalic systems merge, however, in the striatal complex for both groups. The reptilian dorsal cortex projects to the DVR but also to the subpallial striate complex, which also receives a substantial input from the DVR. Similarly, the most important input of the mammalian extrastriate cortex comes from the striate visual cortex, but both – striate and extrastriate – send connections to basal nuclei.

In summary, reptiles and mammals have two systems for visual processing: the first, marked with green lines in Figure 8.1, is tectal and most probably serves to cause rapid and simple innate orienting responses to significant environmental stimuli through what has been called the command releasing system of the reticular formation, which activates the various behavioral programs (Kupfermann & Weiss, 1978). The second system, marked with red lines, is telencephalic and is devoted to analyzing complex characteristics of the visual field (Sewards & Sewards, 2002). Both systems, however, must work together in mammals. Because they have a well-defined somatotopic organization, the connection between them (marked with blue line in Figure 8.1) probably serves to align the two topographic maps (Harvey & Wortington, 1990; Lui, Giolli, Blanks, et al., 1994).

The mammalian visual telencephalic system has a direct motor output through the corticospinal tract. The output of the striatal complex, however, is always essential for the selection and coordination of motor responses. The striatal output is conveyed to the substantia nigra through dopaminergic fibers, merging here with the tectal output in the previously described command releasing system.

However, the telencephalic system has also the capacity to inhibit the tectal output. This was shown after the discovery of the Sprague effect (Sprague, 1966). It was observed that the hemianopia produced in cats after the unilateral destruction of the striate cortex seemed to improve after an additional lesion was produced in the opposite OT. Later studies demonstrated that this improvement could be produced by cutting only the part of the intercollicular commisure containing the nigrotectal GABAergic inhibitory lines (Wallace, Rosenquist, & Sprague, 1990). The Sprague effect was explained as being due to the suppression of the strong inhibitory effect normally produced by the telencephalic output. In other words, the blind hemifield of a cat could be replaced by the homolateral tectum, but only after the disinhibition produced by the second lesion, which allowed for the partial recovery of the lost function. The cortical vision was used for one half of the visual field and the tectal vision for the second half. The opposition between telencephalic and tectal processing has also been shown in the behavioral responses of tree shrews (Jane, Lewey, & Carlson, 1972), but it is particularly evident in the ontogenetic maturation of the brain, which receives particular attention in embryological studies in this review.

To summarize, the Sprague effect provides functional evidence of the double visual system; most interestingly, however, it shows that the two systems are complementary and normally work in opposition. The tectal system serves to recognize key stimuli that cause innate behavioral responses, while the telencephalic system serves to analyze complex visual images. It is worth remembering, however, the reduced extension of the reptilian telencephalic system in comparison with the tectal one. On the contrary, the mammalian isocortex suffered a huge development, in contrast with the reduced tectal extension.

Although the preceding description refers only to the visual system, there is a rather similar one for the somatosensory system. The anatomical features of the auditory system are different, as the acoustic pathways are exclusively collothalamic. However, they are rather similar in function, with mesencephalic (torus semicircularis for reptiles, posterior colliculi for mammals) and pallial–subpallial telencephallic processing. Hence double sensory processing is an essential property of the reptilian and mammalian brains.

Neurological signs of wakefulness in mammals and reptiles

In accordance with the preceding summary of the anatomical and functional basis of sensory analysis in mammals and reptiles, we observe many similarities between the two groups, but also some differences. The important question is this: Are the similarities sufficient to consider that reptilian and mammalian waking systems are homologous or, on the contrary, are the differences so important as to support analogy instead? The empirical results should provide enough evidence to decide this issue. The following discussion compares some neurological traits of reptiles and mammals.

The mammalian electroencephalographic (EEG) arousal pattern has been described as consisting of reductions in amplitude and synchrony, called the desynchronization reaction. Today it has been recognized that the supposed loss

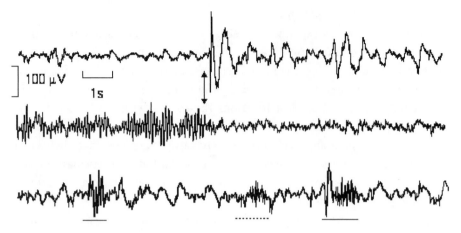

Figure 8.2. EEG arousal in *Gallotia galloti* lizards produced after a handclap (*upper record*) and after eye opening in mammals (*middle record*). In the lower record, the underlined fragments represent the response to a moving stimulus perceived in the visual field of a full waking lizard. The dashed underline corresponds to a spontaneous spindle, probably respiratory.

of synchrony corresponds in fact to a low-amplitude but clear synchronization in the gamma range, from 20 to 50 hertz (Llinás & Rivary, 1993), and the term "EEG activation" has replaced the old terminology (Steriade, 2000).

Mammalian EEG activation depends directly on two main systems, the cholinergic basal telencephalon and the serotonergic raphe, although other neurotransmitters also promote indirect activation through the two basic mechanisms (Dringenberg & Vanderwolf, 1998). Early reports stated that cortical arousal and responsiveness to sensory input were independent (Feldman & Waller, 1962). However, modern reports show that cortical dysfunctions in which the EEG slows down reduce or even block cognitive capacity (Dringenberg & Diavolitsis, 2002; Llinás & Ribary, 1993), thereby emphasizing the importance of cortical high-frequency activation for full waking. Thus, it may be that a full analysis of sensory input can be performed only in a cortex showing an activated EEG.

The cholinergic and serotonergic cellular groups of the basal telencephalon and raphe, respectively, have also been observed in reptiles (Bruce & Butler, 1984; Lohman & VanVoerden-Verkley, 1978; Medina, Smeets, Hoogland, et al., 1993). In general, the main brain regions involved in the control of sleep have been identified for a long time (Broughton, 1972), contrasting with the absence of their neurological signs. Two forms of EEG activation have been recorded in reptiles. The most evident response to sensory stimulation of a wakeful reptile consists in an increase in amplitude, which is particularly salient in the low-frequency range (Nicolau et al., 2000; Susic, 1972; Vasilescu, 1970) (see Figures 8.2 and 8.3).

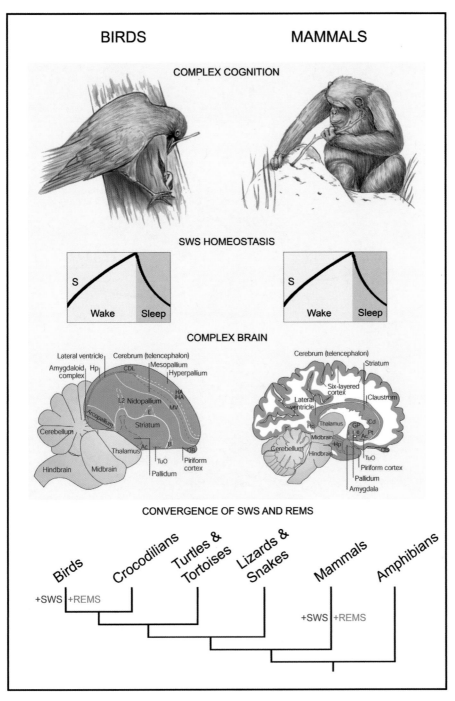

Plate 1. The phylogenetic tree (bottom figure) shows the convergent evolution of slow-wave sleep and rapid-eye-movement sleep in birds and mammals. See Figure 7.6.

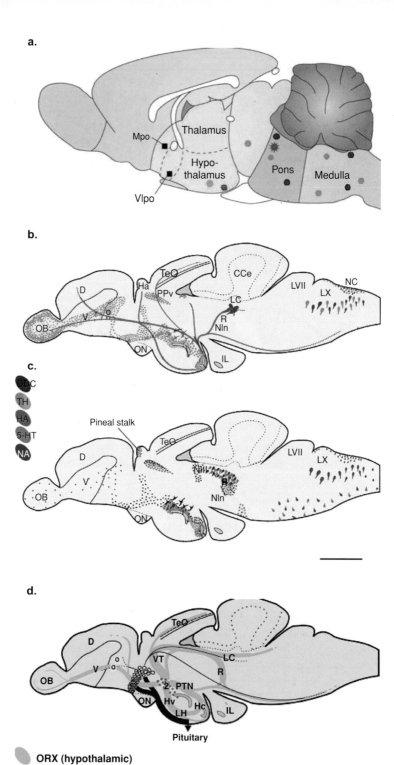

a.

Mpo

Thalamus

Hypo-
thalamus

Vlpo

Pons

Medulla

b.

TeO

CCe

Ha

PPv

LVII

LX

NC

D

LC

V

o

R

Nln

OB

ON

IL

c.

DDC

TH

HA

5-HT

NA

Pineal stalk

TeO

D

NIV

LVII

LX

V

R

OB

Nln

ON

d.

TeO

D

VT

LC

V

o

R

OB

2 PTN

ON

Hv

Hc

IL

LH

Pituitary

ORX (hypothalamic)

ORX (preoptic)

Plate 2. A Schematic Sagittal overview of the mammalian (rat) (a) and zebrafish
(b to d) brain. See Figure 11.1.

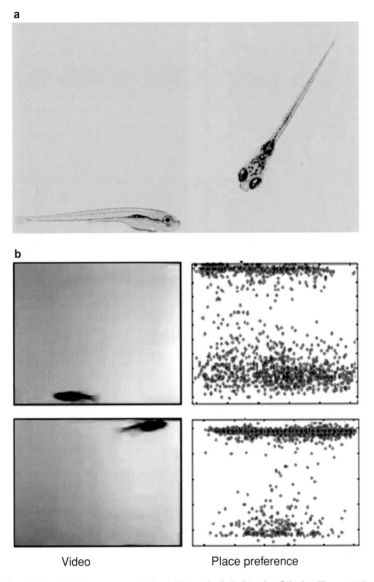

Video   Place preference

Plate 3. Typical sleep postures in larval (a) and adult (b) zebrafish. See Figure 11.3.

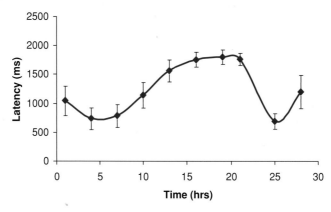

Plate 4. Dirunal variation in latency to visually evoked behavior in zebrafish larvae. See Figure 11.11

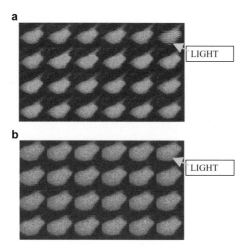

Plate 5. Melatonin attenuates neuronal response to light in MeLc neuron of the nMLF cluster. See Figure 11.12.

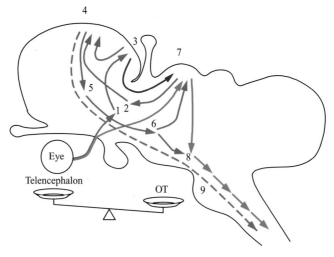

Plate 6. Schematic drawing of the amniote brain showing the collothalamic and lemnothalamic organization of the visual pathways in amniotes. See Figure 8.1.

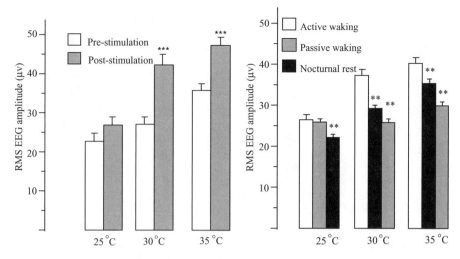

Figure 8.3. The left panel shows the RMS amplitude of the EEG recorded in *Gallotia galloti* lizards before and after sensory (auditive) stimulation. The right panel shows the amplitude of the spontaneous EEG in function of the behavioral state. In both cases, the dependence on body temperature, following the Q_{10} law, was highly significant. (After Gamundi, Akaârir, & Rial, unpublished data.)

Although the slow-wave EEG has been thoroughly looked for in reptiles, most results were negative. It should be remarked, however, that these studies were looking for slow waves during resting time. On the other hand, reptiles are absolutely dependent on external heat sources for their behavioral ectothermic thermoregulation, with the result that their body temperature is at its lowest during rest. Following the Q_{10} law, the EEG amplitude is also minimal during rest, and the negative results obtained should have been expected: reptiles have no slow-wave EEG during rest. Other studies, however, have recognized the effects of both body temperature and activation, so that the EEG amplitude is maximal during waking in warm animals submitted to sensory stimulation, with the highest power always found at the low-frequency end (Figures 8.2 and 8.3) (Bullock, 2003; Bullock & Basar, 1988; de Vera, González, & Rial, 1994; González, Vera, García-Cruz, et al., 1978). In other words, the amplitude of the reptilian EEG follows a pattern opposite that of the mammalian one (Nicolau et al., 2000; Susic, 1972; Vasilescu, 1970) (Figure 8.4).

Another type of response to visual stimulation consists in the production of 12- to 25-hertz spindles (de Vera et al., 1994; González & Rial, 1977; González et al., 1978; Prechtl, 1994; Servit & Strejčkova, 1972; Servit, Strejčkova, & Volanschi, 1971) (Figure 8.2, lower record).

Some researchers have stated that the spindles observed in the reptilian EEG were produced in the olfactory bulb in response to respiratory activity. However,

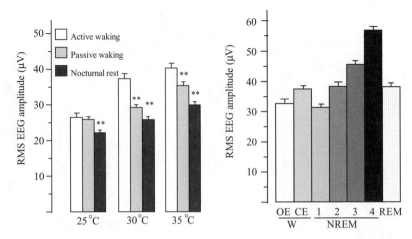

Figure 8.4. EEG amplitude in reptiles (*left panel*) and in mammals (*right panel*) recorded during different behavioral states. The reptile's maximal amplitude occurs in warm and sensory activated animals. The mammals' maximal amplitude is recorded during phase 4 of NREM sleep, whereas the lowest amplitude corresponds to the most activated state. Abbreviations: W, waking; OE, open eyes; CE, closed eyes. (Gamundi, Akaârir, & Rial, unpublished data.)

respiratory and nonrespiratory spindles (Figure 8.2) have been distinguished in the reptilian telencephalic EEG (Gaztelu, García-Austt, & Bullock, 1991). The nonrespiratory ones could be evoked after sensory stimulation and share an extreme similarity with the sleep spindles, as they were recorded simultaneously in the thalamus and in the cortex and depend, as in mammals, on GABAergic activity (Servit & Strejčkova, 1972; Servit et al., 1971). More recent studies (Prechtl, 1994; Prechtl, Cohen, Pesaran, et al., 1997) have demonstrated that the telencephalic spindles of turtles were the result of the telencephalic activation after either patterned visual stimulation or gaze shifts. Moreover, the reptilian spindles could be transformed into high-voltage paroxysmal spikes, as in mammals (Nobili, Ferillo, Baglietto, et al., 1999; van Luijtelaar, 1997, for mammals; Gómez, Bolaños, López, et al., 1990; Rial & González, 1978, for reptiles). Therefore few doubts can be cast on the homology between them.

The mammalian sleep spindles, K complexes, and delta EEG (1 to 4 hertz) depend on rhythmic slow oscillations (<1 hertz) generated intracortically during NREM sleep (Amizca & Steriade, 2002; Steriade & Amizca, 1998). The thalamic cells are in an excitable state and fire in tonic mode during waking, allowing the arrival of sensory inputs to the cortex. After the beginning of sleep, the cells are increasingly hyperpolarized, entering into burst firing mode, which causes the production of spindles and delta EEG (depending on the degree of hyperpolarization attained) (Steriade, 2000) and block the sensory input to the cortex (Coenen, 1995). Early studies showed that sleep spindles were dependent on collicular input (Bremer, 1935). Modern studies have extended those results to the continuous thalamic inhibition produced as a consequence of a reduction in the firing rate of reticular and mesopontine cholinergic nuclei (McCarley, 2007; Steriade, Oakson, & Ropert, 1982). It is important to note that the spindles appear during the transitional state from waking to sleep, with a medium level of hyperpolarization in thalamic reticular neurons (Steriade, 2000), and are substituted by the delta waves with increased hyperpolarization levels. This means that spindles and delta waves cannot be simultaneously generated in a given cell (Steriade, Curro-Dossi, & Nuñez, 1991).

All in all, the neuronal mechanisms causing spindles and slow waves are well understood. Unfortunately, the functional consequences are not so clear. The available data show that both contribute to cortical inhibition, blocking the arrival of sensory input to the cortex. However, taking into account their opposition, some difference should exist, perhaps related to differential activities in various pallial regions. For instance, sleep spindles can be recorded not only in the cortex but also in the hippocampus (Malow, Carney, Kushwaha, et al., 1999), where they may be involved in the transfer of information between hippocampal and neocortical cell assemblies (Sirota, Csicsvari, Buhl, et al., 2003). This would agree with a double interpretation of the functional significance of the spindles, as it has been suggested that they serve to protect sleep (blocking the neocortex) but they can also be indicative of activation (Bowersox, Kaitin, & Dement, 1985; Church, Johnson, & Seales, 1978; Jankel & Niedermeyer, 1985), which would occur in other nonneocortical (paleopallial) regions. It should be recalled here that the medial cortex of reptiles is generally acknowledged to be homologous with the mammalian hippocampus, with well developed anatomical and functional connections with the dorsal cortex, the region from which the isocortex probably developed (Aboitiz et al., 2003).

The meaning of differences in the EEGs of reptiles and mammals

Given the cortical inhibition caused by thalamic neurons, the most parsimonious explanation for the slow-wave EEG recorded in active reptiles should assume a similar blocking function. If the slow EEG waves appear in mammals during sleep but during activity in reptiles, it follows that cortical processing must be normally interrupted in waking reptiles. This conclusion should not be surprising in view of the previously described antagonism between tectal and telencephalic processes. Whereas telencephalic blocking in mammals is necessarily linked to NREM sleep – that is, to unconsciousness – the same blocking in reptiles should be considered necessary to allow essential tectal analysis during the waking state.

On the other hand, if the slow-wave EEG is indicative of telencephalic inhibition, the spindles evoked in reptiles as a response to patterned stimuli should have an opposite meaning. Hence an active reptile may experience continuous oscillations between telencephalic (paleopallial) and subtelencephalic (tectal) activation modes, interspersing the slow-wave EEG with spindles, depending on the nature of the perceived stimulus and the need of the corresponding analysis. This situation is impossible for a mammal if one considers (1) the low processing capacity of the tectal system and (2) that during cortical inactivation (i.e., sleep) the eyes are closed, imposing absolute inactivity for tectal visual analysis as well. Interestingly, the case could be different for somatosensory tectal processing, which could remain active during slow-wave sleep, providing a basis for the well-known sleeping thermoregulatory behavior (Parmeggiani, 2000).

Conclusion: Is the waking of reptiles and mammals homologous?

The preceding paragraphs have provided strong anatomical evidence supporting the existence of similar structures controlling the sensory systems of mammals and reptiles. Moreover, the equivalence between collothalamic and lemnothalamic, telencephalic and tectal sensory processing has also been evidenced in the two groups. Therefore the homology of the nervous structures providing the sensory analysis is supported.

The fine differences in the balance between telencephalic and tectal working modes of the two systems, however, have determined important modifications in the general activity of the reptilian and mammalian brain. Summarizing, mammalian wakefulness is strictly dependent on the telencephalic activity, with a rather meager (and normally inhibited) tectal contribution. On the other hand, mammalian sleep – that is, unconsciousness – unavoidably occurs after telencephalic inhibition. During the first stages of sleep, however, EEG spindles may appear, either spontaneously or after sensory stimulation; these have been attributed to sensory processes in paleocortical structures. The tectal processing could remain active for some sensory modalities during sleep but not for vision, as a consequence of eye closure. The reptilian telencephalic inhibition, heralded by the slow-wave EEG, has few consequences and is even necessary to allow tectal processing, the fundamental one in premammals. On the other hand, the (paleo) telencephalic activation attained with the EEG spindles allows for the analysis, probably learning-related, of complex features of the visual images.

These conclusions show that the mammalian waking state is not equivalent to the reptilian one in functional terms, in spite of the anatomical similarities between the two brains. Thus it must be concluded that they are only analogous states, providing the same function but organized in quite different patterns. A mammal depends on cortical activation to be awake, while a reptile must have a fully functional OT, the state of the telencephalon being less important. Hence the two types of waking may not in fact be homologous. On the contrary, the homology between mammalian slow-wave sleep and the reptilian waking is evident, as shown by the similar EEGs, with slow waves, spindles, and common controlling centers.

How mammalian waking appeared

It has been proposed that the evolutionary transformation from reptiles to mammals was a consequence of the prolongation of diurnal activity, typical of the reptilian machinery, to the early nighttime (Carroll, 1988; Jerison, 1973; Kemp, 1982; Sagan, 1977). The evolutionary pressure to fill the unoccupied nocturnal ecological niche is evident and promoted two types of adaptations. First, the extension to the nocturnal lifestyle was facilitated by the production of endothermic metabolism and the consequent increase in body temperature (Crompton, Taylor, Jagger et al., 1978).

In addition, nocturnal life also caused extensive brain reorganizations. Obviously, the visual system has a low efficiency during times of darkness, while olfactory processes are essential (Jerison, 1973, 1990; Kemp, 1982). Enhanced olfactory processing presumably promoted the increase in telencephalic activation at the expense of tectal activity. According to Sagan (1977) and Lynch (1986), the development of associative networks between the dorsal cortex and the olfactory system via the hippocampus became increasingly important to develop multisensory maps of space and behavior (Eichelbaum, 1998) in which specific odors labeled particular places and routes. This prompted the need to develop olfactory–hippocampal associative networks in primitive mammals; this may have caused an increased production of progenitor cells in the dorsal pallium and other brain regions, leading to the expansion of the dorsal cortex as recipient of visual thalamofugal, tectofugal, and auditory projections, which may have become essential in the early mammalian brain (Aboitiz, 1992; Aboitiz et al., 2003).

When the capability for endothermic thermoregulation appeared, the first mammals could have been able to maintain round-the-clock activity, a state of affairs that was opposed by the imperative need to maintain rest–activity cycles, as postulated by Aschoff (1964). Obviously, natural selection should have favored those individuals able to suppress the newly developed cortical activity during the illuminated part of the circadian cycle. This was not difficult, as reptiles were already able to block the telencephalic processing with slow EEG waves to permit tectal activation. It was only necessary to enhance this capability during light time. In this way, mammalian sleep was born, conserving a number of electrophysiological properties of the reptilian waking state.

Embryological studies

Behavior does not fossilize; thus a direct proof of the evolutionary process cannot be obtained. It can only be deduced from comparative studies, analysis of the anatomy and physiology of related groups, embryological analysis, and logical reasoning. The present report has shown a plausible anatomical and physiological scenario not only for the evolution of waking but also for the advent of mammalian sleep. Additional evidence may be obtained through the analysis of embryologic development, testing the eventual accommodation of the proposed scenario to the recapitulation law of Von Bauer and Haeckel. Indeed, the validity of the recapitulation law has been hotly discussed. However, after the redefinition of Garstang (1922) ("changes in ontogeny create phylogeny" [p. 81]) and after developing the concept of heterochrony (changes in the speed of development), the causes of the success and failure of Haeckel's law are well understood. There are neotenic or pedomorphic processes in which development slows down and the law does not hold, while peramorphic or terminal additions occur when developmental speed increases and new stages are added, a case in which the law is followed. This has been discussed with respect to many aspects of the vertebrate nervous system and, in particular, to the evolution of the mammalian cortex (Montagnini & Treves, 2003) and of sleep and waking (Esteban et al., 2005).

In a first approximation, altricial newborn mammals, with an undeveloped cortex, should continuously show the mammalian counterpart of the reptilian wakefulness – that is, sleep – if the proposed hypotheses are correct. However, we know that newborn mammals show wakefulness, even if only during short periods. Nevertheless, there are interesting reports showing that the contradiction is only apparent. If we concentrate only on human studies, the retinocollicular visual system is well developed at birth, while the lemnothalamic system begins to be functional only 2 months after birth (Braddick, Wattam-Bell, & Atkinson, 1986; Finlay, Hersman, & Darlington, 1998). Hence wakefulness with a reptilian aspect might appear, in early developmental stages, to be substituted by the mammalian type after the completion of cortical development. This is exactly what has been found to happen (Sewards & Sewards, 2002). For instance, it has been shown that human neonates follow the movement of a schematic face (but not a scrambled face or other blank stimuli) a few minutes after birth (Goren, Sarty, & Wu, 1975), a process named "CONSPEC" (conspecifics) (Morton & Johnson, 1991), which is the mechanism to direct the attention of the newborn to faces. This process decays in the second month of life, to be replaced by "CONLEARN," the mechanism allowing the learned recognition of familiar faces.

A similar decline in several other sensorimotor reflexes has been observed, followed by a new development in which the ability to analyze other properties of the stimuli increases. For instance, the orientation reflexes to sound sources declines in 2-month-old infants, emerging again at the end of the fourth month (Johnson, 1990; Muir, Clifton, & Clarkson, 1989). These reports strongly support the idea that the "built in" orienting responses depend on primitive activity (probably tectal), which is later inhibited and replaced after the emergence of cortical activity (Pascalis, de Haan, Nelson, et al., 1998), which blocks tectal processing. Therefore the wakefulness shown immediately after birth in altricial mammals, organized from the OT, could easily be qualified as belonging to the reptilian type, while mammalian wakefulness is shown only after cortical maturation. In conclusion, the ontogenetic law is fully satisfied.

The vital function of sleep: Should it exist?

This chapter began by presenting some doubts as to the existence of a vital function of sleep, and these doubts can be substantiated at the end of the chapter. In 1966, G. C. Williams proposed several rules of thumb that should be used to reject the need of adaptation for a given trait. Among them, it is evident that a trait must not be deemed adaptive when it is the side effect of another truly adaptive trait. For instance, the navel may not have adaptive value, as it is only a scar due to its severance from the placenta, a structure whose adaptiveness nobody can doubt. Applying the rule to the traits distinguishing sleep from rest, it is not difficult to understand that they are scars of a sort, by-product or remnants from the development of the new wakefulness, as described in the foregoing paragraphs. Therefore no adaptive value should be claimed for sleep. Most probably, Mother Nature never selected the bizarre traits distinguishing mammalian sleep from the reptilian resting state, and the functionless sleep is not a mistake of natural selection. The particular traits of reptilian wakefulness were simply relegated to the resting period, which in fact may be considered as a junkyard full of unnecessary remnants of earlier states. The sleep traits are still observed because they are transparent for natural selection, which cannot select traits without phenotypic consequences. Mother Nature has no polygraphic recorders to reject animals without slow-wave EEG, eye movements, or complex neural activity; it had plenty of power to reject animals with continuous wakefulness, cortically based during the dark half of the cycle and tectal during the second half. Natural selection allowed for the survival only of those able to block behavioral activity during the part of the cycle in which their efficiency was lower.

This does not mean that sleep could not have developed secondary functions, perhaps different in different animals. Evolutionary remnants have been reused on many occasions, as the example of the middle ear ossicles shows. Similarly, evolution could have reused some sleep traits for unexpected adaptations. In these cases, however, a firm proof of necessity and sufficiency should be always demanded.

The end of the described evolutionary process producing sleep as a rather unimportant by-product had an enormously positive side. While mammals were developing sleep, they were also developing a uniquely complex structure, the telencephalic cortex. This allowed the emergence of an unparalleled grade of consciousness, which, according to several authors, may occur only in mammals as a result not only of their huge increase in size but also of the development of multiple cortical reentrant connections (Edelmann & Tononi, 2000; Edelman, Baars, & Seth, 2005). Thus, the evolutionary process did not made a mistake either big or even minute. Instead, it allowed for the production of a new phenomenon, unique in the course of evolution. Could Mother Nature plead guilty?

References

Aboitiz, F. (1992). The evolutionary origin of the mammalian cerebral cortex. *Biological Research*, *25*, 41–49.

Aboitiz, F., Morales, D., & Montiel, J. (2003). The evolutionary origin of the mammalian isocortex: Towards an integrated developmental and functional approach. *Behavioral Brain Sciences*, *26*, 535–586.

Abrams, P. A. (2000). The evolution of predator-prey interactions: Theory and evidence. *Annual Review of Ecology and Systematics*, *31*, 79–105.

Allison, T., & Cicchetti, D. V. (1976). Sleep in mammals: Ecological and constitutional correlates. *Science*, *194*, 732–734.

Allison, T., & Van Twyver, H. (1970). The evolution of sleep. *Natural History, 79*, 57–65.

Amizca, F., & Steriade, M. (2002). The functional significance of K-complexes. *Sleep Medicine Reviews*, *6*(2), 139–149.

Aschoff, J. (1964). Survival value of diurnal rhythms. *Symposiums of the Zoological Society of London*, *13*, 79–98.

Benington, J. H. (2000). Sleep homeostasis and the function of sleep. *Sleep*, *23*(7), 959–966.

Bowersox, S. S., Kaitin, K. I., & Dement, W. C. (1985). EEG spindle activity as a function of age: Relationships to sleep continuity. *Brain Research*, *334*, 303–308.

Braddick, J., Wattam-Bell, J., & Atkinson, J. (1986). Orientation specific cortical responses in early infancy. *Nature*, *320*, 617–619.

Bremer, F. (1935). Cerveau "isolé" et physiologie du sommeil ["Insulated" cerebrum and physiology of sleep]. *Comptes Rendues Societé de Biologie (Paris)*, *118*, 1235–1241.

Broughton, R. (1972). Phylogenetic evolution of sleep systems. In M. H. Chase (Ed.), *The sleeping brain: Proceedings of the symposia of the first international congress of the Association for the Physiological Study of Sleep* (pp. 2–7). Los Angeles: Brain Research Institute.

Bruce, L. L., & Butler, A. B. (1984). Telencephalic connections in lizards. I. Projections to cortex. *Journal of Comparative Neurology*, *229*, 585–561.

Bullock, T. H. (2003). Have brain dynamics evolved? Should we look for unique dynamics in the sapient species? *Neural Computation*, *15*, 2013–2027.

Bullock, T. H., & Basar, E. (1988). Comparison of ongoing compound field potentials in the brain of invertebrates and vertebrates. *Brain Research Reviews*, *13*, 57–75.

Butler, A. B. (1974). The evolution of the dorsal pallium in the telencephalon of amniotes: Cladistic analysis and a new hypothesis. *Brain Research Reviews*, *19*, 66–101.

Butler, A. B., & Hodos, W. (2005). *Comparative vertebrate neuroanatomy. Evolution and adaptation* (4th ed.). New York: Wiley Liss.

Campbell, C. B. G., & Hodos, W. (1970). The concept of homology and the evolution of the nervous system. *Brain Behavior and Evolution*, *3*, 353–367.

Carroll, R. I. (1988). *Vertebrate paleontology and evolution*. New York: Freeman Press.

Chalupa, L. M. (1991). Visual function of the pulvinar. In A. G. Leventhal (Ed.), *Vision and visual dysfunction: The neuronal basis of visual function* (Vol. 4, pp. 140–159). London: MacMillan.

Church, M. W., Johnson, L. C., & Seales, D. M. (1978). Evoked K-complexes and cardiovascular responses to spindle-synchronous and spindle-asynchronous stimulus clicks during NREM sleep. *Electroencephalography and Clinical Neurophysiology*, *45*, 443–453.

Coenen, A. (1995). Neuronal activities underlying the electroencephalogram and evoked potentials in sleeping and waking: Implications for information processing. *Neuroscience and Biobehavioral Process*, *19*, 447–463.

Coenen, A., & Van Luijtelaar, E. L. J. M. (1985). Stress induced by three procedures of deprivation of paradoxical sleep. *Physiology and Behavior*, *35*, 501–504.

Crompton, A. W., Taylor C., & Jagger, J. A. (1978) Evolution of homeothermy in mammals. *Nature*, *272*, 333–336.

Dawkins, R. (1986). *The blind watchmaker*. London: Penguin.

de la Iglesia, J. A. L., & López-García, C. (1997). Neuronal circuitry in the medial cerebral cortex of lizards. In J. Mira, R. Moreno-Díaz, & J. Cabestany (Eds.), *Proceedings of the international work-conference on artificial and natural neural networks: Biological and artificial computation: From neuroscience to technology* (pp. 61–71). London: Springer-Verlag.

de Vera, L., González, J., & Rial, R. V. (1994). Reptilian waking EEG: Slow waves, spindles and evoked potentials. *Electroencephalographic and Clinical Neurophysiology*, *90*, 298–303.

Dennett, D. C. (1995). *Darwin's dangerous idea*. London: Penguin.

Dringenberg, H. C., & Diavolitsis, P. (2002). Electroencephalographic activation by fluoxetine in rats: Role of 5-HT1A receptors and enhancement of concurrent acetylcholinesterase inhibitor treatment. *Neuropharmacology*, *42*, 154–161.

Dringenberg, H. C., & Vanderwolf, C. H. (1998). Involvement of direct and indirect pathways in electrocorticographic activation. *Neuroscience and Biobehavioral Reviews*, *22*(2), 243–257.

Edelmann, G. M., & Tononi, G. (2000). *A universe of consciousness*. New York: Basic Books.

Edelman, D. B., Baars, B. J., & Seth A. K. (2005). Identifying hallmarks of consciousness in nonmammalian species. *Consciousness and Cognition*, *14*, 169–187.

Eichelbaum, H. (1998). Using olfaction to study memory. *Annals of the New York Academy of Sciences*, *855*, 657–669.

Esteban, S., Nicolau, M. C., Gamundí, A., Akaârir, M., & Rial, R. V. (2005) Animal sleep: Phylogenetic correlations. In P. L. Parmeggiani & R. Velluti (Eds.), *The Physiologic nature of sleep* (pp. 207–246). London: Imperial College Press.

Feldman, S. M., & Waller, H. J. (1962). Dissociation of electrocortical activation and behavioural arousal. *Nature*, *196*, 1320.

Finlay, B. L., Hersman, M. N., & Darlington, R. B. (1998). Patterns of vertebrate neurogenesis and the paths of vertebrate evolution. *Brain Behavior and Evolution, 52,* 232–242.

Garstang, W. (1922). The theory of recapitulation: A critical restatement of the biogenetic law. *Zoological Journal of the Linnean Society London, 35,* 81–101.

Gaztelu, J. M., García-Austt, E., & Bullock, T. (1991). Electrocorticograms of hippocampal and dorsal cortex of two reptiles: Comparison with possible mammalian homologs. *Brain Behavior and Evolution, 37,* 144–160.

Gómez, T., Bolaños, A., López, J. A., Nicolau, M. C., & Rial, R. (1990). A case report of spontaneous electrographic epilepsy in reptiles (*Gallotia galloti*). *Comparative Biochemistry and Physiology, Series C, 97*(2), 257–258.

Gonzalez, A., & Ruschen, F. T. (1988). Connections of the basal ganglia in the lizard *Gecko gecko*. In W. K. Schwerdtfeger & W. J. A. Smeets (Eds.), *The forebrain of reptiles* (pp. 50–59). Basel: Karger.

González, J., & Rial, R. V. (1977). Electrofisiología de la corteza telencefálica de reptiles (*Lacerta galloti*): EEG y potenciales evocados [Electrophysiology of the telencephalic cortex of reptiles (*Lacerta galloti*): EEG and evoked potentials. *Revista Española de Fisiología, 33,* 239–248.

González, J., Vera, L. M., García-Cruz, C. M., & Rial, R. V. (1978). Efectos de la temperatura en el electroencefalograma y los potenciales evocados de los reptiles (*Lacerta galloti*) [Effects of the temperature in the electroencephalogram and the evoked potentials of the reptiles (*Lacerta galloti*)]. *Revista Española de Fisiología, 34,* 153–158.

Goren, C. C., Sarty M., & Wu, P. Y. (1975). Visual following and pattern discrimination of face-like stimuli by newborn infants. *Pediatrics, 56,* 544–549.

Gould, S. J., & Levontin, R. C. (1979). The spandrels of San Marco and the Panglossian paradigm: A critique of the adaptationist programme. *Proceedings of the Royal Society London, Series B, 205,* 581–598.

Guirado, S., Dávila, J. C., Real, M. A., & Medina, L. (2000). Light and electron microscopic evidence for projections from the thalamic nucleus rotundus to targets in the basal ganglia, the dorsal ventricular ridge, and the amygdaloid complex in a lizard. *Journal of Comparative Neurology, 424*(2), 216–232.

Habaguchi, T., Takakusaki, K., Sitoh K., Sugimoto, J., & Sakamoto, T. (2002). Medullary reticulospinal tract mediating the generalized motor inhibition in cats: II. Functional organization within the medullary reticular formation with respect to postsynaptic inhibition of forelimb and hind-limb motoneurons. *Neuroscience, 113*(1), 65–77.

Hartse, K. M. (1994). Sleep in insects and nonmammalian vertebrates. In M. H. Kryger, T. Roth, & W. C. Dement (Eds.), *Principles and practice of sleep medicine* (pp. 95–104). London: Saunders.

Harvey, A. R., & Wortington, D. R. (1990). The projection from different visual cortical areas to the rat superior colliculus. *Journal of Comparative Neurology, 298,* 281–292.

Hennig, W. (1966). *Phylogenetic systematics.* Urbana, IL: University of Illinois Press.

Hoogland, P. V., & Vermeulen Van Der Zee, E. (1989). Efferent connections of the dorsal cortex in the lizard *Gecko gecko*, studied with *Phaseolus vulgaris* leucoagglutinin. *Journal of Comparative Neurology, 285,* 289–303.

Jane, J. A., Lewey, N., & Carlson, H. J. (1972). Tectal and cortical function in vision. *Experimental Neurology, 35,* 61–77.

Jankel, W. R., & Niedermeyer, E. (1985). Sleep spindles. *Journal of Clinical Neurophysiology, 37,* 538–548.

Jerison, H. J. (1973). *The evolution of the brain and intelligence*. New York: Academic Press.

Jerison, H. J. (1990). Fossil evidence on the evolution of the neocortex. In E. G. Jones & A. Peters (Eds.), *Cerebral cortex* (pp. 285–309). New York: Plenum Press.

Johnson, M. H. (1990). Cortical maturation and the development of visual attention in early infancy. *Journal of Cognitive Neuroscience, 2*, 81–95.

Karmanova, I. G. (1982). *Evolution of sleep: Stages of the formation of the wakefulness-sleep cycle in vertebrates*. Basel: Karger.

Kemp, T. S. (1982). *Mammal-like reptiles and the origin of mammals*. New York: Academic Press.

Kimura, M. (1983). *The neutral theory of molecular evolution*. Cambridge, UK: Cambridge University Press.

Kupfermann, I., & Weiss, K. R. (1978). The command neuron concept. *Behavioral Brain Sciences, 1*, 3–39.

Llinás, R., & Ribary, U. (1993). Coherent 40-Hz oscillation characterizes dream state in humans. *Proceedings of the National Academy of Sciences of the United States of America, 90*, 2078–2081.

Lohman, A. H. M., & VanVoerden-Verkley, I. (1978). Ascending connections to the forebrain in the tegu lizard. *Journal of Comparative Neurology, 182*, 555–594.

Lui, F., Giolli, R. A., Blanks, R. H., & Tom, E. M. (1994). Pattern of striate cortical projections to the pretectal complex in the guinea pig. *Journal of Comparative Neurology, 344*, 598–609.

Lynch, G. (1986). *Synapses, circuits, and the beginnings of memory*. Cambridge, MA: MIT Press.

Malow, B. A., Carney, P. R., Kushwaha, R., & Bowes, R. J. (1999). Hippocampal sleep spindles revisited: Physiologic or epileptic activity? *Clinical Neurophysiology, 110*, 687–693.

McCarley, R. W. (2007). Neurobiology of REM and NREM sleep. *Sleep Medicine, 8*, 302–330.

Meddis, R. (1983). The evolution of sleep. In A. Mayes (Ed.), *Sleep mechanisms in humans and animals: An evolutionary perspective* (pp. 57–106). London: Van Nostrand Reinhold.

Medina, L., Smeets, W. J. A. J., Hoogland, P. V., & Puelles, L. (1993). Distribution of choline acetyltransferase immunoreactivity in the brain of the lizard Gallotia galloti. *Journal of Comparative Neurology, 331*, 261–285.

Monnier, M. (1980). Comparative electrophysiology of sleep in some vertebrates. *Experientia, 36*, 16–19.

Montagnini, A., & Treves, A. (2003). The evolution of mammalian cortex from lamination to arealization. *Brain Research Bulletin, 60*, 387–393.

Montero, V. (1993). Retinotopy of cortical connections between the striate cortex and extrastriate visual areas in the rat. *Experimental Brain Research, 94*, 1–15.

Morton, J., & Johnson, M. H. (1991). CONSPEC and CONLEARN: A two-process theory of infant face recognition. *Psychological Reviews, 98*, 164–181.

Muir, D. W., Clifton, R. K., & Clarkson, M. G. (1989). The development of a human auditory localization response: A U-shaped function. *Canadian Journal of Psychology, 43*, 199–216.

Nicolau, M. C., Akaârir, M., Gamundí, A., González, J., & Rial, R. V. (2000). Why we sleep: The evolutionary pathway to the mammalian sleep. *Progress in Neurobiology, 62*, 379–406.

Nobili, L., Ferrillo, F., Baglietto, M. G., Beelke, M., de Carli, F., de Negri, E., et al. (1999). Relationship of sleep interictal epileptiform discharges to sigma activity (12–16 Hz) in benign epilepsy of childhood with rolandic spikes. *Clinical Neurophysiology, 110*(1), 39–46.

Parmeggiani, P. L. (2000). Physiological regulation in sleep. In M. H. Kryger, T. Roth & W. C. Dement (Eds.), *Principles and practice of sleep medicine* (pp. 169–178). Philadelphia: W. B. Saunders.

Pascalis, O., de Haan, M., Nelson, C. A., & de Schonen, S. (1998). Long-term recognition memory for faces assessed by visual paired comparison in 3- and 6-month old infants. *Journal of Experimental Psychology: Learning, Memory, and Cognition, 24*, 249–260.

Peterson, E. (1980). Behavioral studies of telencephalic function in reptiles. In S. O. E. Ebbeson (Ed.), *Comparative neurology of the telencephalon* (pp. 343–388). New York: Plenum.

Prechtl, J. C. (1994). Visual motion induces synchronous oscillations in turtle visual cortex. *Proceedings of the National Academy of Sciences of the United States of America, 91*, 12467–12471.

Prechtl, J. C., Cohen, L. B., Pesaran, B., Mitra, P. P., & Kleinfeld, D. (1997). Visual stimuli induce waves of electrical activity in turtle cortex. *Proceedings of the National Academy of Sciences of the United States of America, 94*, 7621–7626.

Prigogine, I., & Nicolis, G. (1977). *Self-organization in non-equilibrium systems: From dissipative structures to order through fluctuations.* New York: John Wiley & Sons.

Ramón Cajal, S. (1909). *Histologie du Sytème Nerveux de l'homme et des vertébrés* Tome premier. Paris: A. Maloine. Reprint: 1952, Madrid: CSIC.

Rechtschaffen, A. (1971). The control of sleep. In W. A. Hunt (Ed.), *Human behaviour and its control* (pp. 75–92). Cambridge, MA: Schenkman.

Rial, R. V., & González, J. (1978). Kindling effect in the reptilian brain: Motor and electrographic manifestations. *Epilepsia, 19*, 581–589.

Rial, R. V., Nicolau, M. C., Gamundi, A., Akaârir, M., Aparicio, S., Garau, C., et al. (2007). The trivial function of sleep. *Sleep Medicine Reviews, 11*, 311–325.

Rosa, M. G. P., & Kubitzer, L. A. (1999). The evolution of visual cortex: Where is V2? *Trends in Neurosciences, 22*, 242–248.

Sagan, C. (1977). *The dragons of Eden: Speculations on the evolution of human intelligence.* New York: Ballantine Books.

Servit, Z., & Strejčkova, A. (1972). Thalamocortical relations and the genesis of epileptic electrographic phenomena in the forebrain of the turtle. *Experimental Neurology, 35*, 50–60.

Servit, Z., Strejčkova, A., & Volanschi, D. (1971). Epileptic focus in the forebrain of the turtle (Testudo graeca). Triggering of focal discharges with different sensory stimuli. *Physiologia Bohemoslovaca, 20*, 221–228.

Sewards, T. V., & Sewards, M. A. (2002). Innate visual object recognition in vertebrates: Some proposed pathways and mechanisms. *Comparative Biochemistry and Physiology, Series A, 132*, 861–891.

Siegel, J. M. (2001). The REM sleep-memory consolidation hypothesis. *Science, 294*, 1058–1063.

Siegel, J. M., & McGinty, D. J. (1977). Pontine reticular neurons: Relationship of discharge to motor activity. *Science, 196*, 678–680.

Sirota, A., Csicsvari, J., Buhl, D., & Buzsáki, G. (2003). Communication between neocortex and hippocampus during sleep in rodents. *Proceedings of the National Academy of Sciences, 100*(4), 2065–2069.

Smeets, W. J. A. J., & Medina, L. (1995). The efferent connections of the nucleus accumbens in the lizard *Gecko gecko*. A combined tract-tracing transmitter-immunohistochemical study. *Anatomical Embryology, 191*, 73–81.

Sprague, J. M. (1966). Interactions of the cortex and superior colliculus in mediation of visually guided behaviour in the cat. *Science*, *153*, 1544–1547.

Steriade, M. (2000). Brain electrical activity and sensory processing during waking and sleeping states. In M. H. Kryger, T. Roth, & W. C. Dement (Eds.), *Principles and practice of sleep medicine* (pp. 93–111). London: Saunders.

Steriade, M., & Amizca, F. (1998). Slow sleep oscillation, rhythmic K-complexes and their paroxismal developments. *Journal of Sleep Research*, *7*(Suppl. 1), 30–35.

Steriade, M., Curro-Dossi, R., & Nuñez, A. (1991). Network modulation of a slow intrinsic oscillation of cat thalamocortical neurons implicated in sleep delta waves: Cortical potentiation and brainstem cholinergic suppression. *Journal of Neuroscience*, *11*, 200–217.

Steriade, M., Oakson, G., & Ropert, N. (1982). Firing rates and patterns of midbrain reticular neurons during steady and transitional states of the sleep-waking cycle. *Experimental Brain Research*, *46*, 37–51.

Susic, V. (1972). Electrographic and behavioral correlations of the rest-activity cycle in the sea turtle, *Caretta caretta* L. (Chelonia). *Journal of Experimental Marine Biology and Ecology*, *10*, 81–87.

Takakusaki, K., Habaguchi T., Ohtinata-Sugimoto, J., Saito, K., & Sakamoto, T. (2003). Basal ganglia efferents to the brainstem centers controlling postural muscle tone and locomotion: A new concept for understanding motor disorders in basal ganglia dysfunction. *Neuroscience*, *119*, 293–308.

Tauber, E. S. (1974). Phylogeny of the sleep. *Advances in Sleep Research*, *1*, 133–172.

ten Donkelaar, H. J. (1998). Reptiles. In R. Nieuwenhuis, H. J. ten Donkelaar, & C. Nicholson (Eds.), *The central nervous system of vertebrates* (Vol. 2, pp. 1315–1524). Berlin: Springer.

ten Donkelaar, H. J, & De Boer-Van Huizen, R. (1981). Basal ganglia projections to the brain stem in the lizard Varanus exantematicus as demonstrated by retrograde transport of horseradish peroxidase. *Neuroscience*, *6*, 1567–1590.

Tumarkin, A. (1948). On the phylogeny of the mammalian auditory ossicles. *Journal of Laryngology and Otology*, *62*, 687–690.

van Luijtelaar, E. L. J. M. (1997). Spike-wave discharges and sleep spindles in rats. *Acta Neurobiologiae Experimentalis*, *57*, 113–121.

Vasilescu, E. (1970). Sleep and wakefulness in the tortoise (*Emys orbicularis*). *Revue Roumaine de Biologie Serie de Zoologie*, *15*(3), 177–179.

Vertes, R. P. (2004). Memory consolidation in sleep: Dream or reality. *Neuron*, *1*, 135–148.

Vertes, R. P., & Eastman, K. E. (2000). The case against memory consolidation in REM sleep. *Behavioral Brain Sciences*, *23*, 1057–1063.

Voneida, T. J., & Sligar, C. M. (1979). Efferent projections of the dorsal ventricular ridge and the striatum in the tegu lizard, *Tupinambis nigropuntatus*. *Journal of Comparative Neurology*, *186*, 43–64.

Wallace, S. F., Rosenquist, A. C., & Sprague, J. M. (1990). Ibotenic acid lesions of the lateral substantia nigra restore visual orientation behavior in the hemianopic cat. *Journal of Comparative Neurology*, *296*(2), 222–252.

Williams, G. C. (1966). *Adaptation and natural selection: A critique of some current evolutionary thought*. Princeton, NJ: Princeton University Press.

Zepelin, H. (1994). Mammalian sleep. In M. H. Kryger, T. Roth, & W. C. Dement (Eds.), *Principles and practice of sleep medicine* (pp. 69–80). Philadelphia: W. B. Saunders.

Zepelin, H., & Rechtschaffen, A. (1974). Mammalian sleep, longevity, and energy metabolism. *Brain Behavior and Evolution*, *10*, 425–470.

Zepelin, H., Siegel, J., & Tobler, I. (2005). Mammalian sleep. In M. H. Kryger, T. Roth, & W. C. Dement (Eds.), *Principles and practice of sleep medicine* (pp. 91–100). Philadelphia: W. B. Saunders.

9

The evolution of REM sleep

MAHESH M. THAKKAR AND SUBIMAL DATTA

Introduction

Since the dawn of civilization, sleep has fascinated humankind. Myriad treatises and reviews, scientific and nonscientific, have been written in an attempt to explain the phenomenon of sleep, yet none has been comprehensive enough to gain general acceptance. It is now well established that sleep is neither a unitary nor a passive process. Intricate neuronal systems via complex mechanisms are responsible for controlling sleep. This chapter focuses on the evolution of rapid-eye-movement (REM) sleep; for detailed information about other behavioral states, the reader is referred to several comprehensive reviews (Datta & Maclean, 2007; Jones, 2003; Mignot, 2004; Siegel, 2004; Steriade & McCarley, 2005). We begin with a brief description of the discovery of REM sleep and then describe the phylogeny and evolution of REM.

Discovery of REM sleep

The discovery of REM sleep, a major breakthrough, revolutionized the field of sleep research. The process that led to this discovery began in Kleitman's laboratory at the University of Chicago Medical School in 1953. Kleitman and his graduate student Eugene Aserinsky noticed rhythms in eye movements during sleep in humans and linked this to dreaming (Aserinsky & Kleitman, 1953, 1955). Subsequently, Dement and Kleitman (1957) characterized the electroencephalographic (EEG) activity during dreaming in humans, and later Dement (1958) recorded rapid eye movements during sleep in animals. These discoveries established the presence of the non-REM–REM sleep cycle. However, it was only after Jouvet's demonstration of muscle atonia (total suppression of muscle tone) and the importance of the pontine reticular formation in REM sleep (which he termed "sommeil paradoxal" or paradoxical sleep [referenced in Dement, 2000; Jones, 1991; Jouvet & Mounier, 1960; and Jouvet, Michel, & Courjon, 1959]) that finally established REM sleep as a distinct state of behavior along with wakefulness and non-REM (NREM) sleep.

The evolution of REM sleep

To understand the evolution of REM sleep, it has been important to conduct phylogenetic studies of REM sleep. Elucidation of the quantitative and qualitative expression of REM sleep in diverse animal species has provided insight into the physiology and function of REM sleep. However, several major considerations must be taken into account in evaluating the results of such studies.

The mammalian class encompasses more than 4500 species. However, systematic REM sleep studies have been conducted in few of them, mostly in domesticated animals like cats and dogs as well as in rats and monkeys. Relatively few studies have been conducted in rabbits and pigs, and a very little information is available for other mammals or for other vertebrate species, including birds, reptiles, and amphibians (Siegel, 1999). Finally, there is almost no information about REM sleep in invertebrates. Thus our knowledge about REM sleep is generally limited to mammals, which represent a miniscule fraction of more than a million animal species. In short, by focusing most research effort on mammals, it becomes difficult to draw any conclusions about the evolution of REM sleep.

The major limitation to the study of REM sleep in new species has been the definition of REM sleep. What we know about it has been acquired in the course of studies in humans and domesticated laboratory animals. Most of our knowledge about the cellular mechanisms responsible for the regulation of REM sleep has been achieved by studying neuronal mechanisms in cats. However, because most behavioral and electrophysiological characteristics that describe REM sleep in cats are also observed in humans and other mammalian species, it is likely that similar cellular regulators control REM sleep in other mammalian species, including humans.

Studies in cats have revealed that the primary neuronal systems responsible for the generation of REM sleep are localized in the brainstem (Datta & Maclean, 2007). There, the REM-generating core consists of two major populations of neurons with REM-selective discharge. The norepinephrine (NE)-containing neurons of the locus ceruleus (LC) and serotonin (5-HT)-containing neurons of the dorsal raphe nucleus (DRN) constitute the "REM-off neurons," which cease their activity during REM sleep. On the other hand, the acetylcholine-containing cholinergic neurons of the mesopontine tegmentum along with GABA- and glutamate-containing neurons of the pontine reticular formation increase their activity during REM sleep and are the "REM-on neurons" (Datta & Maclean, 2007; McCarley, Greene, Rainnie, et al., 1995; Siegel, 1995). Activation/inhibition of these neuronal systems is responsible for the behavioral and electrophysiological signatures of REM sleep, including a desynchronized, low-amplitude EEG, hippocampal theta activity, postural muscle atonia (complete loss of activity in antigravity muscles), frequent bursts of eye movements, muscle twitches, and ponto-geniculo-occipital (PGO) waves coupled with a total absence of awareness of the environment as well as blockade of sensory inputs and an elevated arousal threshold (McCarley et al., 1995; Siegel, 1995). Another important characteristic of REM sleep is the presence of REM sleep rebound following REM sleep deprivation. Thus our defining criteria for REM sleep are diverse and complex; they may seem difficult or even impossible to achieve in diverse animal species. For example, it can be difficult to observe muscle atonia in poikilotherms, because they may exhibit general hypotonia during their rest cycle. Even mammalian newborns (especially humans, rats, and cats) do not exhibit a desynchronized EEG (a prominent characteristic of REM sleep in adults), and REM sleep (also known as active sleep) is identified primarily on the basis of muscle twitching – a manifestation of phasic motor activation (Blumberg & Lucas, 1996; Frank & Heller, 1997; Harper, Leake, Miyahara, et al., 1981; McGinty, Stevenson, Hoppenbrouwers, et al., 1977; Siegel, 1999). Active sleep in neonates is generally accepted as an underdeveloped REM sleep state, on the assumption that a "developmental continuity" exists between active sleep in neonates and full-blown REM sleep in adults. Both active sleep in neonates and REM sleep in adult mammals can be identified purely on the basis of behavioral characteristics. The point that we would like to make is that it is possible to identify REM sleep in a newly examined animal species purely on the basis of a few behavioral characteristics.

The cessation of neuronal activity in the monoaminergic neurons along with phasic activation of the motor systems during REM sleep are unique characteristics that readily differentiate REM sleep from two other states of behavior, including wakefulness and NREM sleep. Thus, ideally and ultimately, if one could demonstrate recurrent cessation of monoaminergic neuronal activity and phasic motor activation along with the established primary characteristics of sleep – including (1) spontaneous assumption of species-specific or stereotypic posture, (2) maintenance of behavioral quiescence, (3) elevated behavioral response threshold, (4) rapid reversibility of state with strong threshold stimulus, and (5) presence of compensatory changes following sleep deprivation – it would be certain that a REM-like state distinct and different from NREM sleep is present in a newly examined animal (Campbell & Tobler, 1984; Flanigan, 1973, 1974; Flanigan, Knight, Hartse, et al., 1974; Flanigan, Wilcox, & Rechtschaffen, 1973).

REM sleep in invertebrates

Invertebrates are among the most ancient forms of life on this planet, and most have simple nervous systems.

Although invertebrates make up about 98% of the entire animal kingdom, relatively few studies have been performed to monitor rest and activity cycles in them. The existence of a sleep state has been described in cephalopods (squids and octopi) and mollusks (aplysia) (Mather, 2008; Strumwasser, 1971; Tobler, 1997). In her review, Mather describes sleep rebound following sleep deprivation, along with other characteristics of a sleep-like quiescent behavioral state in octopi, including narrowing of pupils and increased arousal threshold. Furthermore, Mather claims that there may be a cephalopod equivalent of mammalian REM sleep in which the animal changes its color. However, further work is necessary to establish REM sleep in cephalopods.

A sleep-like state of quiescence that satisfies several behavioral criteria of sleep has been reported in insects and arachnids, including cockroaches, bees, and scorpions (reviewed in Hartse, 1994; see also Chapter 2 in this volume), although it has not been claimed that REM sleep exists in insects. Rest in *Drosophila melanogaster* appears to satisfy all the behavioral criteria of sleep (Hendricks, Finn, Panckeri, et al., 2000; Shaw, Cirelli, Greenspan, et al., 2000); however, it is unclear whether the *Drosophila* rest state is similar to the sleep state experienced by humans. Circadian and vigilance changes in sensory response thresholds have been described in honeybees (Frank, 1999; Kaiser & Steiner-Kaiser, 1983). Increased arousal threshold along with compensatory increases in rest following rest deprivation has been reported in crayfish (Ramon, Hernandez-Falcon, Nguyen, et al., 2004). Circadian changes in periods of rest and activity have been observed in cockroaches, although rest deprivation did not produce a major compensatory increase in rest (Tobler & Neuner-Jehle, 1992); however, there was an increase in the metabolic rate (Siegel, 2008; Stephenson, Chu, & Lee, 2007).

REM sleep in vertebrates

Vertebrates are members of the subphylum Vertebrata within the phylum Chordata. The vertebrates are characterized by the presence of backbones or spinal columns. Vertebrata is the largest subphylum of chordates, with more than 50,000 species. They are classified into five groups on the basis of their skin covering, how they reproduce, how they maintain body temperature, and characteristics of their limbs (arms and legs or their equivalents, such as wings or fins). The five classes are fishes, amphibians, reptiles, birds, and mammals. Although a sleep-like state has been demonstrated in some species from each of the five classes, REM sleep has been observed only in mammals and birds.

REM sleep in fishes

The class fishes consists of more than 25,000 recognized species and is the most diverse vertebrate group, comprising about half of all known vertebrates, among which less than 10 species have been examined for sleep (Siegel, 2008; see also Chapters 3 and 11 in this volume).

A substantial number of behavioral studies suggest that fishes may exhibit sleep-like states (Reebs, 2007). Behavioral identification of a sleep-like state with eye movement in several species of Bermuda reef fish was reported by Tauber (Frank, 1999; Tauber, Weitzman, & Korey, 1969). Subsequently, Marshall (1972) and Shapiro and Hepburn (1976) reported the presence of sleep-like states in various species. Although, these earlier studies did not use all of the Flanigan–Tobler criteria (Campbell & Tobler, 1984) to identify sleep, most studies did find states of prolonged inactivity, resting posture, 24-hour rhythmicity, and increased arousal thresholds. Tobler and Borbély (1985) reported a rest–activity rhythm with activity predominating during the light period in two fish species (*Cichlosoma nigrofasciatum* and *Carassius auratus*). Rest deprivation of the perch (*Cichlosoma nigrofasciatum*) by constant light conditions during normally inactive periods produced an increase in rest behavior during the subsequent 12-hour period. No evidence for a compensatory sleep rebound response was reported. Recently, a rest/sleep-like state has been characterized in zebrafish (*Danio rerio*) (Yokogawa et al., 2007; see also Chapter 11 in this volume). The rest/sleep-like state in zebrafish showed circadian variations in responsiveness and activity and reduced responses to stimuli after deprivation of the sleep-like state. There was, however, complete absence of a rest/sleep-like state with long periods of light, with no evidence of subsequent rebound.

Although there are no convincing reports of REM sleep in fishes, eye movement during the period of rest (Tauber, Rojas-Ramírez, & Hernández Peón, 1968) has been recorded in some species and the EEG spikes and slow waves have been recorded from the mid- and forebrain of the catfish (*Ictalurus nebulosus*).

REM sleep in amphibians

Amphibians are believed to have evolved from air-breathing freshwater fish approximately 400 million years ago, during the mid-Devonian period. There are more than 5000 species of amphibians, all of which are members of one of three main groups: frogs and toads (order Anura), salamanders (order Caudata), and caecilians or limbless amphibians (order Gymnophiona). Amphibians were the first animals with backbones to adapt to life on land. They are the ancestors of reptiles, which in turn gave rise to mammals and birds. Thus it would be interesting to study sleep in this class of vertebrates. The majority of sleep studies have been performed in the order Anuras (Hobson, 1967; Hobson, Goin, & Goin, 1968; Huntley, Donnely, & Cohen, 1969; Karmanova & Lazarev, 1979; Segura & De Juan, 1966). Sleep-like states have been reported in diurnally active tree frogs (genus *Hyla*). There was no evidence of REM sleep in these species. Sleep-like states in tree frogs were associated with stereotyped sleep postures, elevated arousal thresholds, and low-amplitude, high frequency (8- to 30-hertz) EEG activity in the forebrain (Hobson et al., 1968). In contrast, the diurnal bullfrog (*Rana catesbeiana*) did not show any sleep-like states (Hobson, 1967). Although its levels of activity varied in a circadian pattern, the animals were *more* responsive during periods of inactivity. Thus, in such periods, the bullfrog maintained a state of resting vigilance (Hobson, 1967). The authors concluded that although these amphibian species are highly vulnerable to predation, they have survived mainly because they can maintain a state of resting without loss of vigilance (Hobson, 1967). Rest/sleep-like states associated with low EEG frequencies (5 to 7 hertz) coupled with reduced amplitude have been observed in toads (*Bufo boreas*) (Huntley et al., 1969). Similar rest/sleep-like states associated with low frequencies (<2 hertz) along with spike-like EEG activity have also been observed in the European frog (*Rana temporaria*). Spectral analysis of the EEG performed in the salamander (*Ambystoma tigrinum*) revealed two distinct patterns of EEG: the quiet, resting state associated with greater power density in the lower-frequency EEG and the alert, active state associated with greater power density in the higher frequencies (Lucas, Sterman, & McGinty, 1969). However, no sleep-like state was reported. Based on these few studies, it appears that some amphibians do show a sleep-like state, although REM sleep is absent in these vertebrates.

REM sleep in reptiles

Fossil studies suggest that reptiles evolved from their amphibian ancestors in the early Carboniferous period, about 340 million years ago. Reptiles surpassed amphibians as the dominant vertebrates on land. There are more than 7000 species of living reptiles, all of which belong to one of four main groups: turtles and tortoises (order Testudines/Chelonia); lizards, worm lizards, and snakes (order Squamata); crocodiles (order Crocodilia); and the lizard-like tuatara (order Rhynchocephalia).

Rest–activity cycles have been studied in various species in the orders Chelonia, Squamata, and Crocodilia, and all studied species showed NREM sleep-like states that satisfied Flanigan–Tobler criteria. The quiescent NREM state in Chelonia – including red-footed tortoise (*Geochelene carbonaria*), box turtle (*Terrapene carolina carolina*), Texas tortoise (*Gopherus berlandieri*), and Bolson tortoise (*Gopherus flavomarginatus*) – was found to be associated with a stereotypic sleep-like posture, increased arousal threshold, compensatory increase in sleep-like state following enforced wakefulness, and telencephalic EEG spikes coupled with sharp waves that disappeared on arousal (Ayala-Guerrero, Calderón, & Pérez, 1988; Ayala-Guerrero, Huitrón-Reséndiz, & Mexicano, 1993; Flanigan, 1974; Flanigan et al., 1974; Frank, 1999). Sleep-like states have also been reported in the European pond turtle (*Emys orbicularis*); however, EEG spikes were absent. Instead, a low-frequency EEG was observed (Frank, 1999; Vasilescu, 1982).

The quiescent NREM sleep–like state similar to that reported in Chelonia and associated with eye closures, increased arousal threshold, and compensatory sleep response to enforced wakefulness has been observed in several reptilian species of the orders Squamata and Crocodilia, including lizards (*Ctenosaura pectinata* and *Iguana iguana*) and crocodiles (*Caiman sclerops*) (Ayala-Guerrero & Huitrón-Reséndiz, 1991; Flanigan, 1973; Flanigan et al., 1973; Warner & Huggins, 1978). Some disagreement exists, however, regarding the type of EEG activity that is present during the NREM sleep–like state. Some authors have reported EEG spikes during NREM sleep in Squamata and Crocodilia (Flanigan, 1973; Flanigan et al., 1973); others have reported an overall decrease in EEG activity (frequency and amplitude) (Ayala-Guerrero & Huitrón-Reséndiz, 1991; Tauber et al., 1968; Warner & Huggins, 1978).

Complete absence of an NREM sleep–like state in some reptilian species – including the tortoise (*Testudo denticulate*), sea turtle (*Caretta caretta*), and American alligator (*Alligator mississipiensis*) – has also been reported (Susic, 1972; Van Twyver, 1973; Walker & Berger, 1973). It is entirely possible that some reptilian species do not have a NREM sleep–like state. Therefore, until further studies can resolve this issue, one must be cautious in drawing conclusions regarding the presence of NREM sleep in reptiles.

A majority of studies did not find any strong evidence of a REM sleep–like state in reptiles, albeit aperiodic saccadic and slow movements of the eyes and/or small movement of the head and mouth not associated with a waking EEG were sometimes observed during reptilian sleep (Flanigan, 1973, 1974; Flanigan et al., 1973, 1974; Frank, 1999; Hartse & Rechtschaffen, 1974, 1982). However, some studies have reported the presence of REM sleep in reptiles.

Tauber and coworkers reported the presence of REM bursts during sleep in the lizard *Ctenosaura pectina* (Tauber et al., 1968) and the chameleons *Chameleo jacksoni* and *Chameleo melleri* (Tauber, Roffwarg, & Weitzman, 1966). Huntley, Friedman, and Cohen (1977; see also Frank, 1999) reported "paradoxical sleep" associated with an irregular cardiorespiratory output, extremely low motor tone, myoclonic bursts, and periodic episodes of theta (5- to 10-hertz) activity in the EEG. Ayala-Guerrero and coworkers reported an "active sleep–like" state characterized by rapid eye movements as well as slight increases in heart rate and phasic muscle activity in lizards (*Iguana iguana, Ctenosaura similis,* and *Ctenosaura pectinata*), the desert tortoise (*Gophe flavomarginalis*), and the turtle *Gopherus berlandieri* (Ayala-Guerrero & Huitrón-Reséndiz, 1991; Ayala-Guerrero & Mexicano, 2008a,b; Ayala-Guerrero et al., 1988; Ayala-Guerrero, Huitrón-Reséndiz, & Mexicano, 1993, 1994). REM sleep–like states associated with REM sleep, muscle atonia, and EEG activation have also been reported in the turtle *Emys orbicular* (Vasilescu, 1983; see also Frank, 1999).

Although there are multiple reports of REM sleep, especially in lizards, these reports should be interpreted cautiously, because brief arousals (by monitoring arousal thresholds) have never been adequately distinguished from REM sleep–like states. In several reports, REM sleep–like states preceded eye opening and wakefulness. Thus it is likely that REM and periodic motor activity in reptiles may represent a preparatory state that normally precedes arousal. Consequently, periods of wake-like EEG during sleep could represent wakefulness and not REM sleep. Finally, Siegel and coworkers recorded discharge activity of brainstem neurons in box turtles (*Terrapene carolina*). There was a marked reduction in brainstem neuronal activity immediately on cessation of waking activity, which was further reduced during extended periods of inactivity. Periodic activation of brainstem neuronal discharge (which would signify the presence of a REM sleep–like state during quiescence) was not present (Eiland, Lyamin, & Seigel, 2001). In summary, it is as yet uncertain whether a REM sleep–like state is present in reptiles.

REM sleep in birds

Birds (class Aves) are bipedal, endothermic (warm-blooded), egg-laying vertebrates. They are particularly interesting for sleep research because both REM sleep state and NREM sleep state are present in them. It is interesting to note that birds and mammals evolved from the same common ancestor, the reptiles. However, unlike the reptiles, which do not have REM sleep–like states, both mammals and birds do have REM sleep, suggesting that this state evolved after the segregation of avian and mammalian lines from their reptilian ancestors (see also Chapter 6 in this volume).

The majority of sleep studies in birds have been performed in pigeons (Newman, Paletz, Rattenborg, et al., 2008; Rattenborg, Obermeyer, Vacha, et al., 2005; Tradardi, 1966; Van Twyver & Allison, 1972; Walker & Berger, 1972) and chickens (Guntheroth, 1979; Mascetti & Vallortigara, 2001; Mascetti, Rugger, & Vallortigara, 1999; Ookawa, 1972; Ookawa & Gotoh, 1965; Schlehuber, Flaming, Lange, et al.,

1974; van Luijtelaar, van der Grinten, Blokhuis, et al., 1987). Sleep studies in other bird species are few (Campbell & Tobler, 1984; Roth, Lesku, Amlaner, et al., 2006). A wide variety of avian species exhibit rest states that satisfy the Flanigan–Tobler criteria for the presence of sleep and are strikingly similar to mammalian NREM and REM sleep. Avian NREM sleep is associated with large-amplitude slow-wave activity similar to that in mammals, although some avian species also show the diphasic EEG spikes associated with reptilian sleep (Amlaner, 1994). The REM-like state similar to mammalian REM sleep is also observed in birds. The avian REM sleep is associated with decreased EEG voltage (desynchronized EEG), changes in cardiorespiratory ouput, increased phasic motor activity, reduced muscle tone, and presence of eye movement (in most species) (Amlaner, 1994; Frank, 1999).

Some major differences exist between avian and mammalian REM sleep. Although muscle tone is reduced during avian REM sleep, the complete motor atonia typical of mammalian REM sleep is rarely observed. Avian REM sleep periods tend to be shorter than those observed in most mammals, and birds spend significantly less time than mammals in REM sleep than mammals. Finally, unlike the case in mammals, in birds there is a complete absence of compensatory increase in sleep following sleep deprivation (Berger & Phillips, 1994; Martinez-Gonzalez, Lesku, & Rattenborg, 2008). However, sleep deprivation does produce an increase in EEG slow-wave activity (0.78 to 2.34 hertz) during NREM sleep along with the time spent in REM sleep (Martinez-Gonzalez et al., 2008). This implicates similar homeostatic mechanisms involved in the control of REM sleep in birds and mammals. Thus it appears that birds do exhibit a REM sleep state that is very similar to that in mammals.

REM sleep in mammals

Mammals are believed to have evolved from mammal-like reptiles called therapsids, which appeared more than 200 million years ago. There are more than 4500 species of mammal. Whereas majority of mammals live on land, there are some that permanently live in water (aquatic mammals) and some that can fly. Mammals are divided into three groups. The monotremes are egg-laying mammals and consist of just three species: the duck-billed platypus and two species of echidnas, or spiny anteaters. The marsupials are pouch-bearing mammalian species. There are more than 200 marsupial species, including kangaroos, koalas, and opossums. The placental include about 4300 species. Within the placental group, one group of aquatic mammals, including whales and dolphins, are grouped together in the order Cetacea.

Many species from all three groups have been studied to ascertain the presence of sleep and REM sleep, and both NREM and REM sleep have been observed in all marsupials and almost all placental species except cetaceans studied to date. There is some controversy regarding REM sleep in monotremes.

REM sleep in mammals has been extensively reviewed by Zepelin (Zepelin, 1994, 2000; Zepelin & Rechtschaffen, 1974); the interested reader is referred to these reviews. Here we review REM-like sleep in primitive mammals in detail and highlight some unresolved issues.

REM sleep in monotremes

The monotremes diverged from placental and marsupial mammals approximately 130 million years ago. Monotremes have changed relatively little since their initial divergence from other mammals and are thought to be closer representatives of the therapsid reptile ancestors than any other extant mammal (Frank, 1999). The monotremes have many typical characteristics of mammals, including the single bone in the lower jaw seen in primitive mammals, three middle ear bones, high metabolic rates, hair, and the production of milk to nourish the young. However, they are very primitive because, like reptiles and birds, they lay eggs rather than having live births. The three extant monotreme species are the short-beaked and long beaked echidnas (*Tachyglossus aculeatus* and *Zaglossus brujini*) and the duck-billed platypus (*Ornithorhynchus anatinus*). The monotremes, therefore, offer a unique glimpse into the evolution of mammalian REM sleep (Frank, 1999; Siegel, 1995, 2008).

Sleep has been studied in two monotremes, the echidna, *Tachyglossus aculeatus*, and the more ancient, amphibious duck-billed platypus, *Ornithorhynchus anatinus* (Allison, Van Twyver, & Goff, 1972; Nicol, Andersen, Phillips, et al., 2000; Siegel, Manger, Nienhuis, et al., 1996, 1999). We begin by reviewing sleep in the duck-billed platypus, followed by a review of sleep in the echidna.

Siegel et al. (1999) conducted the first and only sleep study in the platypus *Ornithorhynchus anatinus*. This platypus displayed a REM sleep–like state that occupied more than 60% of total sleep time and was characterized by the presence of muscle atonia concomitant with REM, muscle twitching, and a heightened arousal threshold. However, the EEG showed moderate or high voltage, similar to NREM sleep in adult mammals. Based on these findings, the authors concluded that REM sleep may have evolved in premammalian reptiles but that the low-voltage EEG is a recently evolved characteristic of REM sleep.

The duck-billed platypus showed an abundance of REM sleep, but there is some controversy regarding the presence of REM sleep in the echidna (*Tachyglossus aculeatus*). The first study conducted by Allison et al. (1972) reported that the echidna had NREM sleep characterized by a high-voltage EEG but that the classic signature of REM sleep, desynchronized EEG, was absent. Although presumptive REM sleep–like states were sometimes noted, these episodes were not associated with heightened arousal thresholds. There were no rapid eye movements, nor was there evidence of phasic motor activation during sleep (Allison et al., 1972). Therefore Allison et al. suggested that REM sleep was absent in monotremes. Subsequently, Siegel et al. (1996) conducted sleep studies in the echidna (*Tachyglossus aculeatus*) and found similar results. However, Siegel et al. found that the active periods preceeded the quiescent periods, as in the REM or paradoxical sleep–like episodes observed by Allison et al., and not the NREM sleep period. Furthermore, Seigel et al. also recorded brainstem neuronal activity in the echidna and reported that this activity in the echidna was higher than the neuronal activity observed during NREM sleep in mammals but lower than neuronal activity observed during mammalian REM sleep. Based on these findings, the authors concluded that the echidna does not have REM sleep and instead exhibits a sleep state intermediate between REM and NREM sleep. According to these authors, echidna sleep represents a primordial mammalian sleep state, preserved in monotremes, that segregated into the distinct states of REM and NREM sleep during the course of mammalian evolution (Siegel, 1995; Siegel, et al., 1996).

Nicol et al. (2000), who repeated sleep recordings of echidna at an ambient temperature of 25°C, reported a REM sleep–like state characterized by a desynchronized EEG with low voltage, a reduced tonic electromyogram, rapid eye movements, and intermittent, occasional muscle twitches. In most cases, REM sleep–like episodes were preceded by NREM sleep and followed by wakefulness. The REM sleep–like state constituted 15.5% of total sleep. Changes in ambient temperature and age affected REM sleep. The authors concluded that "Manifestations of REMS in the phylogenetically ancient echidna are similar to those in all investigated mammalian and avian species" (Nicol et al., 2000, p. 52).

REM sleep in primitive marsupial and placental mammals

The presence of a REM sleep–like state has been observed in many marsupials and in placental mammals. The marsupial opossum (*Didelphis marsupialis*) displays a REM-like state that is characterized by the presence of a desynchronized EEG, reduced muscle tone, phasic motor activation, and elevated arousal thresholds. The opossum spends approximately 18 hours of the day sleeping, of which 5 hours (about 30% of sleep time) is spent in REM sleep (Allison & Van Twyver, 1970; Zepelin, 1994). The presence of REM sleep–like states has also been reported in the primitive placental mammal armadillo (*Dasypus novemcinctus, Chaetophractus villosus*) (Affanni, Cervino, & Marcos, 2001; Prudom & Klemm, 1973). Prudom and Kelm reported that the armadillo (*Dasypus novemcinctus*) spends approximately 8.9% to 21.5% of its sleep time in REM sleep, characterized by the presence of a desynchronized EEG and muscle atonia. However, this REM sleep–like state constituted only about 7.5% of the armadillo's total sleep time. Affanni et al. (2001) reported that the REM-like state in the armadillo (*Chaetophractus villosus*) is characterized by the presence of a desynchronized EEG, REMs, muscle atonia, irregular respiration, muscle twitches, and movements of the vibrissae and that this animal spends approximately 22% of its total sleep time in REM sleep. In addition, Affanni et al. did not observe penile erections during REM sleep in the armadillo (*Chaetophractus villosus*); instead, penile erections were observed during NREM sleep. In humans and rats, penile erections are typically observed during REM sleep (Hirshkowitz & Schmidt, 2005; Schmidt, Valatx, Schmidt, et al., 1994).

The hedgehog (*Erinaceus europaeus*) also displays a REM sleep–like state characterized by a desynchronized EEG, absence of muscle tone, muscle twitches, REMs, and an increased threshold to arousal (Frank, 1999; Monnier, 1980). The hedgehog spends more than 25% of its sleep time in REM sleep (Zepelin, 1989). Similar REM sleep–like states have also been reported in moles (*Scalopus aquaticus* and *Condylura cristata*) and shrews (*Suncus murinus, Blarina brevicauda,* and *Cryptotis parva*) (Allison & Van Twyver, 1970; Allison, Gerber, Breedlove, et al., 1977). The shrews spent approximately 18% and the moles approximately 25% of their sleep time in REM sleep.

REM sleep in marine mammals

Marine mammals are a diverse group of roughly 120 species of mammal that are primarily ocean-dwelling or depend on the ocean for food. They include the cetaceans (whales, dolphins, and porpoises), the sirenians (manatees and dugongs), the pinnipeds (true seals, eared seals, and walruses), and several otters (the sea otter and marine otter). The polar bear is also usually grouped with the marine mammals.

Behavioral observations in the early 1960s indicated that the cetaceans exhibit unilateral eye closure during periods of rest (Hediger, 1969; Lilly, 1964; McCormick, 1969; Shurley, Serafetinides, & Brooks, 1969). Subsequent studies revealed that pilot whales (*Globicephala scammoni*), harbor porpoises (*Phocoena phocoena*), bottlenose dolphins (*Tursiops truncates*), and Amazon dolphins (*Inia geoffrenis*) have a unique pattern of unihemispheric sleep. While one hemisphere displays the slow-wave EEG activity characteristic of NREM sleep, the other displays the low-voltage, high-frequency EEG characteristics of wakefulness (Mukhametov, 1987; Mukhametov, Supin, & Polyakova, 1977; Serafetinides, Shurley, & Brooks, 1972). These animals never show any high-voltage waves bilaterally (Siegel, 2008). This unique unihemispheric sleep allows the marine mammals to swim continuously even when asleep (Mukhametov, 1987; Mukhametov et al., 1977; Serafetinides et al., 1972). Visual observations of resting behavior have revealed signs of phasic motor activation, including body jerks, occasionally twitches, and eye movements resembling those of REM sleep in bottlenose dolphins (*Tursiops truncates*), Amazon dolphins (*Inia geoffrenis*), beluga whales (*Delphinapterus leucas*), and gray whale (*Eschrichtius robustus*) (Lyamin, Manger, Mukhametov, et al., 2000; Lyamin, Shpak, Nazarenko, et al., 2002; Mukhametov & Lyamin, 1994; Oleksenko, Chetyrbok, Polyakova, et al., 1994). Although one report suggests that pilot whales (*Globicephala scammoni*) display a small amount of REM sleep (one 6-minute episode in 3 days) characterized by a marked loss of trunk muscle tone and nonconjugate eye movement (Serafetinides et al., 1972; see also Lyamin et al., 2002), subsequent polygraphic studies in dolphins did not find any EEG features of REM sleep, suggesting that these marine mammals may not display a REM sleep–like state (Mukhametov, 1987; Mukhametov et al., 1977).

One study that conducted unihemispheric sleep deprivation in bottlenose dolphins produced mixed results, with little or no relation between the quantity of slow waves lost in each hemisphere and the quantity of slow waves recovered when the animals were subsequently left undisturbed (Oleksenko, 1992). In another study, it was shown that dolphins were able to maintain continuous vigilance with no decline in accuracy and high levels of target detection for 5 days. At the end of this period, there were no signs of sleep rebound, and the dolphins were able to perform without any detectable decrease in activity or evidence of inattention, although response time was significantly slower during the night (Ridgway, Carder, Finneran, et al., 2006).

Based on the studies described here, it appears that the cetaceans have either no or little REM sleep, even though the cetacean brain has retained the features of primitive placental species like the hedgehog and has neocortical development that is comparable to that of the primate brain (Frank, 1999).

The fur seal, sea lion, and manatee are the other marine mammals that have been studied with respect to sleep–wakefulness. The Amazonian manatee (*Trichechus inunguis*) is an aquatic mammal belonging to the order Sirenia. The Amazonian manatee also displayed interhemispheric asynchrony of EEG slow-wave activity. Although REM sleep was present, it occupied only about 3% of total sleep time (Mukhametov, Lyamin, Chetyrbok, et al., 1992). In contrast, the fur seal, which belongs to the order Otariidae, has displayed a unique pattern of sleep. When fur seals *(Callorhinus ursinus)* are on land, they generally displayed sleep patterns resembling those of most terrestrial mammals, including eyes closure, increased arousal threshold, bilateral synchronized EEG, and NREM–REM cycles (Siegel, 2008). However, in the water, fur seals displayed unihemispheric EEG synchronization with a profound reduction in the amount of time spent in REM sleep. Even after several weeks in the water, there was no REM rebound when the fur seal returned to land.

Reduced or complete absence of REM sleep in almost all marine mammals studied to date, especially when they are in water, may be related to the constraints of the aquatic environment in which these air-breathing mammals live. Complete muscle atonia and irregularities in the cardiovascular and respiratory output typically observed in terrestrial mammals may interfere with surfacing, which is necessary for respiration in this species (Frank, 1999).

What does this mean?

There is no evidence suggesting that REM sleep is present in invertebrates. Within the vertebrates, there is no evidence supporting the presence of REM sleep in fishes or amphibians. There is some evidence, albeit weak, suggesting the presence of REM sleep in reptiles. However, further detailed studies are necessary before we can conclude with any confidence that REM sleep is present in reptiles.

REM sleep is definitely found only in birds and mammals. However, major differences exist between avian and mammalian REM sleep. Birds display briefer REM bouts, and the total amount of time spent in REM sleep is much less than that in mammals. Birds do not display REM sleep rebound following sleep curtailment, suggesting a lack of REM homeostasis in birds. In contrast to mammals, birds do not display total muscle atonia during REM sleep, although muscle tone is greatly reduced. Sleep deprivation does not elicit rebound increases in avian REM sleep, suggesting that the birds lack homeostatic control of REM sleep.

There is convincing evidence of the existence of REM sleep in most or all mammals studied to date. However, detailed, systematic studies of mammalian REM sleep have been conducted mainly in laboratory environments and in domesticated/laboratory species including rats, mice, cats, dogs, and monkeys (Lesku, Roth, Amlaner, et al., 2006; Siegel, 2008). In fact, the definition of REM sleep as we know it today has been based on sleep studies conducted in these laboratory species. REM sleep in most other mammalian species has been identified purely on the basis of behavioral observations. Most of these studies have been performed when the animal has been in captivity. In contrast to the animals in the naturalistic environments, the animals in captivity do not have to struggle for food, and there are no threats from predators (Siegel, 2008). Does this easy lifestyle in a captive environment affect their REM sleep? For example, wild brown-throated three-toed sloths (*Bradypus variegatus*) in captivity spent approximately 70% of their time in sleep. In contrast, in the wild, the sloths spent approximately 40% of their time in sleep (de Moura Filho, Huggins, & Lines, 1983; Rattenborg, Voirin, Vyssotski, et al., 2008). Similarly, giraffes and elephants in captivity (in zoos) spend approximately 5 hours in sleep (Tobler, 1992; Tobler & Schwierin, 1996). These animals migrate for large distances over periods of weeks in the wild (Siegel, 2008). Do they spend the same amount of time in REM sleep (as observed in captivity) during periods of migration? Do they adapt to the environment and alter their REM sleep pattern? Indeed, the marine mammals adapt to reduce the amount of time spent in REM sleep when they are in water, and they do not display any REM sleep rebound. Beyond these definite expression of REM sleep in terrestrial mammals, there are clearly many questions needing further study concerning its evolution.

References

Affanni, J. M., Cervino, C. O., & Marcos, H. J. (2001). Absence of penile erections during paradoxical sleep. Peculiar penile events during wakefulness and slow-wave sleep in the armadillo. *Journal of Sleep Research*, *10*, 219–228.

Allison, T., Gerber, S. D., Breedlove, S. M., & Dryden, G. L. (1977). A behavioral and polygraphic study of sleep in the shrews *Suncus murinus*, *Blarina brevicauda*, and *Cryptotis parva*. *Behavioral Biology*, *20*, 354–366.

Allison, T., & Van Twyver, H. (1970). Sleep in the moles, *Scalopus aquaticus* and *Condylura cristata*. *Experimental Neurology*, *27*, 564–578.

Allison, T., Van Twyver, H., & Goff, W. R. (1972). Electrophysiological studies of the echidna, *Tachyglossus aculeatus*. I. Waking and sleep. *Archives Italiennes de Biologie*, 110, 145–184.

Amlaner, C. J. (1994). Avian sleep. In M. H. Kryger, T. Roth, & W. C. Dement (Eds.), *Principles and practice of sleep medicine* (2nd ed., pp. 81–94). Philadelphia: W. B. Saunders.

Aserinsky, E., & Kleitman, N. (1953). Regularly occurring periods of eye motility and concomitant phenomenon during sleep. *Science*, *118*, 273–274.

Aserinsky, E., & Kleitman, N. (1955). Two types of ocular motility occurring in sleep. *Journal of Applied Physiology*, *8*, 1–10.

Ayala-Guerrero, F., Calderón, A., & Pérez, M. C. (1988). Sleep patterns in a chelonian reptile *(Gopherus flavomarginatus)*. *Physiology and Behavior*, *44*, 333–337.

Ayala-Guerrero, F., & Huitrón-Reséndiz, S. (1991). Sleep patterns in the lizard *Ctenosaura pectinata*. *Physiology and Behavior*, *49*, 1305–1307.

Ayala-Guerrero, F., Huitrón-Reséndiz, S., & Mexicano, G. (1993). Effect of reserpine on sleep patterns in a chelonian reptile *(Gopherous berlandieri)*. *Proceedings of the Western Pharmacology Society*, *36*, 227–231.

Ayala-Guerrero, F., Huitrón-Reséndiz, S., & Mexicano, G. (1994). Effect of parachlorophenyl-alanine on sleep spikes in the iguanid lizard *(Ctenosaura pectinata)*. *Proceedings of the Western Pharmacology Society*, *37*, 149–152.

Ayala-Guerrero, F., & Mexicano, G. (2008a). Sleep and wakefulness in the green iguanid lizard *(Iguana iguana)*. *Comparative Biochemistry and Physiology, Part A: Molecular and Integrative Physiology*, *151*(3), 305–312.

Ayala-Guerrero, F., & Mexicano, G. (2008b). Topographical distribution of the locus coeruleus and raphe nuclei in the lizard *Ctenosaura pectinata*: Functional implications on sleep. *Comparative Biochemistry and Physiology, Part A: Molecular and Integrative Physiology*, *149*, 137–141.

Berger, R. J., & Phillips, N. H. (1994). Constant light suppresses sleep and circadian rhythms in pigeons without consequent sleep rebound in darkness. *American Journal of Physiology*, *267*, R945–R952.

Blumberg, M. S., & Lucas, D. E. (1996). A developmental and component analysis of active sleep. *Developmental Psychobiology, 29*, 1–22.

Campbell, S. S., & Tobler, I. (1984). Animal sleep: A review of sleep duration across phylogeny. *Neuroscience and Biobehavioral Reviews, 8*, 269–300.

Datta, S., & Maclean, R. R. (2007). Neurobiological mechanisms for the regulation of mammalian sleep-wake behavior: Reinterpretation of historical evidence and inclusion of contemporary cellular and molecular evidence. *Neuroscience and Biobehavioral Reviews, 31*, 775–824.

de Moura Filho, A. G., Huggins, S. E., & Lines, S. G. (1983). Sleep and waking in the three-toed sloth, *Bradypus tridactylus. Comparative Biochemistry and Physiology, Part A: Molecular and Integrative Physiology, 76*, 345–355.

Dement, W. (1958). The occurrence of low voltage, fast, electroencephalogram patterns during behavioral sleep in the cat. *Electroencephalography and Clinical Neurophysiology, 10*(2), 291–296.

Dement, W. C. (2000). History of sleep physiology and medicine. In M. H. Kryger, T. Roth, & W. C. Dement (Eds.), *Principles and practice of sleep medicine* (3rd ed., pp. 1–14). Philadelphia: W. B. Saunders.

Dement, W., & Kleitman, N. (1957). Cyclic variation in EEG during sleep and their relation to eye movements, body motility, and dreaming. *Electroencephalography and Clinical Neurophysiology, 9*, 673–690.

Eiland, M. M., Lyamin, O. I., & Siegel, J. M. (2001). State-related discharge of neurons in the brainstem of freely moving box turtles, *Terrapene Carolina major. Archives Italiennes de Biologie, 139*, 23–36.

Flanigan, W. F. Jr. (1973). Sleep and wakefulness in iguanid lizards, *Ctenosaura pectinata* and *Iguana iguana. Brain, Behavior, and Evolution, 8*, 401–436.

Flanigan, W. F. Jr. (1974). Sleep and wakefulness in chelonian reptiles. II. The red-footed tortoise, *Geochelone carbonaria. Archives Italiennes de Biologie, 112*, 253–277.

Flanigan, W. F. Jr., Knight, C. P., Hartse, K. M., & Rechtschaffen, A. (1974). Sleep and wakefulness in chelonian reptiles. I. The box turtle, *Terrapene carolina. Archives Italiennes de Biologie, 112*, 227–252.

Flanigan, W. F. Jr., Wilcox, R. H., & Rechtschaffen, A. (1973). The EEG and behavioral continuum of the crocodilian, *Caiman sclerops. Electroencephalography and Clinical Neurophysiology, 34*, 521–538.

Frank, M. G. (1999). Phylogeny and evolution of REM sleep. In B. N. Mallick & S. Inouse (Eds.), *Rapid eye movement sleep* (pp. 17–38). New Delhi: Narosa Publishing House.

Frank, M. G., & Heller, H. C. (1997). Development of REM and slow-wave sleep in the rat. *American Journal of Physiology, 272*, R1792–R1799.

Guntheroth, W. G. (1979). Cardiopulmonary changes in kittens during sleep. *Science, 205*, 1040–1041.

Harper, R. M., Leake, B., Miyahara, L., Hoppenbrouwers, T., Sterman, M. B., & Hodgman, J. (1981). Development of ultradian periodicity and coalescence at 1 cycle per hour in electroencephalographic activity. *Experimental Neurology, 73*, 127–143.

Hartse, K. (1994). Sleep in insects and nonmammalian vertebrates. In M. H. Kryger, T. Roth, & W. C. Dement (Eds.), *Principles and practice of sleep medicine* (2nd ed., pp. 95–104). Philadelphia: W. B. Saunders.

Hartse, K. M., & Rechtschaffen, A. (1974). Effect of atropine sulfate on the sleep-related EEG spike activity of the tortoise, *Geochelone carbonaria*. *Brain, Behavior, and Evolution, 9*, 81–94.

Hartse, K. M., & Rechtschaffen, A. (1982). The effect of amphetamine, nembutal, alpha-methyltyrosine, and parachlorophenylalanine on the sleep-related spike activity of the tortoise, *Geochelone carbonaria*, and on the cat ventral hippocampus spike. *Brain, Behavior, and Evolution, 21*, 199–222.

Hediger, H. (1969). Comparative observations on sleep. *Proceedings of the Royal Society of Medicine, 62*, 153–156.

Hendricks, J. C., Finn, S. M., Panckeri, K. A., Chavkin, J., Williams, J. A., Sehgal A, et al. (2000). Rest in *Drosophila* is a sleep-like state. *Neuron, 25*, 129–138.

Hirshkowitz, M., & Schmidt, M. H. (2005). Sleep-related erections: Clinical perspectives and neural mechanisms. *Sleep Medicine Reviews, 9*, 311–329.

Hobson, J. A. (1967). Electrographic correlates of behavior in the frog with special reference to sleep. *Electroencephalography and Clinical Neurophysiology, 22*, 113–121.

Hobson, J. A., Goin, O. B., & Goin, C. J. (1968). Electrographic correlates of behaviour in tree frogs. *Nature, 220*, 386–387.

Huntley, A., Donnely, M., & Cohen, H. B. (1969). Sleep in western toad *Bufo boreas*. *Journal of Sleep Research, 7*, 141.

Huntley, A., Friedman, J., & Cohen, H. B. (1977). Sleep in iguanid lizard *Dipsosaurus dorsalis*. *Journal of Sleep Research, 6*, 104.

Jones, B. E. (1991). Paradoxical sleep and its chemical/structural substrates in the brain. *Neuroscience, 40*, 637–656.

Jones, B. E. (2003). Arousal systems. *Frontiers in Bioscience, 8*, S438–S451.

Jouvet, M., & Mounier, D. (1960). Effects des lesions de la formation reticularaire pontique sur le sommeil du chat [Effects of lesions on the "pontique" reticular formation on cat sleep]. *Comptes Rendus des Séances et Mémoires de la Société de Biologie et des ses Filiales, 154*, 2301–2305.

Jouvet, M., Michel, F., & Courjon, J. (1959). Sur un stade d'activite electrique cerebrale rapide au tours du sommeil physiologique [On rapid cerebral electrical activity around physiological sleep]. *Comptes Rendus des Séances et Mémoires de la Société de Biologie et des ses Filiales, 153*, 1024–1028.

Kaiser, W., & Steiner-Kaiser, J. (1983). Neuronal correlates of sleep, wakefulness, and arousal in a diurnal insect. *Nature, 301*, 707–709.

Karmanova, I. G., & Lazarev, S. G. (1979). Stages of sleep evolution (facts and hypotheses). *Waking and Sleeping, 3*, 137–147.

Lesku, J. A., Roth, T. C., Amlaner, C. J., & Lima, S. L. (2006). A phylogenetic analysis of sleep architecture in mammals: The integration of anatomy, physiology, and ecology. *The American Naturalist, 168*, 441–453.

Lilly, C. (1964). Animals in aquatic environments: Adaptation of mammals to the ocean. In D. B. Dill, E. F Adolf, & C. G. Wilbur (Eds.), *Handbook of physiology: Adaptation to the environment* (pp. 741–747). Washington, D.C.: American Physiology Society.

Lucas, E., Sterman, M. B., & McGinty, D. J. (1969). The salamander EEG: A model of primitive sleep and wakefulness. *Psychophysiology, 6*, 230.

Lyamin, O. I., Manger, P. R., Mukhametov, L. M., Siegel, J. M., & Shpak, O. V. (2000). Rest and activity states in a gray whale. *Journal of Sleep Research, 9*, 261–267.

Lyamin, O. I., Shpak, O. V., Nazarenko, E. A., & Mukhametov, L. M. (2002). Muscle jerks during behavioral sleep in a beluga whale (*Delphinapterus leucas L.*). *Physiology and Behavior, 76,* 265–270.

Marshall, N. B. (1972). Sleep in fishes. *Proceedings of the Royal Society of Medicine, 65,* 177.

Martinez-Gonzalez, D., Lesku, J. A., & Rattenborg, N. C. (2008). Increased EEG spectral power density during sleep following short-term sleep deprivation in pigeons (*Columba livia*): Evidence for avian sleep homeostasis. *Journal of Sleep Research, 17,* 140–153.

Mascetti, G. G., Rugger, M., & Vallortigara, G. (1999). Visual lateralization and monocular sleep in the domestic chick. *Cognitive Brain Research, 7,* 451–463.

Mascetti, G. G., & Vallortigara, G. (2001). Why do birds sleep with one eye open? Light exposure of the chick embryo as a determinant of monocular sleep. *Current Biology, 11,* 971–974.

Mather, J. A. (2008). Cephalopod consciousness: Behavioural evidence. *Consciousness and Cognition, 17,* 37–48.

McCarley, R. W., Greene, R. W., Rainnie, D. G., & Portas, C. M. (1995). Brainstem neuromodulation and REM sleep. *Seminars in Neurosciences, 7,* 341–354.

McCormick, J. G. (1969). Relationship of sleep, respiration, and anesthesia in the porpoise: A preliminary report. *Proceedings of the National Academy of Sciences of the United States of America, 62,* 697–703.

McGinty, D. J., Stevenson, M., Hoppenbrouwers, T., Harper, R. M., Sterman, M. B., & Hodgman, J. (1977). Polygraphic studies of kitten development: Sleep state patterns. *Developmental Psychobiology, 10,* 455–469.

Mignot, E. (2004). Sleep, sleep disorders, and hypocretin (orexin). *Sleep Medicine, 5(1),* S2–S8.

Monnier, M. (1980). Comparative electrophysiology of sleep in some vertebrates. *Experientia, 36,* 16–19.

Mukhametov, L. M. (1987). Unihemispheric slow-wave sleep in the Amazonian dolphin, *Inia geoffrensis. Neuroscience Letters, 79,* 128–132.

Mukhametov, L. M., & Lyamin, O. I. (1994). Rest and active states in bottlenose dolphins (*Tursiops truncatus*). *Journal of Sleep Research, 3,* 174.

Mukhametov, L. M., Lyamin, O. I., Chetyrbok, I. S., Vassilyev, A. A., & Diaz, R. P. (1992). Sleep in an Amazonian manatee, *Trichechus inunguis. Experientia, 48,* 417–419.

Mukhametov, L. M., Supin, A. Y., & Polyakova, I. G. (1977). Interhemispheric asymmetry of the electroencephalographic sleep patterns in dolphins. *Brain Research, 134,* 581–584.

Newman, S. M., Paletz, E. M., Rattenborg, N. C., Obermeyer, W. H., & Benca, R. M. (2008). Sleep deprivation in the pigeon using the disk-over-water method. *Physiology and Behavior, 93,* 50–58.

Nicol, S. C., Andersen, N. A., Phillips, N. H., & Berger, R. J. (2000). The echidna manifests typical characteristics of rapid eye movement sleep. *Neuroscience Letters, 283,* 49–52.

Oleksenko, A. I. (1992). Unihemispheric sleep deprivation in bottlenose dolphins. *Journal of Sleep Research, 1,* 40–44.

Oleksenko, A. I., Chetyrbok, I. S., Polyakova, I. G., & Mukhametov, L. M. (1994). Rest and active states in Amazonian dolphins. *Journal of Sleep Research, 3,* 185.

Ookawa, T. (1972). Avian wakefulness and sleep on the basis of recent electroencephalographic observations. *Poultry Science, 51,* 1565–1574.

Ookawa, T., & Gotoh, J. (1965). Electroencephalogram of the chicken recorded from the skull under various conditions. *Journal of Comparative Neurology, 124,* 1–14.

Prudom, A. E., & Klemm, W. R. (1973). Electrographic correlates of sleep behavior in a primitive mammal, the armadillo *Dasypus novemcinctus*. *Physiology and Behavior*, *10*, 275–282.

Ramon, F., Hernandez-Falcon, J., Nguyen, B., & Bullock, T. H. (2004). Slow-wave sleep in crayfish. *Proceedings of the National Academy of Sciences of the United States of America*, *101*, 11857–11861.

Rattenborg, N. C., Obermeyer, W. H., Vacha, E., & Benca, R. M. (2005). Acute effects of light and darkness on sleep in the pigeon *(Columba livia)*. *Physiology and Behavior*, *84*, 635–640.

Rattenborg, N. C., Voirin, B., Vyssotski, A. L., Kays, R. W., Spoelstra, K., Kuemmeth, F., et al. (2008). Sleeping outside the box: Electroencephalographic measures of sleep in sloths inhabiting a rainforest. *Biology Letters*, *4*(4), 402–405.

Reebs, S. G. (2007). Sleep in fishes. Unpublished manuscript, Université de Moncton, Moncton, New Brunswick, Canada.

Ridgway, S., Carder, D., Finneran, J., Keogh, M., Kamolnick, T., Todd, M., et al. (2006). Dolphin continuous auditory vigilance for five days. *Journal of Experimental Biology*, *209*, 3621–3628.

Roth, T. C., Lesku, J. A., Amlaner, C. J., & Lima, S. L. (2006). A phylogenetic analysis of the correlates of sleep in birds. *Journal of Sleep Research*, *15*, 395–402.

Schlehuber, C. J., Flaming, D. G., Lange, G. D., & Spooner, C. E. (1974). Paradoxical sleep in the chick *(Gallus domesticus)*. *Behavioral Biology*, *11*, 537–546.

Schmidt, M. H., Valatx, J. L., Schmidt, H. S., Wauquier, A., & Jouvet, M. (1994). Experimental evidence of penile erections during paradoxical sleep in the rat. *NeuroReport*, *5*, 561–564.

Segura, E. T., & De Juan, A. (1966). Electroencephalographic studies in toads. *Electroencephalography and Clinical Neurophysiology*, *21*, 373–380.

Serafetinides, E. A., Shurley, J. T., & Brooks, R. E. (1972). Electroencephalogram of the pilot whale, *Globicephala scammoni*, in wakefulness and sleep: Lateralization aspects. *International Journal of Psychobiology*, *2*, 129–135.

Shapiro, C. M., & Hepburn, H. R. (1976). Sleep in a schooling fish, *Tilapia mossambica*. *Physiology and Behavior*, *16*, 613–615.

Shaw, P. J., Cirelli, C., Greenspan, R. J., & Tononi, G. (2000). Correlates of sleep and waking in *Drosophila melanogaster*. *Science*, *287*, 1834–1837.

Shurley, J. T., Serafetinides, E. A., & Brooks, R. E. (1969). Sleep in cetaceans: I. The pilot whale, *Globicephala scammoni*. *Psychophysiology*, *6*, 230.

Siegel, J. M. (1995). Phylogeny and the function of REM sleep. *Behavioral Brain Research*, *69*, 29–34.

Siegel, J. M. (1999). The evolution of REM sleep. In R. Lydic & H. Bagdoyan (Eds.), *Handbook of behavioral state control* (pp. 87–100). Boca Raton, FL: CRC Press.

Siegel, J. M. (2004). The neurotransmitters of sleep. *Journal of Clinical Psychiatry*, *65*(16), 4–7.

Siegel, J. M. (2008). Do all animals sleep? *Trends in Neuroscience*, *31*, 208–213.

Siegel, J. M., Manger, P. R., Nienhuis, R., Fahringer, H. M., & Pettigrew, J. D. (1996). The echidna *Tachyglossus aculeatus* combines REM and non-REM aspects in a single sleep state: Implications for the evolution of sleep. *Journal of Neuroscience*, *16*, 3500–3506.

Siegel, J. M., Manger, P. R., Nienhuis, R., Fahringer, H. M., Shalita, T., & Pettigrew, J. D. (1999). Sleep in the platypus. *Neuroscience*, *91*, 391–400.

Stephenson, R., Chu, K. M., & Lee, J. (2007). Prolonged deprivation of sleep-like rest raises metabolic rate in the Pacific beetle cockroach, *Diploptera punctata* (Eschscholtz). *Journal of Experimental Biology*, *210*, 2540–2547.

Steriade, M., & McCarley, R. W. (2005). *Brain control of sleep and wakefulness*. New York: Kluwer-Elsevier.

Strumwasser, F. (1971). The cellular basis of behavior in Aplysia. *Journal of Psychiatric Research*, *8*, 237–257.

Susic, V. (1972). Electrographic and behavioral correlation of rest-activity cycle in the sea turtle *Caretta caretta*. *Journal of Experimental Marine Biology and Ecology*, *10*, 81–87.

Tauber, E. S., Roffwarg, H. P., & Weitzman, E. D. (1966). Eye movements and electroencephalogram activity during sleep in diurnal lizards. *Nature*, *212*, 1612–1613.

Tauber, E. S., Rojas-Ramírez, J., & Hernández Peón, R. (1968). Electrophysiological and behavioral correlates of wakefulness and sleep in the lizard, *Ctenosaura pectinata*. *Electroencephalography and Clinical Neurophysiology*, *24*(5), 424–433.

Tauber, E., Weitzman, E., & Korey, S. (1969). Eye movements during behavioral inactivity in certain Bermuda reef fish. *Communications in Behavioral Biology, Part A*, *3*, 131–135.

Tobler, I. (1992). Behavioral sleep in the Asian elephant in captivity. *Sleep*, *15*, 1–12.

Tobler, I. (1997, September 15). What do we know about the evolution of sleep–when it arose and why? Bacteria surely don't sleep, do they? *Scientific American*. Retrieved November 5, 2008, from http://www.sciam.com/article.cfm?id=what-do-we-know-about-the-1997-09-15.

Tobler, I., & Borbély, A. A. (1985). Effect of rest deprivation on motor activity of fish. *Journal of Comparative Physiology, Series A*, *157*, 817–822.

Tobler, I., & Neuner-Jehle, M. (1992). 24-h variation of vigilance in the cockroach *Blaberus giganteus*. *Journal of Sleep Research*, *1*, 231–239.

Tobler, I., & Schwierin, B. (1996). Behavioural sleep in the giraffe *(Giraffa camelopardalis)* in a zoological garden. *Journal of Sleep Research*, *5*, 21–32.

Tradardi, V. (1966). Sleep in the pigeon. *Archives italiennes de biologie*, *104*, 516–521.

van Luijtelaar, E. L. J. M., van der Grinten, C. P. M., Blokhuis, H. J., & Coenen, A. M. L. (1987). Sleep in the domestic hen *(Gallus domesticus)*. *Physiology and Behavior*, *41*, 409–414.

Van Twyver, H. (1973). Polygraphic studies of the American alligator. *Sleep Research*, *2*, 87.

Van Twyver, H., & Allison, T. (1972). A polygraphic and behavioral study of sleep in the pigeon *(Columba livia)*. *Experimental Neurology*, *35*, 138–153.

Vasilescu, E. (1982). Sleep induced by electrical stimulation of the optic nerves in tortoise *(Emys orbicularis)*. *Neurologie et Psychiatrie*, *20*, 119–124.

Vasilescu, E. (1983). Phylogenetic and general remarks on sleep. *Physiologie*, *20*, 17–25.

Walker, J. M., & Berger, R. J. (1972). Sleep in the domestic pigeon *(Columba livia)*. *Behavioral Biology*, *7*, 195–203.

Walker, J. M., & Berger, R. J. (1973). A polygraphic study of the tortoise *(Testudo denticulata)*. Absence of electrophysiological signs of sleep. *Brain, Behavior, and Evolution*, *8*, 453–467.

Warner, B., & Huggins, S. (1978). An electrographic study of sleep in young caimans in a colony. *Comparative Biochemistry and Physiology, Part A*, *59*, 139–144.

Yokogawa, T., Marin, W., Faraco, J., Pezeron, G., Appelbaum, L., Zhang, J., et al. (2007). Characterization of sleep in zebrafish and insomnia in hypocretin receptor mutants. *Public Library of Science Biology*, *5*, 2379–2397.

Zepelin, H. (1989). Mammalian sleep. In M. H. Kryger, T. Roth, & W. C. Dement (Eds.), *Principles and practice of sleep medicine* (1st ed., pp. 30–49). Philadelphia: W. B. Saunders.

Zepelin, H. (1994). Mammalian sleep. In M. H. Kryger, T. Roth, & W. C. Dement (Eds.), *Principles and practice of sleep medicine* (2nd ed., pp. 69–80). Philadelphia: W. B. Saunders.

Zepelin, H. (2000). Mammalian sleep. In M. H. Kryger, T. Roth, & W. C. Dement (Eds.), *Principles and practice of sleep medicine* (3rd ed., pp. 69–80). Philadelphia: W. B. Saunders.

Zepelin, H., & Rechtschaffen, A. (1974). Mammalian sleep, longevity, and energy metabolism. *Brain, Behavior, and Evolution*, *10*, 425–470.

10

Toward an understanding of the function of sleep: New insights from mouse genetics

VALTER TUCCI AND PATRICK M. NOLAN

Whether all species sleep or meet the common definition of sleep has recently been questioned (Siegel, 2008). In the majority of species that do sleep, however, the evolutionary conservation of DNA elements regulating sleep and its features highlights the physiological importance of this behavior. From an "adaptation" point of view, we would like to think of sleep as solving a problem, just as we do for traits such as eating, drinking, and so on. In such a perspective, the perpetuation of particular sleep genes would have occurred through improved fitness of the individuals with those genes. Clear scientific evidence on this matter, however, is still missing. Historically, the science of sleep has evolved from a key technological innovation: the development of electrophysiological instruments that allow the recording of changes in electrical activity in brain and muscles. Such a phenomenological approach has been successful in providing a practical framework for understanding "how" we sleep, but it has not contributed to solving the question of "why" we sleep.

The year 1953 was an important year for two important research fields: sleep and genetics. The discovery of rapid-eye-movement (REM) sleep at the University of Chicago, announced in *Science* (Aserinsky & Kleitman, 1953), laid the foundation for modern research on sleep. That same year, from the Cavendish laboratory in Cambridge, UK, Crick and Watson sent their proposal of a structural model of DNA to *Nature* (Watson & Crick, 1953b). The discovery of the double helical structure of DNA had a major impact in biology (Watson & Crick, 1953a), but it was many years before sleep scientists began to appreciate it fully, probably because they were fascinated by the revolutionary notion that now sleep could be dissected.

Today the complete DNA sequences of many organisms, including human and mouse, have been determined, but the function of many genes remains unknown. This gap has raised a lot of interest within the scientific community, because many

218

phenotypic traits are searching for a genotype. However, the phenotype problem appears to be more complicated than initially suspected. Indeed, "the" one gene, one disease approach can no longer be supported by modern concepts of genetics in all its complexity. Evidence does exist that genetics is involved in normal sleep function. For example, some of the classic sleep disorders have been associated with single gene mutations (see Kimura & Winkelmann, 2007, for an overview), and several studies have identified genomic regions that contain allelic variations affecting quantitative sleep phenotypes (Tafti, 2007; Tafti, Maret, & Dauvilliers, 2005), commonly referred to as quantitative trait loci (QTLs). It is evident that the study of the genetics of sleep, focusing the analysis at the molecular level, has the potential to refine our understanding of the processes underlying this phenomenon. In evolutionary terms, this will shift the investigation to the "why" question, asking "why did sleep come about?"

To advance rapidly and successfully, sleep genetics needs the support of an appropriate animal model, such as the mouse. The mouse is a pre-eminent model in both the mapping of mammalian genes and determining the function of these genes (functional genomics). This species, among others, is widely recognized as an ideal laboratory model for studying and comparing gene function in animals and humans. Mice are small and are prolific breeders; for this reason, they are much appreciated by geneticists. A female has 5 to 10 litters per year, giving birth to over 100 mice during this time. Mice share many genes with humans, and they also share a number of behaviors that make them a suitable model for studying aspects of human sleep. For example, mice and humans have similar sleep architectures. Indeed, wakefulness, non-rapid-eye movement (NREM), and rapid-eye-movement (REM) states in mice have been widely documented (Figure 10.1). However, mice are nocturnal animals; they sleep during the light phase of the day, and when sleeping, move close to one another for warmth (Figure 10.2). In the early 1970s, Valatx, Bugat, and Jouvet (1972) and Friedmann (1974) proposed that differences in sleep parameters among inbred strains of mice could provide evidence for the genetic basis of sleep. These pioneering experiments initiated the scientific study of sleep in mice.

To date, sleep genetics has collected a vast body of data; however, for the purpose of this chapter, we begin with a brief overview of the actual methods for studying sleep in mice. We then focus on four major areas where the exploitation of mouse models promises to contribute significantly to elucidating the problem of sleep function. First, we describe how it became evident that genetic determinants affect sleep architecture in mice, contributing to the increased use of mice for the genetic investigation of sleep. Second, we discuss the problem of dealing with the increasing quantity of data generated in gene expression profiling experiments carried out on the traits of sleep and wakefulness. The next section focuses on the

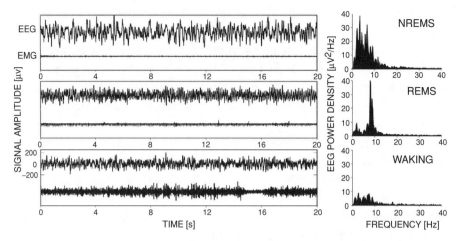

Figure 10.1. Classic EEG and EMG recordings of a mouse during NREM sleep, REM sleep, and wakefulness. (This picture was kindly provided by Dr. Paul Franken.)

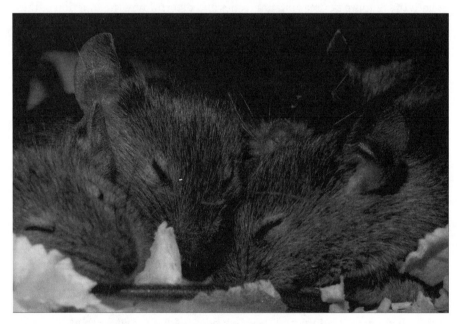

Figure 10.2. Sleeping mice. (This picture was kindly provided by Sara Wells and Russell Joynson at the Mary Lyon Centre, Harwell, UK.)

relation between circadian mechanisms and sleep, in particular looking at the role of circadian genes in sleep homeostasis. In the last part of this chapter, we comment on the imprinting hypothesis of sleep. An increasing body of evidence points to the hypothesis that genomic imprinting has a key role to play in the regulation/evolution of sleep. A deeper understanding of the relation between imprinted genes and sleep has the potential to revolutionize our understanding of sleep and sleep disorders, opening up new theoretical and clinical perspectives.

Methods and technologies in mouse sleep research

There are at least two sets of methods to assess sleep in mice: those based on the monitoring of electrophysiological parameters and those based on the assessment of activity across several days. Standard protocols for the first method involve the continuous recording of EEG and EMG signals; for the second method, any device that can detect variations in activity in the cage can be used (e.g., videos, infrared detectors, etc.).

Electrophysiology in mice

Common protocols to assess sleep in mice, based on EEG and EMG signals, always involve surgery for the implantation of electrodes, followed by a period of recovery. A 2-week postsurgery period of recovery is generally recommended to ensure that the mice have recovered completely and their sleep architecture is fully restored. Indeed, REM sleep is sensitive to prior stress (Sanford, Yang, & Tang, 2003). Conventional systems utilize a cable-recording apparatus, whereas new wireless monitoring systems (telemetry) consist of implantable transmitters that measure physiological parameters and transmit them to a receiver under the home cage. An obvious advantage in using telemetry in mice is that animals are allowed to move freely, with no restrictions from a cable attached to the scalp. Thus telemetry is potentially less stressful for the animal and mice are generally tolerant and adaptable to the implantation of a transmitter (Tang & Sanford, 2002).

Several studies to date (Tang & Sanford, 2002; Sanford et al., 2003) have assessed sleep in mice using telemetric systems. Telemetric recordings of EEG in inbred mouse strains are similar to those of cable recordings systems (Tang & Sanford, 2002). Polysomnography – a quantitative tool of EEG and EMG signals, which has proven to be a standard method to monitor the heterogeneity of sleep (distinguishing sleep states such as NREM versus REM) in humans – has been successfully applied in mice.

High-throughput sleep phenotyping in the postgenomic era

We are now entering an era of genomic exploration. There are currently several large-scale projects to create collections of mutant and knockout mice. All of these projects will be producing extensive archives of mouse lines held as ES cells, frozen sperm, or embryos, which will be available for use by the wider scientific community. However, the next step, which will add immense value to these collections, will be the determination of the phenotype of each of these models. Thus, comprehensive phenotyping of thousands of mutant mouse strains will involve a great effort for many laboratories across the world. For the process of sleep phenotyping to be feasible for large cohorts of mice, the process will have to utilize high-throughput technical methodologies. Automated high-throughput sleep phenotyping within home cage environments will be the key for the success of any screening effort. Current EEG and EMG methods, which are generally time-consuming, are not suitable for large-scale projects such as those using chemical mutagenesis (e.g., see Nolan, Peters, Strivens, et al., 2000).

A recent attempt to resolve this issue has been the development of a new high-throughput technique that estimates sleep and wakefulness in mice based on their lack of activity bouts over 24 hours (Pack, Galante, Maislin, et al., 2007). These authors have based their screen on the principle that the longer a mouse is inactive, the more likely it is to be sleeping (Pack et al., 2007). By using a 40-second minimum threshold duration of inactivity for scoring a sleep episode, they obtained an 88% to 94% agreement with EEG/EMG sleep/wakefulness scoring in the C57BL/6J inbred strain (Pack et al., 2007). This method is very promising, in particular if it is included in a hierarchical screening program, where it can be used as a primary screen for selecting a small cohort of mice to undergo EEG/EMG analysis at a second stage. Similar approaches have the potential to add economies of scale, enhance automated data capture and analysis, increase throughput, and reduce confounding experimental factors induced by surgical implants.

Genetic variation determines the architecture of sleep

The recording of electrical signals from the cortex is a classic method of monitoring the activity of the central nervous system (CNS). Sleep and sleep states are characterized by various changes in oscillatory activity across different frequencies and brain regions, reinforcing the idea that "sleep is about the brain" (Hobson, 2005). Genetics accounts for the phenomenological architecture of sleep in both humans (Linkowski et al., 1991) and mice (Franken, Malafosse, & Tafti, 1998, 1999; Tafti, Franken, Kitahama, et al., 1997). Briefly, during wakefulness, the EEG displays typical activity characterized by high frequencies and low amplitude. For example, in a cognitive effort, as during an attentional task, the EEG signal varies between 20 and 40 hertz (gamma waves). During quiet wakefulness (with eyes closed), the range decreases to 8 to 13 hertz (alpha waves). As sleep begins, the EEG frequencies drop dramatically and the amplitude increases. Slow-wave sleep (SWS) is characterized by 1 to 4 hertz (delta waves), classic rhythmic activity, during which it is possible to observe brief oscillations between 11 and 15 hertz (called spindles). During the course of sleep, a strange event occurs in that the EEG may increase up to 12 hertz (theta waves) and the amplitude decreases.

What distinguishes this electrical state from activity in wakefulness is a transient state of paralysis in the major voluntary musculature. This state is called REM sleep because of the characteristic rapid eye movements (REMs) that signal entry into it. In 1997, the first QTL involved in the expression of REM sleep was reported (Tafti et al., 1997) using a small set of recombinant inbred lines (BALB/cBy x C57BL/6By). This first attempt identified a series of loci on chromosomes 5, 7, 12, and 17 (Tafti et al., 1997) associated with the vigilance phenotype in mice. In a following study by the same group (Tafti, Chollet, Valatx, et al., 1999), the role for a particular QTL on chromosome 5 in the vigilance states was discussed. Franken, Malafosse, and Tafti (1998) presented also a study aimed at characterizing differences in EEG parameters in inbred mouse strains. This was a pivotal study highlighting the advantages of using multiple inbred mouse strains to study the effects of genetic background on phenotypic traits. These authors reported a number of differences between sleep states but also significant genotype-specific variations. Their results not only provided evidence that EEG parameters are under genetic control but also confirmed that the mouse has come to the fore as a model organism in sleep science and that the systematic characterization of new mouse mutant models for sleep abnormalities promises to reveal the genetic mechanisms of sleep itself.

In their study, Franken and colleagues (1998) observed a series of genotype-dependent differences within the theta and delta range of frequencies across a series of inbred mouse strains (AKR/J, BALB/cByJ, C57BL/6J, and DBA/2J). These two sleep EEG waves are associated with different brain structures and receive excitatory input from distinct areas. Theta waves are observed mainly in limbic structures, and are thought to be generated from the septum (Vertes & Kocsis, 1997; Vinogradova, 1995). An endogenous pacemaker in the septum maintains the theta activity in the hippocampus. However, this pacemaker is under the control of the brainstem's reticular formation (RF). As the RF excitatory input increases, the septum-theta pacemaker increases its activity. In contrast, delta waves are generated in the thalamocortical network (Steriade, McCormick, & Sejnowski, 1993). The synchronization of the EEG during SWS is an effect of reduced excitatory input from different brain areas. When the excitatory input decreases, the firing pattern in the thalamocortical neurons becomes rhythmic, and this generates the characteristic SWS EEG pattern.

In the sleeping mouse, significant strain-dependent differences in theta frequency during REM state have been observed (Franken et al., 1998). The authors reasoned that because they did not observe differences in theta waves during wakefulness, the phenotypic differences observed in REM sleep must be related to genotype-specific changes in brainstem activity. Indeed, there is a reduction in the processing of information from external sources during REM sleep. Delta waves also varied with the genotype of the mice. One strain (DBA/2J), which showed low SWS, also displayed low delta and high sigma power. Conversely, two other strains (BALB/cByJ and AKR/J) with a high amount of sleep time presented high delta and low sigma power (Franken et al., 1998). Delta power is the sleep parameter that most consistently reflects the duration of the preceding period of wakefulness. For this reason it is generally considered an accurate marker of sleep homeostasis and provides a measure of SWS need. Indeed, periods of sleep deprivation are followed by rebound increases in SWS and delta power. In mice, it has been shown that changes in delta power that respond to prior sleep deprivation are mathematically predictable (Franken, Chollet, & Tafti, 2001; Huber, Deboer, & Tobler, 2000). From the study of these variations in inbred mouse strains, however, it appears that only the rate of increase of delta power in the absence of SWS varies between genotype, and not the exponential decrease of SWS need (Franken et al., 2001). By screening sleep parameters in inbred mouse strains and using single-nucleotide polymorphism (SNP) genetic maps, it was possible to speculate that the presence of allelic variations may account for basic aspects of sleep regulation (recently estimated to account for between 40% and 60% of the overall sleep variance) (Tafti, 2007). For example, the mechanisms that regulate the amount of sleep were associated with candidate loci lying on chromosomes 4, 5, 9, and 15, whereas particular aspects of REM sleep were associated with loci on chromosomes 1, 17, and 19 (see Tafti, 2007, for an overview). Moreover, a significant QTL on mouse chromosome 13 (*Dps1*: delta power in slow-wave sleep 1) has been associated with changes in the delta power trait (Franken et al., 2001).

Gene expression profiling during sleep and wakefulness

It is an illusion to suppose that simple facts have themselves the power to constitute a theory. It is only the inference based upon them that will advance our viewpoints. (Hess, 1965, pp. 3–8)

The use of rodents and increasingly of mice to investigate gene expression changes across wakefulness and sleep states in several tissues, particularly in the brain, is providing new insights into the function of sleep. Pregenome attempts to solve this issue using gene-targeting approaches were able to identify few sleep-related genes (e.g., Wisor, O'Hara, Terao, 2002). The postgenome revolution expanded the study of gene expression to thousands of transcripts. To date, microarray-based approaches have been successful in identifying transcripts that exhibit circadian rhythms (Akhtar, Reddy, Maywood, et al., 2002; Panda, Hogenesch, & Kay, 2002) and/or sleep-dependent variation (Cirelli, Gutierrez, & Tononi, 2004; Cirelli & Tononi, 1998; Mackiewicz, Shockley, Romer, et al., 2007; Terao, Wisor, Peyron, et al., 2006).

Initial sleep–wake studies have observed an interesting differential expression between the two states for immediate early genes, or IEGs (Cirelli, 2002; Cirelli & Tononi, 1998; Pompeiano, Cirelli, & Tononi, 1994). The expression of IEGs tends to increase during wakefulness in most brain regions, and this was predicted to be related to an arousal state need (O'Hara, Ding, Bernat, et al., 2007). The refinement of microarray techniques and the computational advantages offered by bioinformatics over the last few years have led to new and extensive studies that have made it possible to hypothesize about the function of sleep. Mackiewicz, Shockley, Romer, and colleagues (2007) recently carried out an extensive microarray study in mice aimed at identifying changes in gene expression depending on the amount or lack of sleep. The authors assayed transcript levels in two brain tissues, cerebral cortex and hypothalamus. By using the GeneChip Mouse Genome 430 2.0 array (Affimetrix), which covers approximately 39,000 transcripts, they identified 2090 genes in the cerebral cortex and 409 genes in the hypothalamus that show an altered steady-state level during sleep when compared to wakefulness. These changes were defined as sleep-specific. Furthermore, 3988 genes in the cerebral cortex and 823 genes in the hypothalamus varied between sleep and sleep deprivation. To examine this, mice were sleep-deprived for 3, 6, 9, and 12 hours and compared with sleeping mice sacrified at the same circadian time. An important and interesting conclusion of this study was that sleep is a stage for biosynthesis of a number of macromolecules (Mackiewicz et al., 2007). Indeed, genes of this class, which were defined within the gene ontology database as involved in biosynthesis and transport, were overrepresented among those genes that show an increased expression with sleep. Although significant, a two (or more) fold increase in expression across the sleep–wake cycle (Cirelli, Gutierrez, & Tonini, 2004; Mackiewicz et al., 2007) can be somewhat moderate (less than 50%).

Many authors currently investigating the genetic properties of sleep by microarray analysis reason that the identification of genes that change their expression during sleep has the potential to unravel the mystery of the function of sleep. Gene expression profiling is thought to help in our understanding of at least two key processes: first, the increased expression of genes that promote sleep and, second, the downregulation of genes that restrict the duration of wakefulness (Mackiewicz et al., 2007). However, these microarray data must be interpreted with caution, as a few issues may still be confounding factors within these gene expression studies (for an overview, see Etter & Ramaswami, 2002; O'Hara, Ding, Bernat, et al., 2007; and Verducci, Melfi, Lin, et al., 2006). Also, gene expression data such as those we have described here are often derived from brain tissues (for example, cortex and hypothalamus), which contain a multitude of cell types. Last but not least, gene expression may vary as a function of vigilance state rather than directly causing changes in vigilance state.

Although not high-throughput, in situ hybridization analysis is more thorough in investigating the molecular and cellular correlates of sleep states. Sleep deprivation–related changes in messenger RNA levels throughout the brain, particularly in the cortex, were reported using an in situ hybridization approach (Franken, Thomason, Heller, et al., 2007). For example, *Per1* expression in C57BL/6J mice was increased in the brain after a 6-hour sleep deprivation period. The highest levels of expression were observed in the cerebral cortex, particularly around the cingulate cortex. A significant increase, however, was also recorded in the cerebellum. Recently, Tafti's group in Lausanne were able to refine the *Dps1* QTL region on chromosome 13 from 38 to 11 Mb. Among the annotated genes within this mouse region, the *Homer1a* gene was previously reported to be upregulated after sleep deprivation (Maret, Dorsaz, Gurcel, et al., 2007). These authors have recently generated *Homer1a*-PABP transgenic mice and subjected them to a sleep deprivation paradigm followed by gene expression profiling analysis. In this study, they have established that *Homer1a* changes in the brain highly reflect the sleep loss, although at least three other genes (*Ptgs2, Jph3,* and *Nptx2*) were also overexpressed after sleep deprivation (Maret et al., 2007). The strong role of the *Homer1a* gene in sleep homeostasis has also been confirmed by another group analyzing the *Dps1* QTL (Mackiewicz, Paigen, Naidoo, et al., 2008). *Homer1a* has been shown to play a role in intracellular calcium homeostasis and also in synaptic remodeling. Its activation after sleep loss would support the hypothesis that sleep has a role not only in coping with intracellular stressors but also in linking sleep to cognition and cognitive disorders.

Circadian mechanisms and sleep

The timing and duration of sleep are presumed to be regulated by the interaction of at least two processes (Borbély, 1982). Process C (circadian) dictates the temporal distribution of many physiological events within the organism, whereas process S (sleep homeostasis) reflects the propensity for sleep that accumulates during wakefulness and decreases during sleep. The main difference between these two processes is that the circadian process is self-sustained, whereas the homeostatic process responds to the amount of prior wakefulness or sleep. The mechanisms underlying these two processes have been widely investigated in rodents. The homeostatic process is often quantified by monitoring the daily distribution of slow-wave activity (delta power), particularly following a sleep deprivation period. Nevertheless, classic sleep deprivation experiments that, as we mentioned before, lead to an immediate increase of delta power in NREM sleep were inconclusive in clarifying the underlying mechanism of this sleep rebound process. This compensatory rebound of delta power has been studied in rodents by using, for example, a 6- to 24-hour sleep deprivation paradigm (Franken, Tobler, & Borbély, 1993; Laposky, Easton, Dugovic, et al., 2005; Tobler & Borbély, 1990). Sleep need following deprivation varies among inbred strains. For example AKR/J mice show a dramatic increase in delta power after 6 hours of sleep deprivation, whereas DBA/2J mice have a mild response to the same experience (Franken, Chollet, Tafti, et al., 2001).

The investigation of the circadian component of sleep has been more successful. Over the past few decades, mouse genetics has elucidated several important mechanisms at the molecular level of the circadian process. Moreover, novel genetic factors that influence circadian output continue to be identified by using forward genetics approaches in the mouse (Bacon, Ooi, Kerr, et al., 2004; Godinho, Maywood, Shaw, et al., 2007; Hofstetter, Svihla-Jones, & Mayeda, 2007; Hofstetter, Trofatter, Kernek, et al., 2003). To date a small set of genes have been identified that generate and maintain transcriptional and translational autoregulatory mechanisms that keep the length of the sleep/wake cycle (period) entrained with the rotation of the earth. Circadian genes deserve special attention within sleep science today, given recent data that these genes may also play a role in sleep homeostasis. This was a surprise, because the two processes develop independently and can easily be dissociated. Indeed, the long-standing idea that the circadian and homeostatic components underlying sleep are independent has been questioned by the investigation of genetically modified mice. In particular, mouse models have provided evidence for a modulation of circadian genes in homeostatic mechanisms.

The investigation of mouse circadian models has shown pivotal evidence in the determination of a link between some of the core or related clock genes and sleep phenotypes. For example, mutations in *Clock*, *Bmal1*, and *Cry1*/*Cry2* have been associated with several sleep abnormalities (Laposky et al., 2005; Naylor, Bergmann, Krauski, et al., 2000; Wisor et al., 2002). Other genes – such as *Npas2*, *Dbp*, and *Prok2* – appeared also to influence sleep homeostasis (Dudley, Erbel-Sieler, Estill, et al., 2003; Franken, Lopez-Molina, Marcacci, et al., 2000; Hu, Li, Zhang, et al., 2007). Specifically, mice lacking one or more of the above clock genes presented a reduced NREM sleep duration and delta power after sleep deprivation. However, there are cases of circadian mutations with no effects on sleep homeostasis. For example, mice that carry mutations of *period* genes (*Per1*/*Per2* double mutants) maintained a regular sleep homeostasis even in a prolonged free-running condition (constant darkness), where circadian rhythms of activity are lost (Shiromani, Xu, Winston, et al., 2004). These clock mutations did not affect homeostatic mechanisms even during a sleep deprivation paradigm (Kopp, Albrecht, Zheng, et al., 2002).

Although sleep homeostasis remains intact in mice deficient in *Per* genes, these genes are nevertheless modulated during homeostatic processes. Very recently, a new study investigated the time course of changes in gene expression that follow a sleep deprivation period (Franken et al., 2007). The authors used three inbred strains (AKR/J, C57BL/6J, and DBA2/J mice) that they previously characterized for their different homeostatic regulation (Franken et al., 2001). They confirmed that the expression of *Per* increases after sleep deprivation and observed a specific role of the genotype in such a process. Also, the authors reasoned that a prolonged expression of *Per2* may affect recovery sleep in mice (Franken et al., 2007). Another observation that the disruption of a gene can affect both circadian parameters and sleep homeostasis comes from the study of a loss-of-function mutation in *Rab3a* (Kapfhamer, Valladares, Sun, et al., 2002). A shortening of the circadian period in *Rab3a* mutants segregates with an increased amount of NREM sleep (Kapfhamer, et al., 2002).

Taking the issue of the relation between circadian and homeostatic mechanisms from a different point of view, time-course analysis of circadian brain transcripts has revealed that only 391 transcripts remain rhythmic after sleep deprivation (Maret et al., 2007). These data support the thesis that most diurnal changes in transcription are sleep/wake-dependent. In light of these recent studies, the idea that circadian mechanisms are restricted to circadian rhythm behaviors can no longer be supported. However, the molecular and cellular basis of sleep homeostasis remains unexplained.

Genomic imprinting modulates sleep expression

"Genomic imprinting" refers to the differential expression of inherited alleles depending on the sex of the parent that transmitted the gene. In many cases, this describes the inheritance of resource-acquisition genes (Constancia, Kelsey, & Reik, 2004). According to the conflict theory (Haig & Westoby, 2006), imprinting genes act in an antagonistic fashion to promote early growth if the gene is paternally derived or to inhibit growth if it is maternally derived. In other words, paternally derived alleles increase the allocation of resources, whereas maternally derived alleles restrain resources by the mother to offspring.

Clinical observations of neurodevelopmental disorders suggest that both NREM and REM sleep may be regulated by separate sets of imprinted genes. McNamara (2004) has pointed out that Prader–Willi syndrome (PWS) and Angelman syndrome (AS), both neurodevelopmental syndromes, exhibit opposing imprinting profiles and opposing sleep phenotypes. He also hypothesized that because sleep is strongly involved in the modulation of growth factors, the outcome of genomic imprinting may extend to the neurobiology of sleep (McNamara, 2004). PWS is associated with maternal additions/paternal deletions of alleles on chromosome 15q11-13 and characterized by temperature control abnormalities and excessive sleepiness. Changes in sleep architecture have also been noted in children and young adults with PWS, most specifically REM sleep abnormalities such as sleep-onset REM

periods, REM fragmentation, intrusion of REM into stage 2 sleep, and short latencies to REM (Hertz, Cataletto, Feinsilver, et al., 1993; Vela-Bueno, Kales, Soldatos,
et al., 1984; Vgontzas, Kales, Seip, et al., 1996). Conversely, AS is associated with
paternal additions/maternal deletions on chromosome 15q11-13 and is characterized by severe mental retardation and reductions in sleep. These children may
sleep as little as 1 to 5 hours a night, with frequent and prolonged night awakenings (Clayton-Smith & Laan, 2003; Zhdanova, Wurtman, & Wagstadd, 1999).
Twenty percent of patients who show clinical symptoms of AS have mutations
in one or more genes in the region 15q11-13 (Lalande, Minassian, DeLorey, et al.,
1999). Within this area there are at least two strong candidate genes: *UBE3A* and
GABRB3.

The loss of expression of *UBE3A* suggests a possible abnormality in ubiquitin-
mediated protein degradation during brain development (Colas, Wagstaff, Fort,
et al., 2005). Preliminary work in mice with a *Ube3a* deletion has shown interesting
sleep phenotypes. *Ube3a* mice are characterized by reduced NREM sleep, deteriorated REM sleep, and an increased frequency of waking during the dark–light transition (Colas et al., 2005). Knockout mice for the β3 subunit of the Gamma-Amino
Butyric Acid-A (GABAA) receptor are a good model for AS. Many children affected by
this disease inherit the deletion from the mother. A recent study of heterozygous
mutants reported that mice inheriting the mutant allele from the mother show
an abnormal increase of 7- to 10- hertz theta frequency EEG bursts, specifically
associated with the REM EEG spectrum (Liljelund, Handforth, Homanies, et al.,
2005).

Recent discoveries that imprinted genes are differentially expressed in brain
regions controlling sleep/wake cycles (e.g., hypothalamus) (Kobayashi, Kohda,
Miyoshi, et al., 1997) as well as in key nuclei of the major modulatory neurotransmitters such as the locus ceruleus (e.g., Nesp55) (Plagge, Isles, Gordon, et al.,
2005) suggest that genes of this class are significant for sleep research. The *Gnas*
locus on chromosome 2 in the mouse was one of the first autosomal regions known
to have imprinting properties (Peters & Williamson, 2007). Within this imprinting region is a mouse QTL associated with increased SWS delta power (Franken
et al., 2001). This region is homologous to human chromosome 20q13.2, which
contains, among others, a gene that mediates a low-voltage EEG trait (Anokhin,
Steinlein, Fischer, et al., 1992) and at least two genes that regulate adenosine
levels, known for mediating EEG during SWS (Benington & Heller, 1995; Porkka-
Heiskanen, Strecker, & McCarley, 2000). Moreover, the paternally expressed transcript *Gnasxl* is expressed in specific brain areas including the locus ceruleus and
cholinergic laterodorsal tegmental nucleus. The activity of neurons in the cholinergic laterodorsal tegmental nucleus is particularly important in the mechanisms
underlying REM sleep (Berridge, 2007; Berridge, Isaac, & España, 2003). Mice with
mutations in paternally derived *Gnasxl* transcripts show phenotypic deficits that

are compatible with the parental conflict hypothesis of imprinting, although they cannot be aged to study consequences on sleep parameters (Plagge, Gordon, Dean, et al., 2004).

In addition to growth and development, there is evidence to support a conflict model with regard to sleep phenotype wherein NREM expression is aligned with matriline genes and REM expression is aligned with patriline genes (Table 10.1). For example, serotonin (5-HT) and GABA play critical roles in sleep. Interestingly, maternally expressed 5HT2A receptors mediate aminergic inhibition of REM-on cells in the parabrachialis lateralis region, but paternally expressed GABA B receptors may mediate inhibition of these aminergic inhibitory effects on REM, thus facilitating REM expression (Amici, Sanford, Kearney, et al., 2004). Also, a recent study of the role of 5HT2A receptors in sleep suggests that these receptors modulate NREM sleep (Morairty, Hedley, Flores, et al., 2008).

Additional evidence supporting the conflict model of sleep can be derived from the results of a recent study conducted by Kozlov, Bogenpohl, Howell, and colleagues (2007). These authors generated mice deficient in *Magel2*, a circadian output gene (Panda, Antoch, Miller, et al., 2002), which maps to an imprinted genomic area. The human ortholog, *MAGEL2*, is located on chromosome 15 in the Prader–Willi/Angelman region q11-13. *MAGEL2* is highly expressed in the hypothalamus (Panda, et al., 2002) and, together with a few other genes within the same region (e.g., *MKRN3*, *ND,* and *SNURF-SNRPN*), encodes a protein with paternal allele–specific expression. *Magel2* mice have been shown to be normally entrained in their light–dark schedule by monitoring wheel-running activity (Kozlov et al., 2007). However, their overall activity on wheels was reduced at night and increased during the day as compared to their wild-type controls. Such an abnormal pattern of activity is congruent, in humans, with sleep disorders characterized by intrusions of sleep episodes during wakefulness and awakenings during sleep. Because activity in mice has a strong correlation with EEG sleep (Pack et al., 2007) we may speculate that the reduced activity observed in *Magel2* mice may be the expression of a sleep abnormality. This idea is also supported by an additional phenotype detected in *Magel2*-deficient mice: the reduction of orexin A and B levels in the hypothalamus. Orexin neuropeptide is known to have a crucial role in narcolepsy, a classic sleep disorder (Nishino, 2007) characterized by severe, irresistible sleepiness during the activity phase of the sleep/wake cycle and the presence of abnormal sleep episodes during this phase, with a deteriorated REM sleep pattern.

Outlook

Sleep regulatory processes have been proposed in sleep research to account for the distribution of wakefulness, NREM, and REM sleep states over

24 hours. Some of these, such as the two-process model (see above), were initially proposed in a qualitative version (Borbély, 1982) and only subsequently corroborated quantitatively (Daan, Beersma, & Borbély, 1984). For example, EEG data were very instrumental in feeding mathematical models of the C and S processes. Although, in the case of the two-process model, the delta power index can model the time course of sleep and to some extent the regulation of NREM sleep, it entirely ignores REM sleep. Conversely, REM regulation can be explained by the reciprocal interaction model of REM regulation proposed by McCarley and Hobson (1975). In this model, REM is predicted by the differential activity of two sets of cells: RemOn cells and RemOff cells, associated with the laterodorsal tegmental (LDT)/pedunculopontine (PPT) nuclei and dorsal raphe (DR)/locus ceruleus (LC) areas, respectively. Briefly, RemOn cells generate REM sleep while RemOff cells account for its inhibition. So far, several key genetic processes account for sleep regulation, but none has yet been used to model sleep/wake and/or NREM/REM cycling. However, genetic evidence seems to favor the idea that while process C can be described by the autoregulatory mechanisms of clock genes, process S (sleep homeostasis) and the REM on/off mechanism could follow an epigenetic maternally/paternally derived gene regulation pattern. The application of sleep genetics studies in the mouse brings the promise of a better understanding of several other basic and clinical aspects of sleep. For example, there is a need in sleep research to start characterizing the sleep profile of genetically altered mice, such as those that carry mutations affecting synaptic plasticity or more general CNS function. For example, sleep disturbances are very common in extrapyramidal diseases, and studies in mice have shown sleep problems associated with motor deficits (Daan et al., 1984). For this reason mouse studies can help to determine the relation between sleep and motor learning in mice. It has been shown in humans that sleep triggers overnight learning, particularly using a finger-tapping task (Walker, Brakefield, Hobson, et al., 2003). This suggests that the sleep-dependent learning process selectively provides maximum benefit to fine-motor-skill procedures. Interestingly, mice hold a complete repertoire of fine motor skills (Tucci, Achilli, Blanco, et al., 2007).

Similarly, it is known that sleep disturbances are common in neurodegenerative diseases. Alzheimer's disease (AD) patients, for example, constantly report alterations in sleep/wake cycling, which worsen as the disease progresses. These patients present a lengthening of wakefulness associated with a distinctive slowing in the EEG during this state (Prinz, Vitaliano, Vitiello, et al., 1982). Their sleep is characterized by a reduction in the amount of the two main sleep states: NREM sleep (Loewenstein, Weingartner, Gillin, et al., 1982) and REM sleep (Montplaisir, Petit, Lorrain, et al., 1995) as well as by other clinical EEG features such as increased latency to the first REM episode. Disturbed sleep is also characteristic of

Table 10.1. *Imprinting regulation of sleep*

Matriline/NREM expression	Patriline/REM expression
Increases total sleep	Reduces total sleep
Promotes sleep homeostasis	Prolongs wakefulness
Interferes with REM sleep	Facilitates REM
Abnormal increase in theta frequency	Paternal expression deficiency
	(e.g., *Magel2* mice) reduces orexin levels

Huntington's disease (HD) (Bates, Harper, & Jones, 2002). In HD, sleep efficiency deteriorates and patients experience frequent nocturnal awakenings accompanied by EEG abnormalities (Silvestri, Raffaele, De Domenico, et al., 1995). The association between sleep/wake disturbances and HD is supported by clinical evidence of neurodegeneration in the hypothalamus of patients (Kassubek, Juengling, Kioschies, et al., 2004).

The identification of genetic factors associated with many neurodegenerative disorders has contributed to the development of several transgenic mouse models for the human disease states. The information gained from sleep profiling in mouse models for neurodegeneration will help to solve basic questions in sleep and cognitive science and will also improve our understanding of the physiology and genetics of many severe CNS disorders. For example, mouse studies have reported circadian alterations of gene expression within core circadian genes in R6/2 mice, a transgenic model of HD (Morton, Wood, Hastings, et al., 2005). Consequently, this approach, by assessing the impact on sleep of the neurodegenerative pathology of specific mouse transgenic models, may advance our knowledge of human sleep and cognitive disorders. Also, such an approach will shed light on the ongoing debate as to whether sleep has a role in cognitive processes.

In this chapter, we have highlighted the importance of understanding the genetic mechanisms of sleep – for example, by identifying functional genes. Current progress in mouse functional genetics promises to increase the rate of discovery of sleep-related genes. From an evolutionary point of view, mouse studies will provide an interesting platform for new prospects, including studies of positive selection (e.g., over evolutionary transitions in activity period); this can be detected by examining ratios of nonsynonymous and synonymous substitutions.

References

Akhtar, R. A., Reddy, A. B., Maywood, E. S., Clayton, J. D., King, V. M., Smith, A., et al. (2002). Circadian cycling of the mouse liver transcriptome, as revealed by cDNA microarray, is driven by the suprachiasmatic nucleus. *Current Biology, 12*(7), 540–550.

Amici, R., Sanford L. D., Kearney K., McInerney, B., Ross, R. J., Horner, R. L., et al. (2004). A serotonergic (5-HT2) receptor mechanism in the laterodorsal tegmental nucleus participates in regulating the pattern of rapid-eye-movement sleep occurrence in the rat. *Brain Research*, *996*(1), 9–18.

Anokhin, A., Steinlein, O., Fischer, C., Mao, Y., Vogt, P., Schalt, E., et al. (1992). A genetic study of the human low-voltage electroencephalogram. *Human Genetics*, *90*(1–2), 99–112.

Aserinsky, E., & Kleitman, N. (1953). Regularly occurring periods of eye motility, and concomitant phenomena, during sleep. *Science*, *118*(3062), 273–274.

Bacon, Y., Ooi, A., Kerr, S., Shaw-Andrews, L., Winchester, L., Breeds, S., et al. (2004). Screening for novel ENU-induced rhythm, entrainment, and activity mutants. *Genes, Brain, and Behavior*, *3*, 196–205.

Bates, G. P., Harper, P. S., & Jones, L. (2002). *Huntington's disease*. Oxford, UK: Oxford University Press.

Benington, J. H., & Heller, H. C. (1995). Restoration of brain energy metabolism as the function of sleep. *Progress in Neurobiology*, *45*(4), 347–360.

Berridge, C. W. (2007). Noradrenergic modulation of arousal. *Brain Research Reviews*, *58*(1), 1–17.

Berridge, C. W., Isaac, S. O., & España, R. A. (2003). Additive wake-promoting actions of medial basal forebrain noradrenergic alpha1- and beta-receptor stimulation. *Behavioral Neuroscience*, *117*(2), 350–359.

Borbély, A. A. (1982). A two-process model of sleep regulation. *Human Neurobiology*, *1*, 195–204.

Cirelli, C. (2002). How sleep deprivation affects gene expression in the brain: A review of recent findings. *Journal of Applied Physiology*, *92*(1), 394–400.

Cirelli, C., Gutierrez, C. M., & Tononi, G. (2004). Extensive and divergent effects of sleep and wakefulness on brain gene expression. *Neuron*, *41*(1), 35–43.

Cirelli, C., & Tononi, G. (1998). Differences in gene expression between sleep and waking as revealed by mRNA differential display. *Brain Research, Molecular Brain Research*, *56*(1–2), 293–305.

Clayton-Smith, J., & Laan, L. (2003). Angelman syndrome: A review of the clinical and genetic aspects. *Journal of Medical Genetics*, *40*(2), 87–95.

Colas, D., Wagstaff, J., Fort, P., Salvert, D., & Sarda, N. (2005). Sleep disturbances in Ube3a maternal deficient mice modeling Angelman syndrome. *Neurobiology of Disease*, *20*, 471–478.

Constancia, M., Kelsey, G., & Reik, W. (2004). Resourceful imprinting. *Nature*, *432*(7013), 53–57.

Daan, S., Beersma, D. G., & Borbély, A. A. (1984). Timing of human sleep: Recovery process gated by a circadian pacemaker. *American Journal of Physiology*, *246*(2 Pt. 2), R161–R183.

Dudley, C. A., Erbel-Sieler, C., Estill, S. J., Reick, M., Franken, P., Pitts, S., et al. (2003). Altered patterns of sleep and behavioral adaptability in NPAS2-deficient mice. *Science*, *301*(5631), 379–383.

Etter, P. D., & Ramaswami, M. (2002). The ups and downs of daily life: Profiling circadian gene expression in *Drosophila*. *Bioessays*, *24*(6), 494–498.

Franken, P., Chollet, D., & Tafti, M. (2001). The homeostatic regulation of sleep need is under genetic control. *Journal of Neuroscience*, *21*, 2610–2621.

Franken, P., Lopez-Molina, L., Marcacci, L., Schibler, U., & Tafti, M. (2000). The transcription factor DBP affects circadian sleep consolidation and rhythmic EEG activity. *Journal of Neuroscience*, *20*(2), 617–625.

Franken, P., Malafosse, A., & Tafti, M. (1998). Genetic variation in EEG activity during sleep in inbred mice. *American Journal of Physiology*, *275*(4 Pt. 2), R1127–R1137.

Franken, P., Malafosse, A., & Tafti, M. (1999). Genetic determinants of sleep regulation in inbred mice. *Sleep*, *22*(2), 155–169.

Franken, P., Thomason, R., Heller, H. C., & O'Hara, B. F. (2007). A non-circadian role for clock-genes in sleep homeostasis: A strain comparison. *BMC Neuroscience*, *8*, 87.

Franken, P., Tobler, I., & Borbély, A. A. (1993). Effects of 12-h sleep deprivation and of 12-h cold exposure on sleep regulation and cortical temperature in the rat. *Physiology and Behavior*, *54*(5), 885–894.

Friedmann, J. K. (1974). A diallel analysis of the genetic underpinning of mouse sleep. *Physiology and Behavior*, *12*, 169–175.

Godinho, S. I. H., Maywood, E. S., Shaw, L., Tucci, V., Barnard, A. R., Busino, L., et al. (2007). The after-hours mutant mouse reveals a role for Fbxl3 in determining mammalian circadian period. *Science*, *316*(5826), 897–900.

Haig, D., & Westoby, M. (2006). An earlier formulation of the genetic conflict hypothesis of genomic imprinting. *Nature Genetics*, *38*(3), 271.

Hertz, G., Cataletto, M., Feinsilver, S. H., & Angulo, M. (1993). Sleep and breathing patterns in patients with Prader Willi syndrome (PWS): Effects of age and gender. *Sleep*, *16*(4), 366–371.

Hess, W. R. (1965). Sleep as a phenomenon of the integral organism. In K. Akert, C. Bally, & J. P. Schade (Eds.), *Progress in brain research. Sleep mechanisms* (pp. 3–8). Amsterdam: Elsevier.

Hobson, J. A. (2005). Sleep is of the brain, by the brain, and for the brain. *Nature*, *437*(7063), 1254–1256.

Hofstetter, J. R., Svihla-Jones, D. A., & Mayeda, A. R. (2007). A QTL on mouse chromosome 12 for the genetic variance in free-running circadian period between inbred strains of mice. *Journal of Circadian Rhythms*, *5*, 7.

Hofstetter, J. R., Trofatter, J. A., Kernek, K. L., Nurnberger, J. I., & Mayeda, A. R. (2003). New quantitative trait loci for the genetic variance in circadian period of locomotor activity between inbred strains of mice. *Journal of Biological Rhythms*, *18*(6), 450–462.

Hu, W. P., Li, J. D., Zhang, C., Boehmer, L., Siegel, J. M., & Zhou, Q. Y. (2007). Altered circadian and homeostatic sleep regulation in prokineticin 2-deficient mice. *Sleep*, *30*(3), 247–256.

Huber, R., Deboer, T., & Tobler, I. (2000). Effects of sleep deprivation on sleep and sleep EEG in three mouse strains: Empirical data and simulations. *Brain Research*, *857*(1–2), 8–19.

Kapfhamer, D., Valladares, O., Sun, Y., Nolan, P. M., Rux, J. J., Arnold, S. E., et al. (2002). Mutations in Rab3a alter circadian period and homeostatic response to sleep loss in the mouse. *Nature Genetics*, *32*(2), 290–295.

Kassubek, J., Juengling, F. D., Kioschies, T., Henkel, K., Karitzky, J., Kramer, B., et al. (2004). Topography of cerebral atrophy in early Huntington's disease: A voxel-based morphometric MRI study. *Journal of Neurology, Neurosurgery, and Psychiatry*, *75*(2), 213–220.

Kimura, M., & Winkelmann, J. (2007). Genetics of sleep and sleep disorders. *Cellular and Molecular Life Sciences*, *64*(10), 1216–1226.

Kobayashi, S., Kohda, T., Miyoshi, N., Kuroiwa, Y., Aisaka, K., Tsutsumi, O., et al. (1997). Human PEG1/MEST, an imprinted gene on chromosome 7. *Human Molecular Genetics*, *6*(5), 781–786.

Kopp, C., Albrecht, U., Zheng, B., & Tobler, I. (2002). Homeostatic sleep regulation is preserved in mPer1 and mPer2 mutant mice. *European Journal of Neuroscience, 16*(6), 1099–1106.

Kozlov, S. V., Bogenpohl, J. W., Howell, M. P., Wevrick, R., Panda, S., Hogenesch, J. B., et al. (2007). The imprinted gene Magel2 regulates normal circadian output. *Nature Genetics, 39*(10), 1266–1272.

Lalande, M., Minassian, B. A., DeLorey, T. M., & Olsen, R. W. (1999). Parental imprinting and Angelman syndrome. *Advances in Neurology, 79*, 421–429.

Laposky, A., Easton, A., Dugovic, C., Walisser, J., Bradfield, C., & Turek, F. (2005). Deletion of the mammalian circadian clock gene BMAL1/Mop3 alters baseline sleep architecture and the response to sleep deprivation. *Sleep, 28*(4), 395–409.

Liljelund, P., Handforth, A., Homanies, G., & Olsen, R. (2005). GABAA receptor beta3 subunit gene-deficient heterozygous mice show parent-of-origin and gender-related differences in beta3 subunit levels, EEG, and behavior. *Developmental Brain Research, 157*(2), 150–161.

Linkowski, P., Kerkhofs, M., Hauspie, R., & Mendlewicz, J. (1991). Genetic determinants of EEG sleep: A study in twins living apart. *Electroencephalography and Clinical Neurophysiology, 79*(2), 114–118.

Loewenstein, R. J., Weingartner, H., Gillin, J. C., Kaye, W., Ebert, M., & Mendelson, W. B. (1982). Disturbances of sleep and cognitive functioning in patients with dementia. *Neurobiology of Aging, 3*(4), 371–377.

Mackiewicz, M., Paigen, B., Naidoo, N., & Pack, I. A. (2008). Analysis of the QTL for sleep homeostasis in mice: Homer1a is a likely candidate. *Physiological Genomics, 33*(1), 91–99.

Mackiewicz, M., Shockley, K. R., Romer, M. A., Galante, R. J., Zimmerman, J. E., Naidoo, N., et al. (2007). Macromolecule biosynthesis: A key function of sleep. *Physiological Genomics, 31*(3), 441–457.

Maret, S., Dorsaz, S., Gurcel, L., Pradervand, S., Petit, B., Pfister, C., et al. (2007). Homer1a is a core brain molecular correlate of sleep loss. *Proceedings of the National Academy of Sciences of the United States of America, 104*(50), 20090–20095.

McCarley, R. W., & Hobson, J. A. (1975). Neuronal excitability modulation over the sleep cycle: A structural and mathematical model. *Science, 189*(4196), 58–60.

McNamara, P. (2004). Genomic imprinting and neurodevelopmental disorders of sleep. *Sleep and Hypnosis, 6*(2), 100–108.

Montplaisir, J., Petit, D., Lorrain, D., Gauthier, S., & Nielsen, T. (1995). Sleep in Alzheimer's disease: Further considerations on the role of brainstem and forebrain cholinergic populations in sleep-wake mechanisms. *Sleep, 18*(3), 145–148.

Morairty, S. R., Hedley, L., Flores, J., Martin, R., & Kilduff, T. S. (2008). Selective 5HT2A and 5HT6 receptor antagonists promote sleep in rats. *Sleep, 31*(1), 34–44.

Morton, A. J., Wood, N. I., Hastings, M. H., Barker, R. A., & Maywood, E. S. (2005). Disintegration of the sleep-wake cycle and circadian timing in Huntington's disease. *Journal of Neuroscience, 25*(1), 157–163.

Naylor, E., Bergmann, B. M., Krauski, K., Zee, P. C., Takahashi, J. S., Hotz Vitaterna, M., et al. (2000). The circadian clock mutation alters sleep homeostasis in the mouse. *Journal of Neuroscience, 20*(21), 8138–8143.

Nishino, S. (2007). The hypothalamic peptidergic system, hypocretin/orexin and vigilance control. *Neuropeptides, 41*(3), 117–133.

Nolan, P. M., Peters, J., Strivens, M., Rogers, D., Hagen, J., Spurr, N., et al. (2000). A systematic, genome-wide, phenotype-driven mutagenesis programme for gene function studies in the mouse. *Nature Genetics*, 25(4), 440–443.

O'Hara, B. F., Ding, J., Bernat, R. L., & Franken, P. (2007). Genomic and proteomic approaches towards an understanding of sleep. *CNS and Neurological Disorders – Drug Targets*, 6(1), 71–81.

Pack, A. I., Galante, R. J., Maislin, G., Cater, J., Metaxas, D., Lu, S., et al. (2007). Novel method for high-throughput phenotyping of sleep in mice. *Physiological Genomics*, 28(2), 232–238.

Panda, S., Antoch, M. P., Miller, B. H., Su, A. I., Schook, A. B., Straume, M., et al. (2002). Coordinated transcription of key pathways in the mouse by the circadian clock. *Cell*, 109(3), 307–320.

Panda, S., Hogenesch, J. B., & Kay, S. A. (2002). Circadian rhythms from flies to human. *Nature*, 417(6886), 329–335.

Peters, J., & Williamson, C. M. (2007). Control of imprinting at the gnas cluster. *Epigenetics*, 2(4), 207–213.

Plagge, A., Gordon, E., Dean, W., Boiani, R., Cinti, S., Peters, J., et al. (2004). The imprinted signaling protein XL alpha s is required for postnatal adaptation to feeding. *Nature Genetics*, 36(8), 818–826.

Plagge, A., Isles, A. R., Gordon, E., Humby, T., Dean, W., Gritsch, S., et al. (2005). Imprinted Nesp55 influences behavioral reactivity to novel environments. *Molecular and Cellular Biology*, 25(8), 3019–3026.

Pompeiano, M., Cirelli, C., & Tononi, G. (1994). Immediate-early genes in spontaneous wakefulness and sleep: Expression of c-fos and NGFI-A mRNA and protein. *Journal of Sleep Research*, 3(2), 80–96.

Porkka-Heiskanen, T., Strecker, R. E., & McCarley, R. W. (2000). Brain site-specificity of extracellular adenosine concentration changes during sleep deprivation and spontaneous sleep: An in vivo microdialysis study. *Neuroscience*, 99(3), 507–517.

Prinz, P. N., Vitaliano, P. P., Vitiello, M. V., Bokan, J., Raskind, M., Peskind, E., et al. (1982). Sleep, EEG and mental function changes in senile dementia of the Alzheimer's type. *Neurobiology of Aging*, 3(4), 361–370.

Sanford, L. D., Yang, L., & Tang, X. (2003). Influence of contextual fear on sleep in mice: A strain comparison. *Sleep*, 26(5), 527–540.

Shiromani, P. J., Xu, M., Winston, E. M., Shiromani, S. N., Gerashchenko, D., & Weaver, D. R. (2004). Sleep rhythmicity and homeostasis in mice with targeted disruption of mPeriod genes. *American Journal of Physiology: Regulatory, Integrative and Comparative Physiology*, 287(1), R47–R57.

Siegel, J. M. (2008). Do all animals sleep? *Trends in Neurosciences*, 31, 2008–2213.

Silvestri, R., Raffaele, M., De Domenico, P., Tisano, A., Mento, G., Casella, C., et al. (1995). Sleep features in Tourette's syndrome, neuroacanthocytosis, and Huntington's chorea. *Neurophysiologie Clininique*, 25(2), 66–77.

Steriade, M., McCormick, D. A., & Sejnowski, T. J. (1993). Thalamocortical oscillations in the sleeping and aroused brain. *Science*, 262(5134), 679–685.

Tafti, M. (2007). Quantitative genetics of sleep in inbred mice. *Dialogues in Clinical Neuroscience*, 9(3), 273–278.

Tafti, M., Chollet, D., Valatx, J.-L., & Franken, P. (1999). Quantitative trait loci approach to the genetics of sleep in recombinant inbred mice. *Journal of Sleep Research*, 8(S1), 37–43.

Tafti, M., Franken, P., Kitahama, K., Malafosse, A., Jouvet, M., & Valatx, J.-L. (1997). Localization of candidate genomic regions influencing paradoxical sleep in mice. *NeuroReport*, *8*(17), 3755–3758.

Tafti, M., Maret, S., & Dauvilliers, Y. (2005). Genes for normal sleep and sleep disorders. *Annals of Medicine*, *37*(8), 580–589.

Tang, X., & Sanford, L. D. (2002). Telemetric recording of sleep and home cage activity in mice. *Sleep*, *25*(6), 691–699.

Terao, A., Wisor, J. P., Peyron, C., Apte-Deshpande, A., Wurts, S. W., Edgar, D. M., et al. (2006). Gene expression in the rat brain during sleep deprivation and recovery sleep: An Affymetrix GeneChip study. *Neuroscience*, *137*(2), 593–605.

Tobler, I., & Borbély, A. A. (1990). The effect of 3-h and 6-h sleep deprivation on sleep and EEG spectra of the rat. *Behavioural Brain Research*, *36*(1–2), 73–78.

Tucci, V., Achilli, F., Blanco, G., Lad, H. V., Wells, S., Godinho, S., et al. (2007). Reaching and grasping phenotypes in the mouse (*Mus musculus*): A characterization of inbred strains and mutant lines. *Neuroscience*, *147*(3), 573–582.

Valatx, J. L., Bugat, R., & Jouvet, M. (1972). Genetic studies of sleep in mice. *Nature*, *238*(5361), 226–227.

Vela-Bueno, A., Kales, A., Soldatos, C. R., Dobladez-Blanco, B., Campos-Castello, J., Espino-Hurtado, P., et al. (1984). Sleep in the Prader-Willi syndrome: Clinical and polygraphic findings. *Archives of Neurology*, *41*(3), 294–296.

Verducci, J. S., Melfi, V. F., Lin, S., Wang, Z., Roy, S., & Sen, C. K. (2006). Microarray analysis of gene expression: Considerations in data mining and statistical treatment. *Physiological Genomics*, *25*(3), 355–363.

Vertes, R. P., & Kocsis, B. (1997). Brainstem-diencephalo-septohippocampal systems controlling the theta rhythm of the hippocampus. *Neuroscience*, *81*(4), 893–926.

Vgontzas, A. N., Kales, A., Seip, J., Mascari, M. J., Bixler, E. O., Myers, D. O., et al. (1996). Relationship of sleep abnormalities to patient genotypes in Prader-Willi syndrome. *American Journal of Medical Genetics Part B: Neuropsychiatric Genetics*, *67*(5), 478–482.

Vinogradova, O. S. (1995). Expression, control, and probable functional significance of the neuronal theta-rhythm. *Progress in Neurobiology*, *45*(6), 523–583.

Walker, M. P., Brakefield, T., Hobson, J. A., & Stickgold, R. (2003). Dissociable stages of human memory consolidation and reconsolidation. *Nature*, *425*(6958), 616–620.

Watson, J. D., & Crick, F. H. C. (1953a). Genetical implications of the structure of deoxyribonucleic acid. *Nature*, *171*(4361), 964–967.

Watson, J. D., & Crick, F. H. C. (1953b). Molecular structure of nucleic acids; A structure for deoxyribose nucleic acid. *Nature*, *171*(4356), 737–738.

Wisor, J. P., O'Hara, B. F., Terao, A., Selby, C. P., Kilduff, T. S., Sancar, A., et al. (2002). A role for cryptochromes in sleep regulation. *BMC Neuroscience*, *3*, 20.

Zhdanova, I. V., Wurtman, R. J., & Wagstadd, J. (1999). Effects of a low dose of melatonin on sleep in children with Angelman syndrome. *Journal of Pediatric Endocrinology and Metabolism*, *12*(1), 57–67.

11

Fishing for sleep

I. V. ZHDANOVA

Fish comprise about half of the known vertebrate species. The vast majority of the extant 30,000 currently known species of fish are bony fishes. They occupy diverse habitats in fresh and salty waters of rivers, lakes, seas, and oceans. The dynamic adaptations of fish to these distinctly different environments – including complex reproductive, migratory, and life-cycle adaptations – are truly remarkable. Their adaptive strategies include periods of rest that, in different fish species, can be spent lying quietly on the sea floor, floating, or swimming.

With fishes as with other phylogenetically earlier animals discussed in this book, the decision as to whether they actually sleep or just rest quietly must be based on a combination of behavioral features, electrophysiological patterns of brain activity, and molecular processes that we associate with sleep in mammals. Sleep is thought to be present when the animal is in a species-specific posture of behavioral quiescence and exhibits elevated arousal thresholds as well as rapid reversibility of behavioral quiescence after appropriate stimulation (Campbell & Tobler, 1984). The majority of fish species thus far studied display these behavioral features of sleep accompanied by physiological quietness, including reduced heart rate and respiration (Karmanova, 1975; Karmanova, Churnosov, & Popova, 1976; Karmanova, Titkov, & Popova, 1976; Peyrethon & Dusan-Peyrethon, 1967; Shapiro & Hepburn, 1976; Tobler & Borbély, 1985). Some fish (e.g., blueheads, Spanish hogfish, and several species of wrasses) were reported to exhibit a major increase in arousal threshold during their daily rest to the extent that, at night, they could be lifted by hand to the surface before "waking up" (Tauber, 1974). Other fish species show elaborate preparations to entering the rest phase. For example, parrot fish wrap themselves up in mucus prior to each rest period to prevent predators from sensing them; then they remain quietly under this safe blanket throughout the night.

Based on their behavioral parameters, it appears that fish can sleep. However, the behavioral signs of sleep or lack of it might be somewhat misleading in fish. Some fish species require constant movement for breathing and, unless other sleep-related parameters are assessed, could be considered complete nonsleepers (Kavanau, 1998). Complex sleep adaptations in aquatic mammals as well as birds, including the phenomenon of unihemispheric sleep that appears to be due to the need for constant swimming/movement, might be conserved in fish. They may also have other adaptive ways to either temporarily circumvent the need for sleep or engage in as yet unknown compensatory mechanisms during prolonged periods of wakefulness.

Fish have obvious anatomical limitations in expressing some electrographic patterns of mammalian sleep, especially those that depend on the presence of neocortex (e.g., generation of slow waves). However, the major neuronal structures and neurotransmitter systems with sleep-specific properties (Figure 11.1a) are contained within the isodendritic core of the brain, extending from the medulla through the brainstem, hypothalamus, and up into the basal forebrain.

The neurochemical architecture of these brain areas – including noradrenergic, serotonergic, cholinergic, histaminergic, and orexigenic neurons and their projections – remains largely conserved in fish (Figure 11.1b to Figure 11.1d). Some of the studies also suggest that there are distinct electrographic patterns during periods of fish rest (Karmanova & Lazarev, 1979; Peyrethon & Dusan-Peyrethon, 1967); this has to be explored further.

The daily rotation of our planet relative to the sun defines near-24-hour (circadian) patterns of activity in fish as in other organisms. Different fish species may adapt to a predominantly diurnal or nocturnal lifestyle (i.e., being consistently more active during the day or at night). Such habitual behavioral patterns can be originally based on food availability or risk of predation, but fish show remarkable flexibility under natural and laboratory conditions (Zhdanova & Reebs, 2006). The choice between day and night activity can also be influenced by competition, ontogeny, and light intensity, inasmuch as these factors affect feeding (Chen, Naruse, & Tabata, 2002; Lague & Reebs, 2000). For example, goldfish, golden shiners, or rainbow trout are diurnal when fed by day but can change the timing of their activity when food becomes more available at night (Aranda, Madrid, & Sanchez-Vazquez, 2001; Chen et al., 2002; Gee, Stephenson, & Wright, 1994; Lague & Reebs, 2000; Sanchez-Vazquez, Aranda, & Madrid, 2001; Spieler, Meier, & Noeske, 1978a, 1978b; Spieler & Noeske, 1981). The Atlantic salmon becomes more nocturnal when water temperatures are low, even when the photoperiod is held constant. This might be because cold fish become more sluggish and thus more vulnerable to attack by predators, so that a nocturnal lifestyle provides relative safety (Fraser, Heggenes, Metcalfe, et al., 1995). The most dramatic changes in rest

a.

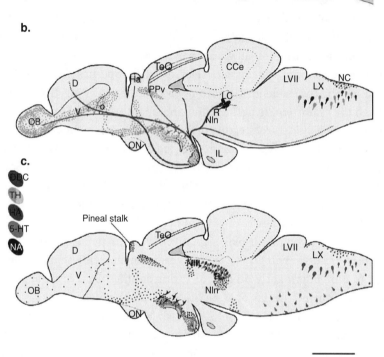

b.

c.

d.

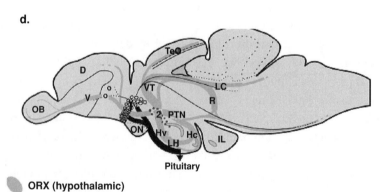

ORX (hypothalamic)

ORX (preoptic)

patterns with significantly reduced or absent sleep state may occur during periods of migration. As is apparently the case with migratory birds, conditions consistent with the environment of migration (temperature and photoperiod) may promote continuous activity patterns in otherwise diurnal wrasse (Olla & Studholme, 1978). Interestingly, such changes in behavior were found only in adult wrasse, not in juveniles, which normally do not migrate on their first fall, thus highlighting ontogenetic differences in sleep behavior within the same species.

More recent studies on the characterization of molecular and behavioral correlates of sleep in fish used zebrafish (*Danio rerio*) as a subject. This is mainly because of the prior and current efficient use of this small teleost in the fields of

Figure 11.1. A schematic sagittal overview of the mammalian (rat) (a) and zebrafish (b to d) brain, showing similarities in structures and neurochemical systems known to be involved in mammalian sleep regulation. (See color Plate 2.) (Overall figure from Zhdanova, 2006, with permission.) Rat: (a) distribution of some key sleep-regulating neuronal populations. (Siegel, 2005.)

The area shaded in gray is both necessary and sufficient for REM sleep generation. The area shaded in yellow is both necessary and sufficient for NREM sleep generation. In the intact animal, both REM sleep and NREM sleep involve interactions between brainstem and forebrain structures.

Circles indicate "REM sleep off" neurons (i.e., low activity in REM relative to wakefulness); orange represents serotonergic neurons (located on the midline), dark blue represents adrenergic or noradrenergic neurons, red represents histaminergic neurons, light blue represents hypocretinergic (orexinergic) neurons. Squares indicate "sleep on" neurons, which are more active during sleep, compared to wakefulness. The green star indicates "REM sleep on" neurons, which are active during REM sleep. Abbreviations: Vlpo, ventrolateral preoptic area; Mpo, median preoptic.

Zebrafish: Major monoaminergic (b), histaminergic (c), and orexigenic (d) cell groups and fiber projections in the adult zebrafish brain. (From Kaslin & Panula, 2001; Kaslin et al., 2004.)

(b): green – distribution of catecholaminergic neurons (tyrosine hydroxylase-immunoreactive [TH-ir]; light green – paraventricular TH-ir neurons; blue – dopamine beta hydroxylase-ir neurons; and red – histaminergic neurons.

(c): orange – serotonin neurons (5-HT); gray – dopa decarboxylase (DDC) – containing neurons.

(d): blue – hypothalamic orexin-producing and orexin-containing neurons and their projections; dark blue – the preoptic putatively orexin-containing and orexin-producing neurons. Abbreviations: D, dorsal telencephalon; Hv, ventral zone of periventricular hypothalamus; IL, inferior hypothalamic lobe; Ha, habenula; HA, histamine; LC, locus ceruleus; LH, lateral hypothalamic nucleus; NA, noradrenalin; NC, commissural nucleus of Cajal; NIn, interpeduncular nucleus; OB, olfactory bulb; ON, optic nerve; PPv, periventricular pretectal nucleus; PTN, posterior tuberal nucleus; R, raphe nuclei; TeO, optic tectum; V, ventral telencephalon; VT, ventral thalamus. (From Zhdanova, 2006, with permission.)

developmental biology and genetics, resulting in a fair amount of accumulated knowledge of its molecular biology as well as the easy availability of diverse mutant and transgenic phenotypes. Thus the rest of this chapter focuses on zebrafish to illustrate several issues of homeostatic and circadian regulation of sleep in fish, the behavioral and molecular mechanisms involved, and their changes during development and aging. We also discuss the effects of endogenous and pharmacological sleep-promoting substances in zebrafish. Such knowledge may help to elucidate the sleep-related processes in other vertebrates and perhaps assist in deciphering and treating human sleep disorders.

Do zebrafish sleep?

Zebrafish are diurnal animals; as such, they are active during the day and rest at night. While we were searching for a genetically well-characterized diurnal vertebrate with which to study the sleep effects of melatonin, this was one of the primary reasons we became interested in the sleep behavior of zebrafish. Melatonin is a phylogenetically ancient molecule that is present in unicellular organisms. In vertebrates, it is produced by the pineal gland and retina and can promote sleepiness. Melatonin is produced at night in both nocturnal and diurnal species; this might explain why it promotes sleep only in those animals that habitually sleep at the time of high circulating melatonin levels (Zhdanova, 2005). While we were able to observe that melatonin significantly reduces activity levels in zebrafish, the remaining question was the extent to which this reduction in behavioral activity was analogous to mammalian sleep. Thus we and then others conducted a series of studies in zebrafish of different ages to characterize their sleep (Prober, Rihel, Onah, et al., 2006; Yokogawa, Marin, Faraco, et al., 2007; Zhdanova, 2006; Zhdanova, Wang, Leclair, et al., 2001).

Zebrafish develop very quickly, hatching from their chorion around 50 hours postfertilization (hpf). They become behaviorally active miniature fish soon thereafter, showing complex behaviors and skillful prey capture by 7 days postfertilization (dpf). The small size of larval zebrafish, which are about 4 to 6 millimeters in length at 5 to 10 days postfertilization (dpf), allows us to conduct high-throughput recordings in many individual animals in parallel by placing them in small, individual wells and documenting their behavior using image-analysis techniques (Cahill, Hurd, & Batchelor, 1998; Prober et al., 2006; Yokogawa et al., 2007; Zhdanova, et al., 2001). The behavior of adult zebrafish can be recorded in a similar way using substantially larger fish tanks (Yokogawa et al., 2007; Zhdanova, 2006; Zhdanova, Yu, Lopez-Patino, et al., 2008). This allows for the study of sleep-related behaviors during the early stages of vertebrate development, through adulthood, and into senescence.

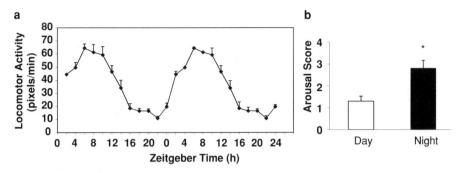

Figure 11.2. Diurnal activity pattern in zebrafish is associated with increased daytime locomotion (a) and reduced arousal threshold (b); *p < 0.05. (From Zhdanova, et al., 2001, with permission.)

The larval, young adult, and aged zebrafish display periods of quietness during the day. However, the duration and frequency of such inactivity bouts is substantially increased at night (Figure 11.2a).

The pattern is preserved after larvae are transferred from a light–dark cycle to constant conditions of dim light or darkness, indicating a true intrinsic regulation of circadian rhythmicity. In conditions of both light–dark and constant darkness, prolonged periods of inactivity (beyond 5 seconds) are accompanied by significant increases in arousal threshold in both larval and adult zebrafish (Figure 11.2b). Thus we and others adopted a 6-second bout of continuous inactivity as a threshold for differentiating between quiet wakefulness and onset of a sleep-like state (Yokogawa et al., 2007; Zhdanova, 2006).

Although complete inactivity may last only a few seconds and is then interrupted by slow movements of fins with or without locomotion, several distinct postures are associated with sleep-like states in zebrafish. Larvae either float with their heads down or stay in a horizontal position close to the bottom of the tank (Figure 11.3a).

Adult fish typically float either in horizontal position or with the head slightly upward, showing occasional small pectoral fin movements. During nighttime rest, some zebrafish alternate between staying at the bottom of the tank and at its surface, especially if maintained in individual tanks (Figure 11.3b).

The reason for such alternation is not yet clear. Such behavior is less frequent when zebrafish are housed in groups, when they tend to distribute throughout the tank or form several groups of two or three fish staying close to each other. These sleep behaviors appear to vary between zebrafish strains and even different social groups and may reflect social relationships within the group. Similar sleep-related behaviors were reported earlier in groups of tilapia, a diurnally active schooling teleost fish (Shapiro & Hepburn, 1976). Throughout the night, the majority of

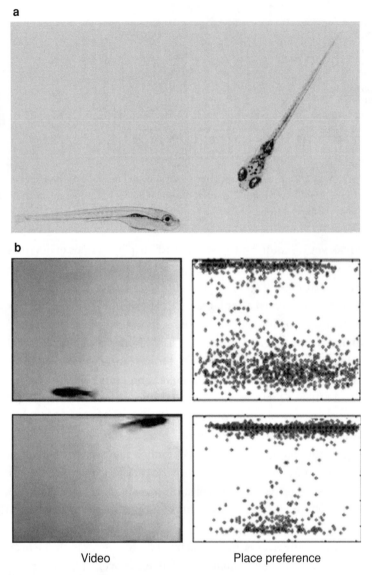

Figure 11.3. Typical sleep postures in larval (a) and adult (b) zebrafish (from Zhdanova et al., 2001, and Yokogawa et al., 2007, with permission). (See color Plate 3.)

tilapia rest on the bottom of the tank, while the minority continue active movement. Toward the end of the dark period, the behavior becomes more variable, with more fish actively swimming around the tank or floating in the middle of the water column.

Unlike daytime rest, nighttime sleep in zebrafish is often associated with a decline in the frequency of mouth and gill movements, reflecting a reduced

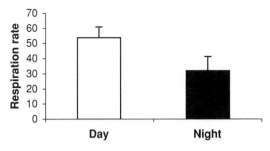

Figure 11.4. Reduction in nighttime respiration rate in larval zebrafish. Data (bursts/min) were collected in the middle of the light and dark periods; expressed as mean ± SEM group number. N = 18. (From Zhdanova, 2006, with permission.)

respiration rate. However, we have observed sporadic augmentation of respiration during sleep in adult zebrafish, without concurrent body movements. Stimulation immediately after such periods leads to an increased arousal threshold relative to the daytime rest state – that is, the fish appear to continue their sleep. We did not observe eye movements in adult fish during such episodes of increased respiration in sleep.

In larvae, the respiration rate is also changed during sleep (Figure 11.4). During the day, zebrafish often have bursts of respiratory movements, with two to five individual movements per burst. The burst rate and the number of movements per burst decline during nighttime sleep, at the time of increased arousal threshold. Whether oxygen absorption through the skin, characteristic of larvae, also changes at that time remains unknown. Prolonged nighttime rest periods in larval zebrafish, especially in the head-down position, can be associated with spontaneous eye movements that may be accompanied by low-amplitude tail movements. Such changes in respiration or occurrence of eye movements may represent microarousals or changes in the quality of sleep. The nature of these events in the context of the sleep-like state in zebrafish is not yet clear.

Homeostatic regulation of sleep and effects of light

To evaluate the homeostatic regulation of sleep in zebrafish, recordings are made either during the light–dark cycle or under conditions of constant illumination (complete darkness or dim light). Rest deprivation can be achieved by different experimental approaches. Repeated pulses of mild vibration or tapping, or continuous slow movement of the perforated partition through which animals have to swim makes it possible to deprive many animals of sleep simultaneously. Another approach is to apply an arousing stimulus only when each animal ceases to move for a certain period of time, which requires individual sleep deprivation

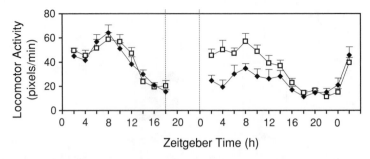

Figure 11.5. Effects of sleep deprivation on subsequent sleep in zebrafish. Reduction in daytime activity level in sleep-deprived (*black diamond*) but not control (*white square*) larval zebrafish. (From Zhdanova et al., 2001, with permission.)

chambers and the use of stimuli that are not perceived by all the animals in the recording system, (e.g., mild electric shock) (Yokogawa et al., 2007).

Both larval and adult zebrafish respond to nighttime sleep deprivation by increasing their subsequent sleep duration (Figure 11.5). This "compensatory sleep" includes longer sleep bouts and increased arousal thresholds, suggesting that sleep in zebrafish is under homeostatic control (Zhdanova, 2006; Zhdanova, et al., 2001), just as it is in mammals. Moreover, the importance of sleep to both juvenile and adult zebrafish is illustrated by the presence of performance deficits in the conditioned place preference (CPP) paradigm following sleep deprivation (Zhdanova, 2006). It remains to be determined whether such performance deficits in zebrafish might be related to memory formation, recall, attention level, visual sensitivity, or a combination of factors. Several methods of sleep deprivation and performance assessment will have to be compared before firm conclusions can be drawn about sensitivity to the cognitive consequences of sleep loss in zebrafish.

Interestingly, strong environmental factors, such as constant bright light, experienced during the sleep deprivation period or at the time of anticipated sleep rebound, can disrupt the homeostatic response to sleep deprivation. If zebrafish are kept under these conditions for several days, they do not show regular signs of sleep, prolonged inactivity periods, or increased arousal thresholds. Moreover, when they are returned to their regular environment (light–dark cycle), they do not show sleep rebound. After being held under these unusual conditions for a week or two, the fish may again display a sleep-like state, which, however, lacks a clear circadian pattern (Yokogawa et al., 2007). Such modulation by environmental illumination is likely to be due to its critical role in the diurnal adaptation of zebrafish to avoiding predators or finding prey. Consistent with this, other fish also show high light-dependence, with rest being suppressed by light in diurnal perch, goldfish, and tilapia (Shapiro & Hepburn, 1976; Tobler & Borbély, 1985) but being promoted in nocturnal tench (Campbell & Tobler, 1984).

Collectively, the presence of characteristic postures, physiological changes, elevated arousal threshold to sensory stimulation during habitual nighttime hours of prolonged rest, and a compensatory rest rebound following rest deprivation under low illumination allows us to classify the rest state in zebrafish as a sleep-like state or sleep and to show that this process is under the control of homeostatic regulation. A peculiar absence of sleep rebound after sleep deprivation in bright light, however, suggests that some additional compensatory sleep mechanisms might be present in zebrafish and potentially other fish species.

Circadian regulation

The complex machinery of the intrinsic circadian clock includes several core clock genes and proteins organized in complex feedback loops to maintain close to 24-hour oscillations and serving as transcription factors (Reppert & Weaver, 2002). As a result, the clock-controlled genes and their products modulate multiple output processes. Moreover, posttranslational mechanisms are also actively involved in clock function. Considering the important role of the circadian factors in sleep regulation in the majority of the species studied, their analysis in zebrafish is of obvious interest. This is especially true because few detailed data are available in other diurnal vertebrates and nocturnal and diurnal species exhibit an inverted temporal relationship between the molecular clock mechanisms and habitual sleep periods.

Like the majority of other species, fish display the circadian rhythms of activity, food intake, sleep, and physiological functions (Zhdanova & Reebs, 2006). Our knowledge of the well-conserved molecular mechanisms of the fish circadian system is based mainly on the original studies conducted in zebrafish (Cahill et al., 1998; Pando & Sassone-Corsi, 2002). Owing to partial genome duplication in teleosts, the zebrafish circadian system has more circadian clock or clock-controlled genes, which are orthologs of those in other species. This might be viewed as a complication – for example, the presence of six melatonin receptors (Reppert, Weaver, Cassone, et al., 1995), but it may also prove to be an advantage in deciphering tissue-specific functions of individual homologs.

Zebrafish have multiple tissues containing autonomic oscillators rhythmically expressing the core clock genes. These oscillators have direct sensitivity to light that can entrain them (Cermakian, Whitmore, Foulkes, et al., 2000; Kaneko, Hernandez-Borsetti, & Cahill, 2006; Whitmore, Foulkes, Strahle, et al., 1998), but they gradually desynchronize in continuous darkness (Carr & Whitmore, 2005; Pando, Pinchak, Cermakian, et al., 2001; Whitmore, Foulkes, & Sassone-Corsi,

2000). These oscillators appear to be similar to recently discovered non–light-sensitive peripheral oscillators in mammals, which can sustain autonomic activity for some time in the absence of the "master clock" but generally require its synchronizing input for normal functioning (Vansteensel, Michel, & Meijer, 2008).

In zebrafish, the eyes and pineal gland are the principal clock structures. They carry out autonomic oscillations, photoreception, and melatonin production. Starting with early embryogenesis (first dpf), melatonin production and melatonin receptor expression provide a unifying neuroendocrine circadian signal through specific melatonin receptors (Cahill et al., 1998; Danilova, Krupnik, Sugden, et al., 2004). Hence, as in the case of mammals, melatonin is the major neurohumoral output of the circadian system in zebrafish. Zebrafish have the suprachiasmatic nucleus of the hypothalamus (SCN), the site of the "master clock" in mammals, but its role in circadian rhythmicity, if any, has not been confirmed (Cahill, 1998). In spite of this, the brain tissue analyzed as a whole rhythmically expresses the core clock and clock-controlled genes (Whitmore, et al., 1998).

We find that in zebrafish, the expression of core genes of the positive limbs of the clock – for example, *bmal1* and *clock1* – is initiated close to the habitual sleep time (Figure 11.6a,c; Zhdanova et al., 2008). The expression of the genes of the negative limb of the clock – for example, *per1* – occurs at the end of the sleep period and continues into the early hours of daily activity phase (Figure 11.6b). Constant light exposure can significantly reduce the amplitude of expression for the clock and clock-controlled genes (Figure 11.7; Shang & Zhdanova, 2007), and this might be, at least in part, responsible for the sleep alterations in bright light described earlier.

Neurochemical mechanisms of sleep regulation in zebrafish

The neuroanatomical structures involved in sleep regulation in mammals are also typical of zebrafish, with corresponding neurotransmitters and their receptors being typically well conserved in this teleost (Figure 11.1) (Arenzana, Clemente, Sanchez-Gonzalez, et al., 2005; Clemente, Arenzana, Sanchez-Gonzalez, et al., 2005; Clemente, Porteros, Weruaga, et al., 2004; Eriksson, Peitsaro, Karlstedt, et al., 1998; Faraco, Appelbaum, Marin, et al., 2006; Kaslin, Nystedt, Ostergard, et al., 2004; Kaslin & Panula, 2001; Panula, Kaarlstedt, Sallmen, et al., 2000; Ruuskanen, Peitsaro, Kaslin, et al., 2005). Studies on the hypnotic effects of drugs and physiological agents targeting the GABAergic, melatonin, histamine, hypocretin, cholinergic, dopamine, and adrenergic signaling pathways confirm that these anatomical similarities translate into the functional ones (Renier, Faraco, Bourgin, et al., 2007; Ruuskanen et al., 2005; Zhdanova, Wang, Leclair, et al., 2001; Zhdanova, et al., 2008).

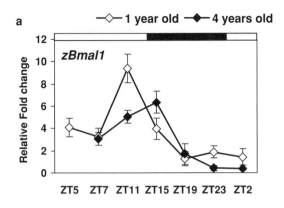

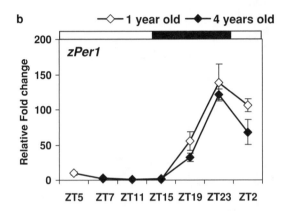

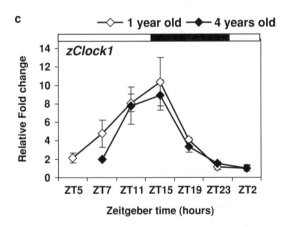

Figure 11.6. Daily pattern of expression of core clock genes, *bmal1*, *per1*, and *clock1*, in young and aged adult zebrafish. (From Yu, Tucci, Kishi, & Zhdanova, 2006, with permission.)

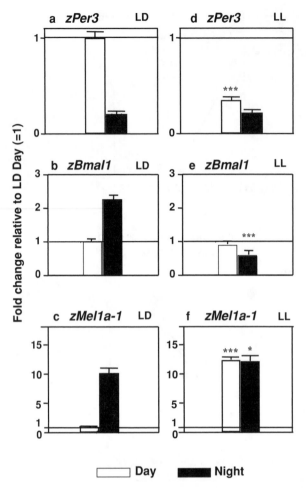

Figure 11.7. Normal diurnal variation in *per3*, *bmal1*, and melatonin receptors is altered in embryos raised under constant light conditions. Daytime increase in *per3* expression in LD (a) is partially preserved in LL (d). Daily variation in *bmal1*, *mel1a-1*, and other melatonin receptors is abolished in LL (b versus e; c versus f). Y axis in all plots: fold change relative to LD day level (=1). N = 3 to 4 samples per time point in LD or LL group, 25 embryos per sample. Mean (SEM), *t*-test, *P < 0.05, **P < 0.01, ***P < 0.001 indicates difference between LD and LL, at day or night, accordingly. (From Shang & Zhdanova, 2007, with permission.)

The inhibitory neurotransmitter gamma-aminobutyric acid (GABA) plays an important role in the physiological regulation of sleep. Moreover, its receptors are the target of the vast majority of sedative hypnotics (barbiturates, benzodiazepines, or nonbenzodiazepine agonists of benzodiazepine receptors), mediating an increase in intracellular chloride levels and leading to neuronal hyperpolarization. Selective targeting of the GABAergic pathway can facilitate sleep onset

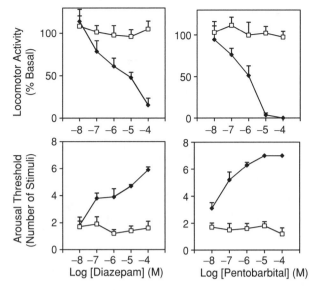

Figure 11.8. Conventional GABAergic sedatives promote rest behavior in larval zebrafish. Diazepam and sodium pentobarbital (barbital) significantly and dose-dependently reduce zebrafish locomotor activity and increase arousal threshold. Each data point represents mean (SEM) group changes in a 2-hour locomotor activity relative to basal activity, measured in each treatment or control group for 2 hours prior to treatment administration. Arousal threshold data are expressed as the mean (SEM) group number of stimuli necessary to initiate locomotion in a resting fish. Closed diamond – treatment, open square – vehicle control; N = 20, each group. (From Zhdanova et al., 2001, with permission.)

and extend sleep duration, but it also appears to be responsible for side effects of available hypnotic substances (e.g., amnesia, ataxia, morning-after sedation, or abuse potential). In large doses, these drugs induce general anesthesia and may cause death.

Zebrafish have well-developed GABAergic neurotransmission (Doldan, Prego, Holmqvist, et al., 1999; Higashijima, Mandel, & Fetcho, 2004) and are highly sensitive to GABAergic hypnotics (Renier et al., 2007; Zhdanova et al., 2001). The dose-dependence of their behavioral effects is quite similar to that in mammals. At low doses of drug, the activity levels decline and arousal thresholds rise, followed by inactivity and lack of spontaneous or stimulated arousal in response to higher doses, consistent with general anesthesia (Figure 11.8).

Finally, if zebrafish are exposed to these drugs for prolonged periods, they do not survive. The latter effect is more pronounced in adult zebrafish than in larvae. This might be explained by apparent reduction in respiratory movements in adult fish treated with barbiturates or benzodiazepines. In larvae, cutaneous respiration may compensate for this effect. Indeed, when larvae are washed out after hours

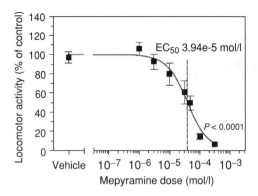

Figure 11.9. Dose–response curve of the effects of mepyramine (pyrilamine) on spontaneous locomotor activity in zebrafish larvae. (From Renier et al., 2007, with permission.)

of benzodiazepine-induced anesthesia, many of them survive and thereafter show normal behavior.

Antagonists of histamine receptors have sedative effects in mammals via direct antagonism of H1 receptors and additional antiadrenergic or antimuscarinic activity (Montoro, Sastre, Bartra, et al., 2006). Similarly, administration of histamine H1 antagonists (Figure 11.9) such as diphenhydramine or mepyramine produce dose-dependent effects in larval zebrafish, ranging from mild sedation to general anesthesia (Renier et al., 2007).

The neuropeptides hypocretins (Hcrt 1 and 2, also known as orexins A and B) were linked to the human sleep disorder narcolepsy, characterized by excessive daytime sleepiness, a fragmented sleep–wake cycle, and a sudden loss of muscle tone during waking, called cataplexy (Zeitzer, Nishino, & Mignot, 2006). Deficiency in these peptides or their G protein–coupled receptors alters alertness and sleep in humans and mammalian models. When Hcrt 1 is injected into the cerebrospinal fluid (CSF), locomotor activity in mammals is typically increased and sleep inhibited. Zebrafish express Hcrt in the neurons of the posterior hypothalamus (Faraco et al., 2006; Kaslin, et al., 2004), like those of other teleost fish (fugu, tetraodon, medaka, and stickleback). These neurons project to monoaminergic and cholinergic nuclei (Kaslin et al., 2004; Prober et al., 2006; Yokogawa et al., 2007). The Hcrt receptors are distributed in the telencephalon, hypothalamus, posterior tuberculum, and hindbrain (Yokogawa et al., 2007).

Two studies on the role of Hcrt in zebrafish produced somewhat discrepant results (Yokogawa et al., 2007). The mutation of the hypocretin receptor is found to disrupt the consolidation of sleep/wake behavior in zebrafish, reminiscent of insomnia in narcoleptic patients. Such mutants do not display a decrease in wake bout length, however, whether in the light or the dark, or sudden episodes of paralysis, analogous to cataplexy. Moreover, intracerebroventricular injection of

Hcrt 1 (but not Hcrt 2) leads to only a mild reduction in zebrafish activity, in contrast to robust activation following such injection in mammals (Zeitzer et al., 2006). The authors suggest that the powerful stimulating effect of light might suppress sleep irrespective of Hcrt deficiency and that the arousing effect of this peptide might be unnecessary in this species.

In contrast, Prober et al. reported increased locomotor activity and decreased sleep in zebrafish larvae with overexpression of hypocretin peptide, suggesting that the role of Hcrt as an arousal-promoting agent is conserved (Prober et al., 2006). These two reports also disagreed on the presence of Hcrt innervation of the locus ceruleus in zebrafish, which is one of the ways Hcrt can promote wakefulness. The Hcrt system in other fish is poorly characterized, although some increase in locomotor activity was found in goldfish injected with mammalian Hcrt 1 (Nakamachi, Matsuda, Maruyama, et al., 2006). Further studies in zebrafish and other teleosts might clarify these issues on the role of Hcrt in fish sleep physiology.

Melatonin, the principal hormone of the circadian system, has been known to have a hypnotic-like effect in humans when used in physiological or pharmacological doses (Zhdanova, 2005). Its administration also promotes sleep in diurnal primates (Zhdanova, Cantor, Leclair, et al., 1998; Zhdanova, Geiger, Schwagerl, et al., 2002) and birds (Aparicio, Garau, Nicolau, et al., 2006; Mintz, Phillips, & Berger, 1998; Paredes, Terron, Valero, et al., 2007). However, the nocturnal species appear to be immune to this effect of melatonin, consistent with nocturnal production of this hormone corresponding to their active period. In larval and adult zebrafish, exposure to melatonin promotes a sleep-like state, reducing their locomotor activity and elevating their arousal threshold (Figure 11.10).

Independent of the dose used, this effect does not induce anesthesia. Rather, following melatonin administration, both locomotor activity and arousal threshold reach a plateau at higher doses, and the resulting behavior remains close to that observed normally at night. Interestingly, this dose-dependence of melatonin's effects on sleep is documented in both humans and nonhuman primates, suggesting that the mechanisms through which this effect is mediated are also likely to be the same in fish and primates. Moreover, zebrafish allowed us for the first time to confirm that the effects of melatonin on sleep are mediated through melatonin receptors, because they can be attenuated by luzindole, a specific melatonin receptor antagonist (Zhdanova et al., 2001).

Neuronal structures involved in zebrafish sleep

Imaging techniques have dramatically changed research approaches and medical practice. The sleep field is no exception, with magnetic resonance imaging and positron emission tomography contributing to our understanding of structures involved in sleep regulation or sleep alterations. Although

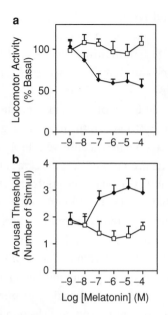

Figure 11.10. Melatonin significantly and dose-dependently reduces locomotor activity (a) and increases arousal threshold (b) in zebrafish. Each data point represents mean (SEM) group changes in a 2-hour locomotor activity relative to basal activity, measured in each treatment or control group for 2 hours prior to treatment administration. Arousal threshold data are expressed as the mean (SEM) group number of stimuli necessary to initiate locomotion in a resting fish. Closed diamond – treatment, open square – vehicle control; N = 20, each group. (From Zhdanova et al., 2001, with permission.)

neurophysiological experiments in adult and especially larval zebrafish pose a challenge, the optical transparency of larval fish allows another approach by the visual observation of their neuronal activity. The larval fish can be embedded in water-saturated agar to restrain them and allow imaging of the same cells for prolonged periods of time, relying on larval cutaneous respiration. Because electrically active vertebrate neurons usually have large increases in calcium, fluorescent calcium indicators represent an excellent tool for monitoring this neuronal activity. Such indicators (e.g., calcium green dextran) are readily transported both retrogradely and anterogradely by neurons and can be used to fill populations of motoneurons and interneurons in different brain regions. As a result, confocal or two-photon imaging methods allow the study of neuronal activity with single-cell resolution in intact zebrafish (Fetcho & O'Malley, 1995; O'Malley, Zhou, & Gahtan, 2003).

These experimental approaches have begun to yield new insights into the brain structures and individual neurons involved in sleep processes. Our studies in larval zebrafish suggest that multiple brain areas are involved in both arousal and sleep-related states. Typically, the latency and threshold to neuronal calcium

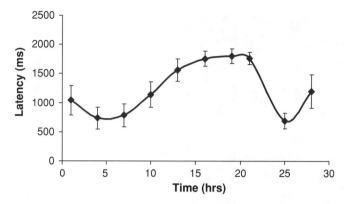

Figure 11.11. Diurnal variation in latency to visually evoked behavior in zebrafish larvae. (See color Plate 4.) The motor response of the tail to blue light pulse (450 nm; 2 sec) in agar-embedded larvae, recorded using a high-speed camera (500 fps) in parallel with monitoring the neuronal calcium responses. X axis: time (hours) elapsed from the beginning of the experiment; y axis: mean latency (SEM) to locomotor response; n = 6 larvae (7 dpf); each fish tested three times at each time. (From Zhdanova, 2006, with permission.)

responses are increased during nighttime sleep, and this correlates with similarly changed behavioral responses recorded in parallel (Figure 11.11).

The well-known involvement of the brainstem in sleep/wake mechanisms, including motoneuron inhibition during sleep, makes this brain region interesting for analyzing sleep-like behavior in zebrafish. The specific neuronal clusters and individual neurons of the brainstem nuclei can be screened for possible involvement in zebrafish sleep regulation following caudal injections of fluorescent indicators. Using confocal calcium imaging in labeled cells (Figure 11.12) and concurrent recording of locomotor activity (Figure 11.11), we have screened multiple reticulospinal neurons and their responses to melatonin treatment at concentrations that induce sleep-like states in zebrafish larvae.

In the majority of these neurons, melatonin did not affect calcium response to sensory stimuli (tapping or light). However, the activity of several neurons of the nucleus of medial longitudinal fasciculus (nucMLF) demonstrated a robust response to light stimulation and was inhibited by melatonin treatment (Figure 11.12; Zhdanova, 2006).

Whereas the effects of melatonin appear to be more localized, the GABA-ergic hypnotic drugs inhibit neuronal activity in the majority of brain areas. Such difference might be due to the wide presence of GABA receptors, mediating chloride-dependent hyperpolarization, in contrast to more discretely localized areas of melatonin receptor expression. Detailed characterization of neuronal representation of sleep-related effects of pharmacological and physiological agents is currently under investigation.

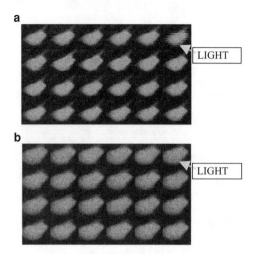

Figure 11.12. Melatonin attenuates neuronal response to light in MeLc neuron of the nMLF cluster. (See color Plate 5.) Framescan recordings, with frames collected at 440-millisecond intervals, show fluorescence response (*red*) after light pulse applied at the 6th frame (a). In the same cell, such response to light is not observed 20 minutes after 10-mM melatonin treatment (b). (From Zhdanova, 2006, with permission.)

Genetically encoded calcium indicators (Higashijima, Masino, Mandel, et al., 2003) are even more promising in monitoring neuronal activity in sleep and wakefulness than the injected ones. So far, they provide less robust signals; but when perfected, they should become a standard technique in studying the role of individual structures in zebrafish sleep. The development of transparent adult zebrafish (White, Sessa, Burke, et al., 2008) should further enhance our ability to analyze localized sleep-related processes in adults. Furthermore, the identification of mutants with alterations in the sleep/wake cycle or sleep homeostasis (Prober et al., 2006; Yokogawa et al., 2007) and crossing them with transgenic fish carrying genetically encoded calcium indicators may help to identify the structures defining specific sleep phenotypes and sleep disorders.

Age-related changes in zebrafish sleep

The zebrafish has recently attracted attention as a promising model for studying aging (Gerhard, 2003; Herrera & Jagadeeswaran, 2004; Kishi, 2004; Zhdanova et al., 2008). Under laboratory conditions, zebrafish mature within 6 months and survive up to 6 years. They experience a gradual senescence and start showing age-dependent changes in multiple physiological and cognitive parameters at around 2 years of age (Kishi, 2004; Tsai, Tucci, Uchiyama, et al., 2007). The distinct interindividual variability in these changes suggests that zebrafish, like mammals, can have successful and unsuccessful aging processes.

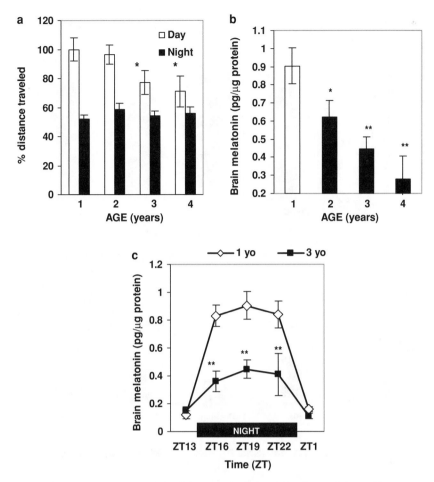

Figure 11.13. Aging in zebrafish results in reduced daytime activity and lower nighttime melatonin production. (a) Relative percent distance traveled in LD during the day (*white*) and at night (*black*) in zebrafish of four age groups ($n = 11$ to 12 per group; same groups at daytime and at night), with daytime distance traveled by 1-year-old fish represented as 100%. (b) Brain melatonin levels (pg/μg protein) in the middle of the dark period in young (*white*) and aged (*black*) zebrafish (n = 5 to 7 fish per group). (c) Comparison of daily patterns of brain melatonin levels (pg/μg protein) in 1-year-old (white diamond) and 3-year-old (*black square*) zebrafish (4 to 7 fish per group per time point). Horizontal black bar represents the night period. Mean (SEM); $^*P < 0.05$, $^{**}P < 0.001$, compared to 1-year-old zebrafish. (From Zhdanova, et al., 2008, with permission.)

The circadian functions are disrupted in aged zebrafish (Zhdanova et al., 2008), including a reduced daily amplitude of locomotor activity and melatonin production (Figure 11.13). Similarly, the amplitude of daily expression of core clock genes is also reduced during zebrafish aging (Figure 11.6). This might explain why

258   Fishing for sleep

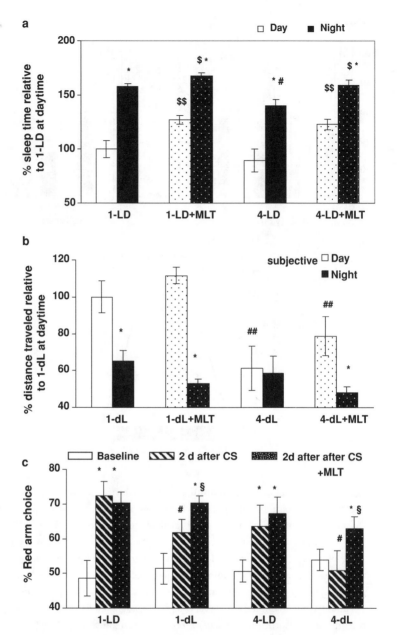

Figure 11.14. Aging in zebrafish is associated with sleep alterations, reduced intrinsic circadian rhythm of activity, and cognitive performance; these effects are counteracted by repeated overnight melatonin administration. (a) Change in percentage of sleep time during the day (ZT 5 to 7) and at night (ZT 18 to 20), with or without 30-minute melatonin (+MLT) pretreatment in 1- and 4-year-old zebrafish, relative to mean sleep time in 1-year-old zebrafish during the day represented as 100%.

the daily activity rhythms in aged zebrafish become more vulnerable to environmental factors. Although both young and aged animals show more robust daily patterns of locomotion in the light–dark cycle, lack of entraining environmental cues under constant dim light conditions reduces both rest and alertness in aged zebrafish and severely disrupts their circadian pattern of activity. Such changes in the circadian system may also, at least in part, underlie age-dependent modifications in zebrafish sleep, manifesting as low nighttime sleep duration (Figure 11.14a) and high sleep fragmentation. At the same time, the basal daytime arousal threshold during wakefulness is increased in aged zebrafish, potentially reflecting a lower level of alertness due to reduced circadian amplitude and/or nighttime sleep deficits.

These observations are further supported by overnight melatonin treatment, promoting sleep and resulting in a reduced daytime arousal threshold in aged fish; this suggests that sensitivity to melatonin typically does not decline with age (Figure 11.14a,b).

These observations are consistent with the expression levels for melatonin receptors remaining similar in young and aged zebrafish. In addition, the ability of melatonin to improve cognitive performance in aged zebrafish following overnight administration under constant dim light conditions (Figure 11.14c) also suggests that sleep-related and circadian abnormalities are part of the reason for cognitive alterations in aged fish.

←——

Figure 11.14. (*continued*). (b) Percentage of distance traveled during the day and at night under constant dim light conditions in 1-year-old (1 dL) and 4-year-old (4 dL) zebrafish at baseline and after melatonin administration (1-dL-MLT and 4-dL-MLT), relative to 1 dL during the day, represented as 100%. In a and b: *$P < 0.05$ relative to the same age and treatment condition in the day; $ $P < 0.05$ and $$ $P < 0.001$, relative to the same age and time in the absence of melatonin treatment; #$p < 0.05$ and ##$p < 0.001$, relative to the same time and treatment condition in the young group. (c) Generalization of conditioned response after exposure to 2 days of different light conditions (LD or dL), with or without overnight melatonin administration in 1- and 4-year-old zebrafish. Percentage of choosing the red arm of the T maze at baseline (*white*), 2 days after the end of conditioning (*diagonal*), and 2 days after the end of conditioning with overnight melatonin treatment (*black with dots*). *$P < 0.0001$ relative to the same age and light condition at baseline; $ $P < 0.01$, relative to the same age and light condition without melatonin treatment, # $P < 0.05$ relative to the same age and treatment condition in LD. Data presented as group mean (SEM); N = 8 to 10 fish per data point; proc mixed for all comparisons. (From Zhdanova, et al., 2008, with permission.)

Conclusion

Sleep function remains an enigma of modern biology. This is especially surprising in view of the substantial time animals and humans spend in this distinct physiological state, major similarities in its behavioral manifestations observed in different species, and typically deleterious effects of sleep deprivation on behavioral, autonomic, and cognitive functions. Although all this attests to sleep being a basic necessity, the question of whether sleep function is single and universal among diverse taxa remains to be determined. To reveal such common function requires in-depth investigation of the sleep processes in phylogenetically distant organisms adapted to different environments. In this respect, fish are interesting for a number of reasons. They represent some of the earliest vertebrates on the planet, with abundant variation in species, habitats, and adaptations to periodically changing environments. Fish have a well-developed brainstem, midbrain, and diencephalon, with sensory, motor, and integrative central nervous circuits that are highly comparable to those in mammals. However, the forebrain in fish is relatively small and its structure is distinctly different from that in mammals, including lack of the regular cerebral cortex. Considering that the cortex in mammals contributes to such distinct sleep phenomena as slow-wave sleep, the study of the sleep process in fish might help to elucidate the extent to which cortical changes affect the intrinsic mechanisms of sleep in humans.

Rest behavior has been evaluated in several species of fish, with the results supporting the idea that prolonged rest in these vertebrates shares principal similarities with mammalian sleep. The behavioral criteria for sleep – such as periodic reductions in activity, increases in arousal threshold, and rebound after sleep deprivation – are common in fish. Similarly, with the notable exception of the cerebral cortex, the principal neuronal structures involved in mammalian sleep are conserved in fish and have a neurochemical composition similar to that in higher vertebrates.

It is well established that sleep in mammals is regulated by both homeostatic and circadian processes, which complement each other and together provide an adaptive balance between the need for sleep and the optimal time to fulfill its physiological functions. The daily and circadian (i.e., under constant conditions) rhythms of sleep, the shorter latency to sleep onset (Shapiro & Hepburn, 1976), and the increased duration of sleep (Zhdanova et al., 2001) after sleep deprivation suggest that both types of sleep regulation are present in fish. Changes in cognitive performance following sleep deprivation also imply that sleep deficit affects brain function in fish (Zhdanova et al., 2008), as also observed in mammals (Stickgold, 2005).

There are, however, many questions in fish sleep physiology that require further investigation. One of them is the remarkable ability of fish to seemingly circumvent sleep need when faced with challenges of constant bright environmental illumination (Shapiro & Hepburn, 1976; Yokogawa et al., 2007). Because constant light exposure is known to suppress or significantly distort the circadian system, including that in fish (Shang & Zhdanova, 2007), such lack of sleep may suggest that the circadian regulation of sleep in fish is more critical than the homeostatic one. However, there might be special adaptive strategies in play during such challenging conditions (e.g., similar to the unihemispheric sleep found in aquatic mammals and birds) (Bobbo, Galvani, Mascetti, et al., 2002; Mascetti & Vallortigara, 2001; Mukhametov, 1987). This issue needs further elucidation.

Studies of endogenous agents that promote or attenuate sleep in both fish and mammals are of special interest. This is because their neurochemical actions and target structures should help to decipher the complex network of sleep mechanisms and to develop drugs to combat human sleep disorders. Melatonin, with its sleep-promoting effect in both humans and zebrafish, is one example of such a physiological agent. Its role in homeostatic and circadian sleep regulation is clearly conserved in diurnal vertebrates of different taxa, including fish (Zhdanova, 2005). The role of another important endogenous regulator of sleep, hypocretin (orexin), which promotes wakefulness in humans and other mammals, requires intense investigation in fish, especially owing to the somewhat controversial results obtained so far (Prober et al., 2006; Yokogawa et al., 2007). Further studies on the neurophysiological correlates of sleep in fish should be of great value and will benefit from new techniques involving electrophysiology or the in vivo imaging of changes in intracellular calcium responses and gene expression. Together, these studies should significantly contribute to our understanding of sleep evolution and its physiological functions.

References

Aparicio, S., Garau, C., Nicolau, M. C., Rial, R. V., & Esteban, S. (2006). Opposite effects of tryptophan intake on motor activity in ring doves (diurnal) and rats (nocturnal). *Comparative Biochemistry and Physiology, Part A: Molecular and Integrative Physiology, 144*(2), 173–179.

Aranda, A., Madrid, J. A., & Sanchez-Vazquez, F. J. (2001). Influence of light on feeding anticipatory activity in goldfish. *Journal of Biological Rhythms, 16*(1), 50–57.

Arenzana, F. J., Clemente, D., Sanchez-Gonzalez, R., Porteros, A., Aijon, J., & Arevalo, R. (2005). Development of the cholinergic system in the brain and retina of the zebrafish. *Brain Research Bulletin, 66*(4–6), 421–425.

Bobbo, D., Galvani, F., Mascetti, G. G., & Vallortigara, G. (2002). Light exposure of the chick embryo influences monocular sleep. *Behavioural Brain Research, 134*(1–2), 447–466.

Cahill, G. M., Hurd, M. W., & Batchelor, M. M. (1998). Circadian rhythmicity in the locomotor activity of larval zebrafish. *NeuroReport, 9*(15), 3445–3449.

Campbell, S. S., & Tobler, I. (1984). Animal sleep: A review of sleep duration across phylogeny. *Neuroscience and Biobehavioral Reviews, 8*(3), 269–300.

Carr, A. J., & Whitmore, D. (2005). Imaging of single light-responsive clock cells reveals fluctuating free-running periods. *Natural Cell Biology, 7*(3), 319–321.

Cermakian, N., Whitmore, D., Foulkes, N. S., & Sassone-Corsi, P. (2000). Asynchronous oscillations of two zebrafish CLOCK partners reveal differential clock control and function. *Proceedings of the National Academy of Sciences of the United States of America, 97*(8), 4339–4344.

Chen, W. M., Naruse, M., & Tabata, M. (2002). The effect of social interactions on circadian self-feeding rhythms in rainbow trout *Oncorhynchus mykiss Walbaum. Physiology and Behavior, 76*(2), 281–287.

Clemente, D., Arenzana, F. J., Sanchez-Gonzalez, R., Porteros, A., Aijon, J., & Arevalo, R. (2005). Comparative analysis of the distribution of choline acetyltransferase in the central nervous system of cyprinids. *Brain Research Bulletin, 66*(4–6), 546–549.

Clemente, D., Porteros, A., Weruaga, E., Alonso, J. R., Arenzana, F. J., Aijon, J., et al. (2004). Cholinergic elements in the zebrafish central nervous system: Histochemical and immunohistochemical analysis. *The Journal of Comparative Neurology, 474*(1), 75–107.

Danilova, N., Krupnik, V. E., Sugden, D., & Zhdanova, I. V. (2004). Melatonin stimulates cell proliferation in zebrafish embryo and accelerates its development. *The FASEB Journal, 18*(6), 751–753.

Doldan, M. J., Prego, B., Holmqvist, B. I., & de Miguel, E. (1999). Distribution of GABA-immunolabeling in the early zebrafish (*Danio rerio*) brain. *The European Journal of Morphology, 37*(2–3), 126–129.

Eriksson, K. S., Peitsaro, N., Karlstedt, K., Kaslin, J., & Panula, P. (1998). Development of the histaminergic neurons and expression of histidine decarboxylase mRNA in the zebrafish brain in the absence of all peripheral histaminergic systems. *The European Journal of Neuroscience, 10*(12), 3799–3812.

Faraco, J. H., Appelbaum, L., Marin, W., Gaus, S. E., Mourrain, P., & Mignot, E. (2006). Regulation of hypocretin (orexin) expression in embryonic zebrafish. *The Journal of Biological Chemistry, 281*(40), 29753–29761.

Fetcho, J. R., & O'Malley, D. M. (1995). Visualization of active neural circuitry in the spinal cord of intact zebrafish. *Journal of Neurophysiology, 73*(1), 399–406.

Fraser, N., Heggenes, J., Metcalfe, N. B., & Thorpe, J. E. (1995). Low summer temperatures cause juvenile Atlantic salmon to become nocturnal. *Canadian Journal of Zoology, 73*, 446–451.

Gee, P., Stephenson, D., & Wright, D. E. (1994). Temporal discrimination learning of operant feeding in goldfish (*Carassius auratus*). *Journal of the Experimental Analysis of Behavior, 62*(1), 1–13.

Gerhard, G. S. (2003). Comparative aspects of zebrafish (*Danio rerio*) as a model for aging research. *Experimental Gerontology, 38*(11–12), 1333–1341.

Herrera, M., & Jagadeeswaran, P. (2004). Annual fish as a genetic model for aging. *The Journals of Gerontology, Series A, Biological Sciences and Medical Sciences, 59*(2), 101–107.

Higashijima, S., Mandel, G., & Fetcho, J. R. (2004). Distribution of prospective glutamatergic, glycinergic, and GABAergic neurons in embryonic and larval zebrafish. *The Journal of Comparative Neurology, 480*(1), 1–18.

Higashijima, S., Masino, M. A., Mandel, G., & Fetcho, J. R. (2003). Imaging neuronal activity during zebrafish behavior with a genetically encoded calcium indicator. *Journal of Neurophysiology, 90*(6), 3986–3997.

Kaneko, M., Hernandez-Borsetti, N., & Cahill, G. M. (2006) Diversity of zebrafish peripheral oscillators revealed by luciferase reporting. *Proceedings of the National Academy of Sciences of the United States of America, 103*(39), 14614–14619.

Karmanova, I. G. (1975). New data on the circadian biorhythm of wakefulness and sleep in vertebrates. *Doklady Akademii Nauk SSSR, 225*(6), 1457–1460.

Karmanova, I. G., Churnosov, E. V., & Popova, D. I. (1976). Daily form of rest in the catfish *Ictalurus nebulosus* and the frog *Rana temporaria. Zhurnal Evoliutsionnoĭ Biokhimii i Fiziologii, 12*(6), 572–578.

Karmanova, I. G., & Lazarev, S. G. (1979). Stages of sleep evolution (facts and hypotheses). *Waking and Sleeping, 3*(2), 137–147.

Karmanova, I. G., Titkov, E. S., & Popova, D. I. (1976). Species characteristics of the diurnal periodicity of rest and activity in Black Sea fish. *Zhurnal Evoliutsionnoĭ Biokhimii i Fiziologii, 12*(5), 486–488.

Kaslin, J., Nystedt, J. M., Ostergard, M., Peitsaro, N., & Panula, P. (2004). The orexin/hypocretin system in zebrafish is connected to the aminergic and cholinergic systems. *The Journal of Neuroscience, 24*(11), 2678–2689.

Kaslin, J., & Panula, P. (2001). Comparative anatomy of the histaminergic and other aminergic systems in zebrafish (*Danio rerio*). *The Journal of Comparative Neurology, 440*(4), 342–377.

Kavanau, J. L. (1998). Vertebrates that never sleep: Implications for sleep's basic function. *Brain Research Bulletin, 46*(4), 269–279.

Kishi, S. (2004). Functional aging and gradual senescence in zebrafish. *Annals of the New York Academy of Sciences, 1019*, 521–526.

Lague, M., & Reebs, S. G. (2000). Phase-shifting the light-dark cycle influences food-anticipatory activity in golden shiners. *Physiology and Behavior, 70*(1–2), 55–59.

Mascetti, G. G., & Vallortigara, G. (2001). Why do birds sleep with one eye open? Light exposure of the chick embryo as a determinant of monocular sleep. *Current Biology, 11*(12), 971–974.

Mintz, E. M., Phillips, N. H., & Berger, R. J. (1998). Daytime melatonin infusions induce sleep in pigeons without altering subsequent amounts of nocturnal sleep. *Neuroscience Letters, 258*(2), 61–64.

Montoro, J., Sastre, J., Bartra, J., del Cuvillo, A., Davila, I., Jauregui, I., et al. (2006). Effect of H1 antihistamines upon the central nervous system. *Journal of Investigational Allergology and Clinical Immunology, 16*(Suppl. 1), 24–28.

Mukhametov, L. M. (1987). Unihemispheric slow-wave sleep in the Amazonian dolphin, *Inia geoffrensis. Neuroscience Letters, 79*(1–2), 128–132.

Nakamachi, T., Matsuda, K., Maruyama, K., Miura, T., Uchiyama, M., Funahashi, H., et al. (2006). Regulation by orexin of feeding behaviour and locomotor activity in the goldfish. *Journal of Neuroendocrinology, 18*(4), 290–297.

Olla, B. L., & Studholme, A. L. (1978). Comparative aspects of the activity rhythms of tautog, *Tautoga onitis,* bluefish, *Pomatomus saltatrix,* and Atlantic mackerel, *Scomber scombrus,* as related to their life habits. In J. E. Thorpe (Ed.), *Rhythmic activity of fishes* (pp. 131–151). London: Academic Press.

O'Malley, D. M., Zhou, Q., & Gahtan, E. (2003). Probing neural circuits in the zebrafish: A suite of optical techniques. *Methods, 30*(1), 49–63.

Pando, M. P., Pinchak, A. B., Cermakian, N., & Sassone-Corsi, P. (2001). A cell-based system that recapitulates the dynamic light-dependent regulation of the vertebrate clock. *Proceedings of the National Academy of Sciences of the United States of America, 98*(18), 10178–10183.

Pando, M. P., & Sassone-Corsi, P. (2002). Unraveling the mechanisms of the vertebrate circadian clock: Zebrafish may light the way. *Bioessays, 24*(5), 419–426.

Panula, P., Karlstedt, K., Sallmen, T., Peitsaro, N., Kaslin, J., Michelsen, K. A., et al. (2000). The histaminergic system in the brain: Structural characteristics and changes in hibernation. *Journal of Chemical Neuroanatomy, 18*(1–2), 65–74.

Paredes, S. D., Terron, M. P., Valero, V., Barriga, C., Reiter, R. J., & Rodriquez, A. B. (2007). Orally administered melatonin improves nocturnal rest in young and old ringdoves (*Streptopelia risoria*). *Basic and Clinical Pharmacology and Toxicology, 100*(4), 258–268.

Peyrethon, J., & Dusan-Peyrethon, D. (1967). Polygraphic study of the wakefulness-sleep cycle of a teleostean (*Tinca tinca*). *Comptes Rendus des Séances de la Société de Biologie et de ses Filiales, 161*(12), 2533–2537.

Prober, D. A., Rihel, J., Onah, A. A., Sung, R. J., & Schier, A. F. (2006). Hypocretin/orexin overexpression induces an insomnia-like phenotype in zebrafish. *The Journal of Neuroscience, 26*(51), 13400–13410.

Renier, C., Faraco, J. H., Bourgin, P., Motley, T., Bonaventure, P., Rosa, F., et al. (2007). Genomic and functional conservation of sedative-hypnotic targets in the zebrafish. *Pharmacogenetics and Genomics, 17*(4), 237–253.

Reppert, S. M., & Weaver, D. R. (2002). Coordination of circadian timing in mammals. *Nature, 418*(6901), 935–941.

Reppert, S. M., Weaver, D. R., Cassone, V. M., Godson, C., & Kolakowski, L. F., Jr. (1995). Melatonin receptors are for the birds: Molecular analysis of two receptor subtypes differentially expressed in chick brain. *Neuron, 15*(5), 1003–1015.

Ruuskanen, J. O., Peitsaro, N., Kaslin, J. V., Panula, P., & Scheinin, M. (2005). Expression and function of alpha-adrenoceptors in zebrafish: Drug effects, mRNA and receptor distributions. *Journal of Neurochemistry, 94*(6), 1559–1569.

Sanchez-Vazquez, F. J., Aranda, A., & Madrid, J. A. (2001). Differential effects of meal size and food energy density on feeding entrainment in goldfish. *Journal of Biological Rhythms, 16*(1), 58–65.

Shang, E. H., & Zhdanova, I. V. (2007). The circadian system is a target and modulator of prenatal cocaine effects. *PLoS ONE, 2*(7), e587.

Shapiro, C. M., & Hepburn, H. R. (1976). Sleep in a schooling fish, Tilapia mossambica. *Physiology and Behavior, 16*(5), 613–615.

Siegel, J. M. (2005). Clues to the functions of mammalian sleep. *Nature, 437*, 1264–1271.

Spieler, R. E., Meier, A. H., & Noeske, T. A. (1978a). Timing of a single daily meal affects daily serum prolactin rhythm in gulf killifish, *Fundulus grandis*. *Life Sciences, 22*(3), 255–258.

Spieler, R. E., Meier, A. H., & Noeske, T. A. (1978b). Temperature-induced phase shift of daily rhythm of serum prolactin in gulf killifish. *Nature, 271*(5644), 469–470.

Spieler, R. E., & Noeske, T. A. (1981). Timing of a single daily meal and diel variations of serum thyroxine, triiodothyronine, and cortisol in goldfish *Carassius auratus*. *Life Sciences, 28*(26), 2939–2944.

Stickgold, R. (2005). Sleep-dependent memory consolidation. *Nature, 437*(7063), 1272–1278.

Tauber, E. S. (1974). The phylogeny of sleep. In E. D. Weitzman (Ed.), *Advances in sleep research* (pp. 133–172). New York: Spectrum Publications.

Tobler, I., & Borbély, A. A. (1985). Effect of rest deprivation on motor activity of fish. *Journal of Comparative Physiology, Part A, 157*(6), 817–822.

Tsai, S. B., Tucci, V., Uchiyama, J., Fabian, N. J., Lin, M. C., Bayliss, P. E., et al. (2007). Differential effects of genotoxic stress on both concurrent body growth and gradual senescence in the adult zebrafish. *Aging Cell, 6*(2), 209–224.

Vansteensel, M. J., Michel, S., & Meijer, J. H. (2008). Organization of cell and tissue circadian pacemakers: A comparison among species. *Brain Research Reviews, 58*(1), 18–47.

White, R. M., Sessa, A., Burke, C., Bowman, T., LeBlanc, J., Ceol, C., et al. (2008). Transparent adult zebrafish as a tool for in vivo transplantation analysis. *Cell Stem Cell, 2*(2), 183–189.

Whitmore, D., Foulkes, N. S., & Sassone-Corsi, P. (2000). Light acts directly on organs and cells in culture to set the vertebrate circadian clock. *Nature, 404*(6773), 87–91.

Whitmore, D., Foulkes, N. S., Strahle, U., & Sassone-Corsi, P. (1998). Zebrafish clock rhythmic expression reveals independent peripheral circadian oscillators. *Nature Neuroscience, 1*(8), 701–707.

Yokogawa, T., Marin, W., Faraco, J., Pezeron, G., Appelbaum, L., Zhang, J., et al. (2007). Characterization of sleep in zebrafish and insomnia in hypocretin receptor mutants. *PLoS Biology, 5*(10), 2379–2397.

Yu, L. , Tucci, V., Kishi, S., & Zhdanova, I. V. (2006). Cognitive aging in zebrafish. *PLoS 1*(1), E14.

Zeitzer, J. M., Nishino, S., & Mignot, E. (2006). The neurobiology of hypocretins (orexins), narcolepsy, and related therapeutic interventions. *Trends in Pharmacological Sciences, 27*(7), 368–374.

Zhdanova, I. V. (2005). Melatonin as a hypnotic: Pro. *Sleep Medicine Reviews, 9*(1), 51–65.

Zhdanova, I. V. (2006). Sleep in zebrafish. *Zebrafish, 3*(2), 215–226.

Zhdanova, I. V., Cantor, M. L., Leclair, O. U., Kartashov, A. I., & Wurtman, R. J. (1998). Behavioral effects of melatonin treatment in nonhuman primates. *Sleep Research Online, 1*(3), 114–118.

Zhdanova, I. V., Geiger, D. A., Schwagerl, A. L., Leclair, O. U., Killiany, R., Taylor, J. A., et al. (2002). Melatonin promotes sleep in three species of diurnal nonhuman primates. *Physiology and Behavior, 75*(4), 523–529.

Zhdanova, I. V., & Reebs, S. G. (2006). Circadian rhythms in fish. In K. Sloman, R. Wilson, S. Balshine (Eds.), *Behavior and physiology of fish* (pp. 197–238). San Diego: Elsevier.

Zhdanova, I. V., Wang, S. Y., Leclair, O. U., & Danilova, N. P. (2001). Melatonin promotes sleep-like state in zebrafish. *Brain Research, 903*(1–2), 263–268.

Zhdanova, I. V., Yu, L., Lopez-Patino, M., Shang, E., Kishi, S., & Guelin, E. (2008). Aging of the circadian system in zebrafish and the effects of melatonin on sleep and cognitive performance. *Brain Research Bulletin, 75*(2–4), 433–441.

Index

acousticolateralis system, in
teleosts, 63–64
schooling functions of,
63–64
AD. *See* Alzheimer's disease
adaptationism, 173, 174
adenosine triphosphate (ATP)
in biochemical regulation
of sleep, 95–96
in organization of sleep, 99
aggressiveness, insomnia
and, 115–116
aging, sleep and
in *Drosophila melanogaster*,
46–47
in honeybees, 40–41
insomnia, 110
in zebrafish, 256–259
albacores. *See* scombrids
Alzheimer's disease (AD), 231
amphibians
evolution of, 201–202
REM sleep in, 201–202
amylase, 51
Angelman syndrome (AS), 228
animal models, of sleep, 145.
See also mouse models,
for sleep
in *Drosophila melanogaster*,
145

mice as, 219–232
circadian mechanisms
and, 226–228
electrophysiology
protocols in, 221
genetics and, 219–232
genomic imprinting in,
228–230
high-throughput
technology protocols
in, 221–222
NREM in, 228–232
REM during, 223,
230–232
in sleep-wake studies,
224–225
wakefulness and,
224–226, 230–232
animals in the wild
future sleep research on,
9–10
REM and, captive animals
v., 210–211
sleep times for, captive
animals v., 14–15
ANS. *See* automatic nervous
system
anxiety, insomnia and, 114
Apis mellifera. *See* honeybees,
sleep in

AS. *See* Angelman syndrome
Aserinsky, Eugene, 197
ATP. *See* adenosine
triphosphate
automatic nervous system
(ANS), 3
avian sleep. *See* birds, sleep in

baboons. *See* nonhuman
primates, sleep in
bears, hibernation and, 8
bees. *See* honeybees, sleep in
behavioral measures, of sleep,
2–3, 35
circadian influences on,
35
homeostatic factors in, 35
in insects, 37
in honeybees, variations
among species, 40
NREM, 2–3
HVSW in, 3
REM, 2
bihemispheric sleep, 7
birds, sleep in, 145–164
brain size and, 157
homeostasis for, 152–155
sleep deprivation and,
152–153
SWA and, 153

267